Springer Geology

Springer Geology Field Guides

FieldGuides

Springer Geology Field Guides is a book series that provides the details of both well-known and little known transects to discover the beauty and knowledge of Geology, worldwide. Springer Geology Field Guides aims to bring geology field trips to professionals, students, and amateurs to find the most interesting geology worldwide. This series includes carefully crafted guidebooks that help generations of geologists explore the terrain with minimum or no guidance. In this series, the audience will also find field methodologies and case studies as examples.

This book series will welcome both authored and edited field guides of all geology disciplines, including structural geology, tectonics, sedimentology, stratigraphy, paleontology, economic geology, among others. Photo-atlases are also welcome.

Soumyajit Mukherjee
Editor

Structural Geology and Tectonics Field Guidebook—Volume 2

Springer

Editor
Soumyajit Mukherjee
Department of Earth Sciences
Indian Institute of Technology Bombay
Mumbai, Maharashtra, India

ISSN 2197-9545 ISSN 2197-9553 (electronic)
Springer Geology
ISSN 2730-7344 ISSN 2730-7352 (electronic)
Springer Geology Field Guides
ISBN 978-3-031-19575-4 ISBN 978-3-031-19576-1 (eBook)
https://doi.org/10.1007/978-3-031-19576-1

This Springer imprint is published by the registered company Springer Nature Switzerland AG
The registered company address is: Gewerbestrasse 11, 6330 Cham, Switzerland

Dedicated to (retired) Prof. P.K. Saraswati
(Department of Earth Sciences, IIT Bombay)

Preface

The purpose of this book is exactly same as that of the prequel volume 1 on the same broad topic (Mukherjee 2023)—this book presents few well-known and several rather unknown transects where exciting structures exist, and field programs can be established.

Cite individual chapters in the following format:

Mukherjee S. (2023) Introduction to Structural Geology and Tectonics Field Guidebook—Volume 2. In: Mukherjee S. (Ed) Structural Geology and Tectonics Field Guidebook—Volume 2. Springer Nature Switzerland AG. Cham. pp. xi–xiv. ISBN: 978-3-031-19575-4.

Cite this book in the following format:

Mukherjee S. (2023) Structural Geology and Tectonics Field Guidebook—Volume 2. Springer Nature Switzerland AG. Cham. pp. 1–418. ISBN: 978-3-031-19575-4.

Mumbai, India
December 2022

Soumyajit Mukherjee
smukherjee@iitb.ac.in

Reference

Mukherjee, S. (2021). Structural geology and tectonics field guidebook (vol. 1, pp. 1–723). Springer Nature Switzerland AG. ISBN: 978-3-030-60142-3.

Acknowledgements

Mohamedharoon A. Shaikh, **Bikramaditya Mandal**, **Shatavisa Chatterjee** (IIT Bombay), **Mohit Kumar Puniya** (National Geotechnical Facility, Dehradun), **Mery Biswas** (Presidency University, Kolkata) and **Ankita Paul** (JIS Group of Educational Initiatives) assisted in preparing this book. The **Springer (proofreading) team—Boopalan Renu**, **Alexis Vizcaino** and **Doerthe Mennecke-Buehler** interacted. I thank the **contributing authors** and the **reviewers** for participation.

Introduction

The book is a sequel of the previous edited volume 1 on the same broad subject (Mukherjee 2023). Indian terrain has been a matter of national and international attention to geoscientists because of its pure and applied geological research issues (e.g., Mukherjee 2015, 2020; Mukherjee et al., 2015, 2017). This edited book consists of **14** chapters contributed by **45** authors and co-authors from **6** countries. Geoheritage in India has been reviewed by Chandrasekharam (2007) and Kaur (2022).

Porcher et al. (2022; Chapter "Field Guide for a Complete Cross-Section of the Central Andes Along Main Roads") present a field guide for a seven-day fieldwork in central Andes. The purpose of this chapter is to familiarize the reader with the structural, morphotectonic, stratigraphic, volcanic and sedimentary features of the orogen. **Pamplona et al. (2022; Chapter** "Structures Associated with the Dynamics of Granitic Rock Emplacement (NW Portugal)") provide structural examples through photographs how granite rock emplacement in NW Portugal has been its cause. An enclave disruption mechanism within the granite body is proposed in this work. **Novakova (2022; Chapter** "Tectonically Significant Fault Zones in Central Europe (Germany, Czech Republic and Poland) and Their Surface and Subsurface Outcrops: Franconian Line, Hronov-Porici Fault, Sudetic Marginal Fault and Lusatian Fault") describes the four major fault zones with NW trend in Central Europe. Understanding these faults will be important since they are active. **Singh et al. (2022; Chapter** "Geological Field Observations Along the Pandoh Syncline: The Mandi-Kataula-Bajura Section of Himachal Pradesh, NW-India") discuss lithologies and structures of the Mandi-Kataula-Bajura section of the Indian Himachal Lesser Himalaya. These workers also add up new metamorphic information into their study. **Ganguli et al. (2022; Chapter** "The Rock Outcrops at Raghunathdi, SE of Ghatsila (Jharkhand, India): a Spectacular Preservation of Polyphase Folding") present Paleo-Proterozoic lithologies and structures from the Indian state Jharkhand. The structures mainly include superposed folding. **Samanta and Kundu (2022; Chapter** "Spectacular Soft-Sediment Deformation Structures in Sedimentary Rock Outcrops of Damodar Valley Basin, West Bengal, India: A Field Guide") describe primary structures in sedimentary rocks including syn-sedimentary deformation structures from the Damodar

Valley in West Bengal (India). **Lohani et al. (2022; Chapter** "Structural Geological Field Guide: Bhuj Area (Gujarat, India)") present in detail structures associated with the active segment of Katrol Hill Range Fault Zone, Kutch area, Gujarat, India. **Sinha et al. (2022; Chapter** "Structural and Sedimentary Field Studies in Angul District, Odisha, India") describe a geological fieldwork with sedimentology and structural geology as focus from the Angul District, Odisha, India. The rock types in this region are of diverse ages—Archean-Proterozoic metamorphics and migmatites, Gondwana Supergroup of sedimentary succession and the overlying Quaternary deposits. **Puniya et al. (2022a; Chapter** "New Structural Geological Input from the Barmer Basin, Rajasthan (India)") provide new field-based structural geologic data from the eastern, western and northern parts of the Barmer basin in terms of brittle faults, brittle shear zones and dykes. A N-S compression was decoded, which could be the result of India–Eurasia collision. In another work, **Puniya et al. (2022b; Chapter** "Structural Geology and Stability Issue of the Giral Lignite Mine, Rajasthan, India") present structural geology from the Giral Lignite Mine, Barmer area. Two normal faults were documented and were correlated with Barmer basin's rifting. **Puniya et al. (2022c; Chapter** "Relationship Between Deformation Structures and Rock Mass Rating: A Case Study of Underground Power House, Andhra Pradesh—India") discuss structural geological studies relevant to an underground powerhouse construction in West and East Godavari Districts, Andhra Pradesh, India. The authors documented four sets of joints from the study area. **Bhu et al. (2022; Chapter** "Microstructures Mimic Meso-Scale Structures") discuss structures from a deformed Precambrian metasedimentary terrane of the Aravalli craton, India. Through mesoscale and microscale studies, the authors prove the established fact that structures can be fractal in nature. **Ansari (2022; Chapter** "Review on Role of Multi-Constellation Global Navigation Satellite System-Reflectometry (GNSS-R) for Real-Time Sea-Level Measurements") reviews how multi-constellation Global Navigation Satellite System Reflectometry can aid a field geologist in coastal tectonics studies. **Haldar et al. (2022; Chapter** "Architecture and Structures of Kiradu Temple (Barmer Region, Rajasthan, India)") discuss in great detail the architectures of the less known Kiradu temple in the Barmer area, India.

Mumbai, India

Soumyajit Mukherjee
smukherjee@iitb.ac.in

Acknowledgements I thank **Springer (proofreading) team** and especially **Boopalan Renu, Alexis Vizcaino** and **Doerthe Mennecke-Buehler** in successful completion of this book.

References

Ansari, K. (2022). Review on role of multi-constellation global navigation satellite system-reflectometry (GNSS-R) for real-time sea-level measurements In: S. Mukherjee (Ed.) Structural geology and tectonics field guidebook (vol. 2, pp. 333–358). Springer Nature Switzerland AG. ISBN: 978-3-031-19575-4.

Bhu, H., Purohit, R., Dutta, R. (2022). Macrostructures mimic Meso-structures. In: S. Mukherjee (Ed.) Structural geology and tectonics field guidebook (vol. 2, pp. 323–332), Springer Nature Switzerland AG. ISBN: 978-3-031-19575-4.

Biswas, M., Gogoi, M.P., Mondal, B., Sivasankar, T., Mukherjee, S., Dasgupta, S. (2022a). Geomorphic assessment of active tectonics in Jaisalmer basin (Western Rajasthan, India). Geocarto International. https://doi.org/10.1080/10106049.2022.2066726.

Biswas, M., Puniya, M.K., Gogoi, M.P., Dasgupta, S., Mukherjee, S., Kar, N.R. (2022b). Morphotectonic analysis of petroliferous Barmer rift basin (Rajasthan, India). *Journal of Earth System Science 131*, 140.

Chandrasekharam, D. (2007). Geo-mythlogy of India. In: L. Piccardi, W.B. Masse (Eds.) Myth and geology (vol. 273, pp. 29–37). Geological Society, London, Special Publications.

Dasgupta, S., Mukherjee, S. (2017). Brittle shear tectonics in a narrow continental rift: Asymmetric nonvolcanic Barmer Basin (Rajasthan, India). *Journal of Geology 125*, 561–591.

Dasgupta, S., Mukherjee, S. (2019). Remote sensing in lineament identification: Examples from western India. In: A. Billi, A. Fagereng (Eds.) Problems and solutions in structural geology and tectonics. Developments in structural geology and tectonics book series (vol. 5, pp. 205–221). Series Editor: S. Mukherjee, Elsevier. ISBN: 9780128140482.

Ganguli, S., Samanta, A., Kundu, A. (2022). The rock outcrops at Raghunathdi, SE of Ghatsila (Jharkhand, India): A spectacular preservation of polyphase folding. In: S. Mukherjee (Ed.) Structural geology and tectonics field guidebook (vol. 2, pp. 203–216). Springer Nature Switzerland AG. ISBN: 978-3-031-19575-4.

Kar, N.K., Mani, D., Mukherjee, S., Dasgupta, S., Puniya, M.K., Kaushik, A.K., Biswas, M., Babu, E.V.S.S.K. (2022). Source rock properties and kerogen decomposition kinetics of Eocene shales from petroliferous Barmer basin, western Rajasthan, India. *Journal of Natural Gas Science and Engineering 100*, 104497.

Kaur, G. (2022). Geodiversity, geoheritage and geoconservation: A global perspective. *Journal of Geological Society of India 91*, 1221–1228.

Lohani, N., Mukherjee, S., Singh, S., Pawar, A., Shaikh, M. (2022). Structural geological field guide: Bhuj area (Gujarat, India). In: S. Mukherjee (Ed.) Structural geology and tectonics field guidebook (vol. 2, pp. 227–250). Springer Nature Switzerland AG. ISBN: 978-3-031-19575-4.

Mukherjee, S. (2015). Petroleum geosciences: Indian contexts. Springer Geology. ISBN 978-3-319-03119-4.

Mukherjee, S. (2020). Tectonics and structural geology: Indian context. Springer. ISBN 978-3-319-99341-6.

Mukherjee, S. (2022). Structural geology and tectonics field guidebook (vol. 1, pp. 1–722). Springer Nature Switzerland AG. ISBN 978-3-030-60143-0.

Mukherjee, S, Carosi, R., van der Beek, P.A., Mukherjee, B.K. Robinson, D.M. (2015). Tectonics of the Himalaya. *Geological Society, London, Special Publications, 412*, 1–3. ISBN 978-1-86239-703-3.

Mukherjee, S., Misra, A.A., Calvès, G., Nemčok, M. (2017). Tectonics of the Deccan Large Igneous Province. *Geological Society, London, Special Publications 445*. http://doi.org/10.1144/SP445.

Novakova, L. (2022). Tectonically significant fault zones in Central Europe (Germany, Czech Republic and Poland) and their surface and subsurface outcrops: Franconian Line, Hronov-Porici Fault, Sudetic Marginal Fault and Lusatian Fault. In: S. Mukherjee (Ed.) Structural geology and tectonics field guidebook (vol. 2, pp. 157–178). Springer Nature Switzerland AG. ISBN: 978-3-031-19575-4.

Pamplona, J., Rodrigues, B.C., Peternell, M., Lorenz, A., Schmidt, A., Mengert, M., Altmeter, T., Köpping, J. (2022). Structures associated with the dynamics of granitic rock emplacement

(NW Portugal). In: S. Mukherjee (Ed.) Structural geology and tectonics field guidebook (vol. 2, pp. 61–156). Springer Nature Switzerland AG. ISBN: 978-3-031-19575-4.

Puniya, M.K., Kaushik, A.K., Mukherjee, S., Dasgupta, S., Kar, N.R., Biswas, M., Choudhary. (2022a). New structural geological input from the Barmer Basin, Rajasthan (India). In: S. Mukherjee (Ed.) Structural geology and tectonics field guidebook (vol. 2, pp. 285–296). Springer Nature Switzerland AG. ISBN: 978-3-031-19575-4.

Puniya, M.K., Kaushik, A.K., Mukherjee, S., Dasgupta, S., Kar, N.R., Biswas, M., Choudhary. (2022b). Structural geology and stability issue of the Giral lignite mine, Rajasthan, India. In: S. Mukherjee (Ed.) Structural geology and tectonics field guidebook (vol. 2, pp. 297–310). Springer Nature Switzerland AG. ISBN: 978-3-031-19575-4.

Puniya, M.K., Kumar, S., Kaushik, A.K., Puniya, N. (2022c). Relationship between deformation structures and rock mass rating: A case study of under ground power house, Andhra Pradesh-India. In: S. Mukherjee (Ed.) Structural geology and tectonics field guidebook (vol. 2, pp. 311–322). Springer Nature Switzerland AG. ISBN: 978-3-031-19575-4.

Porcher, C.C., Gomes, M.E.B., Martini, A., De Toni, G.B., Machado, J.P. (2022). Field guide for a complete cross-section of the Central Andes along main roads. In: S. Mukherjee (Ed.) Structural geology and tectonics field guidebook (vol. 2, pp. 1–60). Springer Nature Switzerland AG. ISBN: 978-3-031-19575-4.

Samanta, A., Kundu, A. (2022). Spectacular soft-sediment deformation structures in sedimentary rock outcrops of Damodar Valley Basin, West Bengal, India: A field guide. In: S. Mukherjee (Ed.) Structural geology and tectonics field guidebook (vol. 2, pp. 217–226). Springer Nature Switzerland AG. ISBN: 978-3-031-19575-4.

Singh, P., Sethy, P.C., Rastogi, H., Singh, M.R., Singh, A.K., Thakur, S.S., Singhal, S. (2022). Geological field observations along the Pandoh Syncline: The Mandi-Kataula-Bajura section of Himachal Pradesh, NW-India. In: S. Mukherjee (Ed.) Structural geology and tectonics field guidebook (vol. 2, pp. 179–202), Springer Nature Switzerland AG. ISBN: 978-3-031-19575-4.

Contents

Field Guide for a Complete Cross-Section of the Central Andes Along Main Roads 1
Carla Cristine Porcher, Márcia Elisa Boscato Gomes, Amós Martini, Giuseppe Betino De Toni, and João Pacífico Machado

Structures Associated with the Dynamics of Granitic Rock Emplacement (NW Portugal) 61
Jorge Pamplona, Benedito C. Rodrigues, Mark Peternell, Alex Lorenz, Alex Schimdt, Melissa Mengert, Thomas Altmeyer, and Jonas Köpping

Tectonically Significant Fault Zones in Central Europe (Germany, Czech Republic and Poland) and Their Surface and Subsurface Outcrops: Franconian Line, Hronov-Porici Fault, Sudetic Marginal Fault and Lusatian Fault 157
Lucie Novakova

Geological Field Observations Along the Pandoh Syncline: The Mandi-Kataula-Bajura Section of Himachal Pradesh, NW-India 179
Paramjeet Singh, Pratap Chandra Sethy, Hrithik Rastogi, M. Rajanikanta Singh, A. Krishnakanta Singh, Satyajit Singh Thakur, and Saurabh Singhal

The Rock Outcrops at Raghunathdi, SE of Ghatsila (Jharkhand, India): a Spectacular Preservation of Polyphase Folding 203
Srinanda Ganguly, Arpita Samanta, and Abhik Kundu

Spectacular Soft-Sediment Deformation Structures in Sedimentary Rock Outcrops of Damodar Valley Basin, West Bengal, India: A Field Guide 217
Arpita Samanta and Abhik Kundu

Structural Geological Field Guide: Bhuj Area (Gujarat, India) 227
Nidhi Lohani, Soumyajit Mukherjee, Seema Singh, Aashu Pawar, and Mohamedharoon Shaikh

Structural and Sedimentary Field Studies in Angul District, Odisha, India 251
Subhajit Sinha, Ananya Ghosh, and Somraj Mishra

New Structural Geological Input from the Barmer Basin, Rajasthan (India) 285
Mohit Kumar Puniya, Ashish Kumar Kaushik, Soumyajit Mukherjee, Swagato Dasgupta, Nihar Ranjan Kar, Mery Biswas, and Ratna Choudhary

Structural Geology and Stability Issue of the Giral Lignite Mine, Rajasthan, India 297
Mohit Kumar Puniya, Ashish Kumar Kaushik, Soumyajit Mukherjee, Nihar Ranjan Kar, Mery Biswas, and Ratna Choudhary

Relationship Between Deformation Structures and Rock Mass Rating: A Case Study of Underground Power House, Andhra Pradesh—India 311
Mohit Kumar Puniya, Sohan Kumar, Ashish Kumar Kaushik, and Nikhil Puniya

Microstructures Mimic Meso-Scale Structures 323
Harsh Bhu, Ritesh Purohit, and Riya Dutta

Review on Role of Multi-Constellation Global Navigation Satellite System-Reflectometry (GNSS-R) for Real-Time Sea-Level Measurements 333
Kutubuddin Ansari

Architecture and Structures of Kiradu Temple (Barmer Region, Rajasthan, India) 359
Paramita Haldar, Mohit Kumar Puniya, Mery Biswas, Soumyajit Mukherjee, Nihar Ranjan Kar, and Ratna Choudhary

Field Guide for a Complete Cross-Section of the Central Andes Along Main Roads

Carla Cristine Porcher, Márcia Elisa Boscato Gomes, Amós Martini, Giuseppe Betino De Toni, and João Pacífico Machado

Abstract This guide presents seven itineraries to investigate the geology of the Central Andes, which can be done in seven days of field work along main roads in northern Argentina and Chile. A total of 32 field stops are organized in a complete cross-section that cuts over 600 km of the mountain range, from the Subandean Zone and Eastern Cordillera, between latitudes 24°–23° 20′ (Purmamarca–Quebrada de Humahuaca), to the Coastal Cordillera, between latitudes 22° 10′–20° 17′ (Tocopilla–Iquique), passing through the domains of North Puna, Western Cordillera, Modern Volcanic Arc, Cordillera de la Sal and Cordillera Domeyko. The proposed transect exhibits diverse structural, morphotectonic, stratigraphic, volcanic and sedimentary features of the magnificent Andean mountain range, which is the main example of a non-collisional orogen in the world. The geological features mentioned in this guide are widely documented and contextualized here according to recent bibliography. The stops were selected from a broader field trip guide, part of the *Geology of the Central Andes* course, held during four years by the Geosciences Institute of the Federal University of Rio Grande do Sul (BR), with groups with up to 50 people, so that the feasibility of the route with large bus was properly tested and is assured. Additionally, travel logistics tips, based on our experience, are provided along this guide.

Keywords Orogeny · Accretionary orogen · Field trip guide · Central Andes · Northern Argentina · Northern Chile

C. C. Porcher (✉) · M. E. B. Gomes
Institute of Geosciences, Federal University of Rio Grande do Sul (UFRGS), Porto Alegre, Brazil
e-mail: carla.porcher@ufrgs.br

A. Martini
Institute of Sciences and Technology, Federal University of Jequitinhonha and Mucuri Valleys (UFVJM), Diamantina, Brazil

G. B. De Toni
Universidade Federal do Pampa (UNIPAMPA), Caçapava do Sul, Brazil

J. P. Machado
Center of Applied Geosciences (CGA), Geological Survey of Brazil (CPRM), Rio de Janeiro, Brazil

S. Mukherjee (ed.), *Structural Geology and Tectonics Field Guidebook–Volume 2*, Springer Geology, https://doi.org/10.1007/978-3-031-19576-1_1

1 Introduction

The Andes mountain range is one of the greatest orogenic systems active on Earth. Cradle of civilizations that had their cultural intelligence forged by the natural challenges of this environment, such as the Incas and Quechuas, and of mineral wealth that spurred the Spanish invasion, this majestic mountain chain has been admired by geologists since Charles Darwin (and before) and meticulously investigated as one of the pillars for the construction of the Plate Tectonics Theory (e.g. Isacks & Molnar, 1971; James, 1971). The intense research on this region allowed the characterization from processes that can be considered endogenous, as the development non-collisional orogenesis, shallow subduction, continental magmatic arcs, and associated metallogenesis, to the understanding of more exogenous processes, as the development of foreland sedimentary basins and the relationship between climate and orogenesis.

This orogenic system, the longest in the world with about 7500 km in length, is divided into the northern, central and southern segments (Fig. 1). The Central Andes segment is the longest one, and results of the interaction between the Nazca and South American plates. This segment displays the largest widths (up to 750 km), highest altitudes (up to 6961 m in Aconcagua Mount, and average altitude of 2500 m—Horton, 2018), and the largest amount of shortening and thickening of the Andes Mountains. As Wörner et al. (2018a) affirms, this segment constitutes a unique tectonic environment, being the only current example of an orogen with an orogenic plateau generated as a result of the convergence between oceanic and continental plates.

In this field guide we present a route for a cross-section in the Central Volcanic Zone (CVZ, Fig. 1) of the Central Andes, between Argentina (Purmamarca), to the east, and Chile (Tocopilla), to the west, that can be done in seven days. In this segment, the tectonic provinces of the Subandean Zone, Eastern Cordillera, N-Puna (orogenic plateau), Western Cordillera (Central Volcanic Zone), Pre-Cordillera (Cordillera Domeyko), Central Depression and Coastal Cordillera are crossed (Fig. 3), being possible to observe some of the main units and structures representing each domain. Despite the high elevations, this Andes segment can be traveled on paved roads and full geological sessions can be visited with large groups on tour buses. Between the years 2016 and 2019 we held a field course named "Geology of the Central Andes" with students from the Institute of Geosciences of the Federal University of Rio Grande do Sul, Brazil, aiming at developing knowledge about the Andean orogeny and providing the opportunity to analyze outcrops in an active orogeny. During the organization of the course, we perceived the difficulty in gathering specific information for the proposed itinerary, as well in choosing outcrops to visit, in particular considering bus access with capacity up to 50 people. We realized, therefore, the need to write this guide, gathering the main information of the geological context, location of the outcrops and information on the features to be observed in the proposed route, as well as the logistics of travel and accommodation for large groups.

Fig. 1 South American digital relief model (Shuttle Radar Topography Mission—NASA) with main divisions of the Andes Cordillera. The red dashed line indicates the position of the transect presented in this work. Main Andean volcanoes are shown as red triangles, while the major segments of the Andes (after Ramos, 1999, 2009) are: BFS—Bucamaranga Flat Slab; NVZ—Northern Volcanic Zone; PeFS—Peruvian Flat Slab; CVZ—Central Volcanic Zone; PaFS—Pampean Flat Slab; SVZ—Southern Volcanic Zone; PVG—Patagonian Volcanic Gap; AVZ—Austral Volcanic Zone

2 Geology of the Central Andes at the Central Volcanic Zone

The Central Andes is a segment of the Andes Cordillera between Guayaquil and Penas gulfs, where the oceanic Nazca Plate is subducting under the South America Plate. While subduction of the oceanic plate occurs along this entire segment, there are two regions of volcanic gap that characterize flat slab subduction (Peruvian and Pampean flat slabs), which flank the Central Volcanic Zone (Fig. 1) (Ramos, 1999, 2009). The CVZ presents the largest crustal widths and thicknesses of the Andes mountain range. For these reasons, this is one of the most investigated segments in the Andes Cordillera and a classic region for studying orogenic processes of non-collisional orogens, being target of numerous recent research and publications (see DeCelles et al., 2011; Horton, 2018; Kearey et al., 2009; Ramos, 2009; Wörner et al., 2018a, 2018b). Other notable features of the Central Andes are the Altiplano–Puna

Orogenic Plateau (the second largest in the world, after the Tibet Plateau in the Himalayas) and the Bolivian Orocline.

High topography is a marked feature of active orogenic belts, which cannot be observed in ancient orogenic belts. Therefore, it is interesting to observe, in this segment of the Andes, the relationships between the high topography and the associated crustal processes: where igninbritic volcanism occurs, how is the passage from the coastal mountain range to the sea, how does the relief controls the quaternary sedimentary deposits, where are the active andesitic volcanism, where are thrusts faults, etc. Important aspects of the section covered in this excursion are the large width of the orogenic zone, the presence of the orogenic plateau and the high mean elevation of the entire area, which indicates an important crustal shortening and thickening. This shortening is accommodated by several thrust faults that can be observed along the entire profile, and which affect Quaternary sedimentary deposits as well. It is important to notice that the Andean topography has evolved through time (Amilibia et al., 2008; Coutand et al., 2001; Schildgen & Hoke, 2018; Walk et al., 2020), as a result of the protracted plate convergence and variations on the subduction geometry along the South America margin.

The oceanic plate convergence and subduction under the continental margin started in the end of the Jurassic period, with the overriding continental plate evolving from a noncompressive/extensional regime to a predominantly compressive one in the Late Cretaceous (Fig. 2) (Kearey et al., 2009; Ramos, 2009 and references therein). From the Late Cretaceous until present time the predominant compressive stress in the overriding plate resulted in the Andean Orogeny (Megard, 1984). The Andean Orogeny evolved with migration of the orogenic front to the east, through deformational phases called Peruvian (107–65 Ma), Incaic (45–35 Ma) and Quechua (20 Ma until present) (Megard, 1984; Gianni et al., 2015, and references therein). Some authors consider the inception of the Andean Orogeny only at 23 Ma, i.e. in the Quechua Phase, which is related to the beginning of subduction of the Nazca Plate, originated from fragmentation of the extinct Farallon Plate (Horton, 2018).

The modern Central Andes segment is related to the Quechua Phase, and can be divided into domains that represent large morphostructural units with specific geological characteristics (Gansser, 1973; Kearey et al., 2009; Mpodozis et al., 2005). The domains crossed in the transects of Central Andes proposed by this guide, from the subduction zone to the interior of the continent, are shown in Fig. 3 and are known as: (i) Coastal Cordillera, (ii) Central Valley, (iii) Domeiko Cordillera (Pre-Cordillera), (iv) Western Cordillera (Cordillera Occidental), (v) Puna, (vi) Eastern Cordillera (Cordillera Oriental), (vii) Sub-Andean Belt and Santa Bárbara System. A brief description of the geological and tectonic evolution of each domain will be presented at the beginning of each session of this guide.

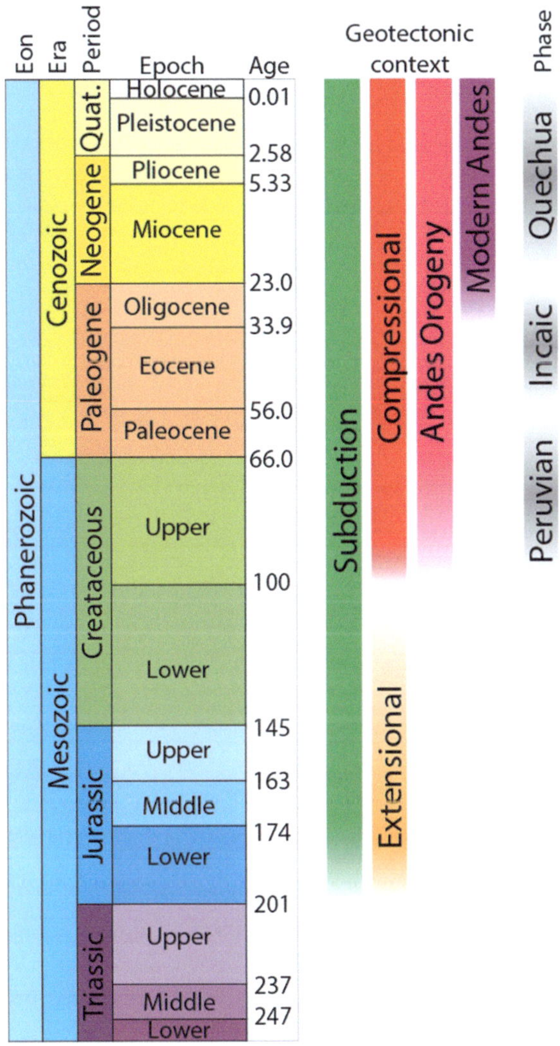

Fig. 2 Synthesis of the tectonic evolution of the Andes Orogeny, with main deformational phases indicated (according to Gianni et al., 2015; Kearey et al., 2009; Megard, 1984; Ramos, 2009)

3 Field Guide

In this field trip we pass through one of the most remote, least populated and arid regions on the planet. We cross peculiar landscapes representative of environmental extremes, landscapes built by the historical exploration of minerals and metals, sceneries that reveal catastrophic events such as earthquakes and tsunamis, and others forged by climatic phenomena, such as El Niño and La Niña, which challenge human societies. These regions are populated by nature species that mark the Andes identity, as llamas and cacti, which are deeply connected with the remarkable culture of the

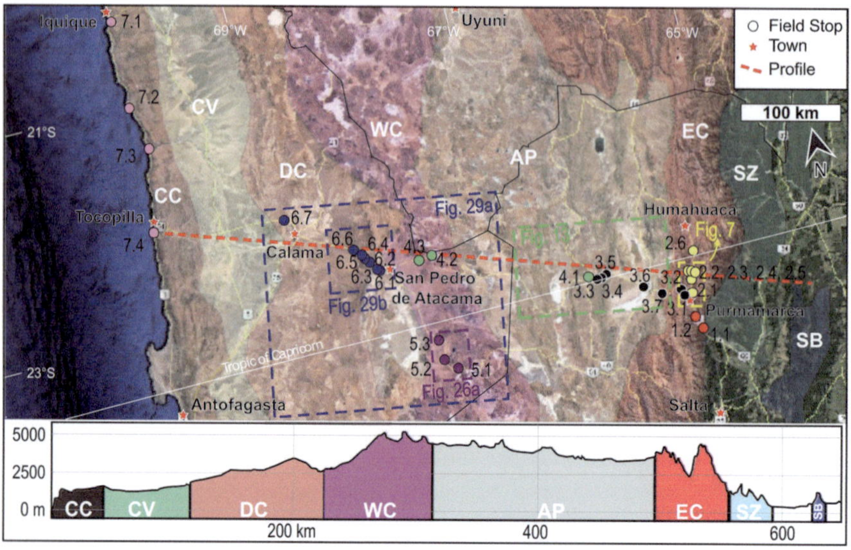

Fig. 3 Google Earth image and elevation profile showing the main divisions of Central Andes, after Kearey et al. (2009). Colored circles represent observation points of this guide, red stars some key towns, black lines international frontiers, yellow lines mains roads, and red dashed line the position of the schematic profile below. CC: Coastal Cordillera; CV: Central Valley; DC: Domeiko Cordillera; WC: Western Cordillera; AP: Altiplano Puna; EC: Eastern Cordillera; SZ: Subandean Zone; SB: Santa Bárbara

Andean people. Due to its peculiarities, some logistic aspects of field travel, hosting infrastructure and local customs will be presented to serve as references in this guide.

The field trip begins in San Salvador de Jujuy (AR), a town located at the edge of the Subandean Zone. Following the Ruta 9 to west, along the Quebrada de Humahuaca, it is possible to observe the significant landscape changes caused by the orogeny, from a fairly flat foreland basin to the high topography of the Eastern Cordillera, easternmost flank of the Altiplano–Puna Orogenic Plateau (Fig. 4). The best period for this field trip is during spring of the southern hemisphere, between October and the beginning of December, avoiding the rigorous winter as well the rainy summer. All the sectors can be traveled by paved or high-quality unpaved roads, suitable for big tour buses. Coordinates of each stop are given in the supplementary data (Table 1). Most stops are at altitudes of 2000 m above sea level, commonly at elevations near 3000 m, peaking at 4700 m, except the ones from the last day, that are along Chile coast. Large parts of the route are in unpopulated areas, and phone signal, gas and other services can be rare. Water scarcity and high altitudes sickness are the main challenges faced during the field trip, so extra attention must be paid to carry enough amount of drinkable water and to the potential altitude sickness (called *The Puna* in Argentina), for which the local tradition indicates chewing coca leaves (or drinking coca tea) as relief, which is easily found in all localities.

Fig. 4 Oblique Google Earth image showing the central and eastern main domains of Central Andes and the observation points of these domains (colored circles) Yellow lines represent the mains roads and red dashed lines the approximated limits of each domain. Note the alternation of N–S trending mountain ranges and the elongated salt flat basins on Altiplano Puna. Vertical exaggeration 3x

The arrival in Argentina for the beginning of the field trip can be done by plane in the town of Perico, which receives international flights and is located about 33 km (20 miles) southeast of San Salvador de Jujuy. Other possibilities are the cities of Salta and San Miguel de Tucumân, further south.

4 Subandean Zone

The Subandean zone has considerably lower elevations (around 1500 m) than the Eastern Cordillera (up to 5000 m; see Figs. 3 and 4). It was uplifted only in the Miocene, when the orogenic front progressively migrated from the Eastern Cordillera (west) towards the foreland (east) (Gianni et al., 2015; Scheuber et al., 2006). The Subandean Zone is considered a fold and thrust belt with thin-skin structures detached in Silurian strata during the Quechua Phase (e.g. Rojas Vera et al., 2019). Northwards, the Subandean Zone is structurally related with the Interandean Zone (thin-skin structures detached in Ordovician strata), which occurs between the Eastern Cordillera and Subandean Zone. The Interandean zone narrows towards the south (Rojas Vera et al., 2019), so it is not significantly exposed along our itinerary. Further south, the thin-skinned tectonics of the Subandean Zone is replaced by the thick-skinned tectonics in the Sierras Santa Bárbara. Subandean Zone basement units are represented mainly by Puncoviscana Formation and Paleozoic rocks, which are affected by anticlines bounded by thrust and backthrust systems.

Table 1 Summary of the location of all field trip stops

Domain	ID	Quad	Long	Lat	Z
Internandes	1–1	20 J	253,524	7,338,251	1625
Internandes	1–2	20 K	249,057	7,349,030	2170
Eastern Cordillera	2–1	20 K	248,129	7,371,900	2235
Eastern Cordillera	2–2	20 K	253,864	7,385,290	2438
Eastern Cordillera	2–3	20 K	254,771	7,390,551	2478
Eastern Cordillera	2–4	20 K	255,725	7,389,593	2560
Eastern Cordillera	2–5	20 K	257,056	7,388,680	2719
Eastern Cordillera	2–6	20 K	259,795	7,409,257	2705
Eastern Cordillera	3–1	20 K	245,732	7,371,522	2320
Eastern Cordillera	3–2	20 K	241,629	7,375,990	2614
N Puna	3–3	19 K	777,061	7,406,213	3791
N Puna	3–4	19 K	780,583	7,405,045	3645
N Puna	3–5	19 K	783,637	7,406,527	3557
N Puna	3–6	20 K	206,154	7,387,163	3412
N Puna	3–7	20 K	222,151	7,376,811	3788
N Puna	4–1	19 K	770,032	7,409,086	3697
N Puna	4–2	19 K	626,816	7,465,227	4749
Western Cordillera	4–3	19 K	615,576	7,464,271	4093
Western Cordillera	5–1	19 K	626,277	7,358,335	4183
Western Cordillera	5–2	19 K	616,606	7,368,039	3816
Western Cordillera	5–3	19 K	615,108	7,388,004	3494
Cordillera de la Sal	6–1	19 K	577,874	7,465,492	2592
Cordillera de la Sal	6–2	19 K	576,835	7,466,209	2616
Cordillera de la Sal	6–3	19 K	567,932	7,475,528	2788
Cordillera de la Sal	6–4	19 K	563,534	7,481,661	3280
Cordillera de la Sal	6–5	19 K	562,096	7,483,046	3380
Cordillera de la Sal	6–6	19 K	558,736	7,488,848	3372
Cordillera Domeiko	6–7	19 K	499,128	7,531,935	3329
Coastal Cordillera	7–1	19 K	382,059	7,755,501	20
Coastal Cordillera	7–2	19 K	380,417	7,670,639	10
Coastal Cordillera	7–3	19 K	390,246	7,630,091	5
Coastal Cordillera	7–4	19 K	374,617	7,553,751	6

First Day—Comprises two stops in the route between the town of San Salvador de Jujuy and the village of Purmamarca, through the Quebrada de Humahuaca (Humahuaca Basin) (Fig. 4). Purmamarca is the base camp for the first three nights of field work. This location was chosen because it is a central point for the route on day 2, to the north by the RN9, and for the route on day 3, to the west by the RN52 (international route to Chile). The town of Tilcara is also a possible lodging option. Purmamarca is a little tourist village with ca. 2000 inhabitants, with good infrastructure for lodging and groceries. The village blends elements of Andean culture with strong marks of Spanish colonization, such as the seventeenth century Jesuit church *Santa Rosa de Purmamarca*, around which the town was developed. There is a native marketplace in the town main square and several typical restaurants, where varieties of corn can be tasted, cereal grain which has cosmic significance for the Andean people *(regalo de los deuses)* and is considered as an essential component of the development of civilizations in South America (Echeverría & Muñoz, 1988). The main attraction of Purmamarca is the "Paseo de los Colorados", a trail to the colored rocks of Salta Group.

Stop 1-1

Feature: Rio Grande valley, Puncoviscana Fm, Quebrada de Humahuaca.
UTM: 20 J, 253,524, 7,338,251, 1625 m.
Directions: Ruta 9, Mirador de León.
Description: Panoramic view of the Rio Grande valley (Fig. 5), just before the Quebrada de Humahuaca. The Rio Grande is an anastomosing river that runs parallel to the Andes Cordillera and represents an intermountain drainage system (Pingel et al., 2013). Northwards from this point occurs the transition between the Subandean sierras and the Eastern Cordillera domains, marked by changes in the topography and vegetation. The Eastern Cordillera corresponds to a orographic barrier to the moisture coming from eastern South America, that causes a higher precipitation frequency in the Subandean/Interandean zones (Alonso et al., 2006; Schildgen & Hoke, 2018), which favour the development of a denser vegetation in the region. Across the road from the Mirador crops out the Puncoviscana Fm, composed here by brown phyllites overlain in an angular unconformity by coarse Quaternary deposits.

Stop 1-2

Feature: Alluvial cone Arroyo del Médio, Quebrada de Humahuaca.
Location: 20 K, 249057, 7349030, 2170 m.
Directions: Ruta 9, about 15 km northwards from stop 1-1.
Description: This point is within the Quebrada de Humahuaca sector. From here the large alluvial cone Arroyo del Médio can be observed (Fig. 6), formed mainly by debris flow associated with seismic activity in 1945, reinforced by the intense precipitation episodes during the rainy summer season (Kay et al., 2008). Another catastrophic landslide occurred in January 2017, which partially covered the Volcan Village a few kilometers to the north. The recurrent seismic activity and catastrophics landslides caused by episodes of intense precipitation

Fig. 5 The Rio Grande valley during the dry season (October), showing the exuberant vegetation typical of the lower ranges of subandean/interandean zones

in the region impose problems for human occupation, causing the deactivation of ferry lines and limiting the growth of towns. However, the gentle relief and abundant water resources favoured the development of the ancient civilizations in the past (Marcato et al., 2009). The adaptation to verticality is the evolutionary basis of the Andean people, settled on the large-scale movements between different ecological zones. Thus, ancestral routes (*quebradas*) are recognized as cultural itineraries. On account of that, the Quebrada de Humahuaca is, since 2003, an UNESCO World heritage site.

5 Eastern Cordillera

The Eastern Cordillera was an orogenic front during the Inca phase, beginning its rise at 40 Ma to 15 Ma (DeCelles et al., 2011; Gianni et al., 2015; Kay et al., 2008). This is an intensely deformed Andean domain, bordering the east limit of the Altiplano–Puna Orogenic Plateau and with crustal thickness reaching a maximum of 74 km behind the Bolivia Orocline (Allmendinger et al., 1997; Kearey et al., 2009). It represents a thick-skin fold-thrust belt, in which the thrust faults, with vergence both to east and west (Allmendiger et al. 1997), affect the regional basement in a typical piggy-back structural style (Kay et al., 2008). The change in thrust fault vergence from eastward only vergence, typical of the Interandean and Subandean zones, to the double vergence, characteristic of the Eastern Cordillera (Allmendinger et al., 1997) is related to reactivations of pre-existing structures of the extensional regime

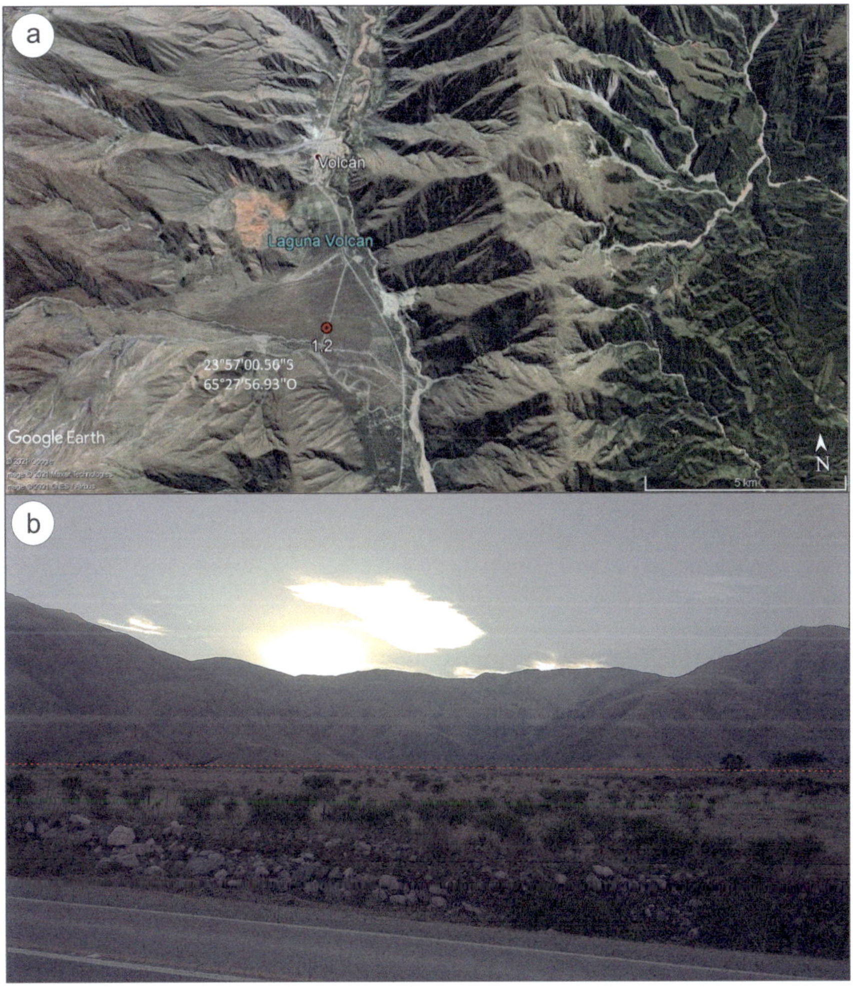

Fig. 6 Volcanic cone deposit. Image at the top shows the dimension of the cone. The location of the stop and direction of observation are marked. In the bottom photo, it is possible to observe a road side with the cone sediments. The dashed line marks the upper limit of the sedimentary package

active during the Late Cretaceous, which led to the formation of rift basins in which the Salta Group was deposited (see Gianni et al., 2015, and references therein).

The topography of the Eastern Cordillera in northern Argentina is marked by high mountains ranges (up to 5800 m) alternated with narrow and deep longitudinal valleys, with range to valley differential of altitudes from 1700 to 4300 m (Steinmetz & Galli, 2015). The uplift of the Eastern Cordillera began with the propagation of the Puna orogenic front during the Incaic phase, starting on the western portion of the mountain range, around 40 Ma, and migrating to the east at 25 Ma (DeCelles et al. 2011; Gianni et al., 2015), with rapid exhumation in the late Eocene and late Miocene

(Carrapa et al., 2014). During the migration of the orogenic wedge from the Domeiko Cordillera to the east, the current region of the Eastern Cordillera changed from a rift basin system (Grupo Salta) to a foreland system (with fordeep/bulge/forebulge), which also progressively migrated eastwards (DeCelles et al., 2011; Horton, 2018).

The Eastern Cordillera domain will be observed in two sectors: within the Quebrada de Humahuaca along road Ruta 9, to the north of Purmamarca; and in route along Quebrada de Purmamarca–Sierra Cobres, on road Ruta 52, to the west of Purmamarca (Fig. 7, stops 2-1 to 2-6 and 3-1 to 3-2). The first one, in the direction of Humahuaca town, beside the homonymous cliff, allows to observe varied structures and geological units of the Eastern Cordillera and of the Humahuaca Basin (Pingel et al., 2013). The second route, towards Susques (Fig. 7), cuts across the Cordillera Oriental towards the Puna Altiplano.

6 Quebrada de Humahuaca

Along the Quebrada de Humahuaca occurs the narrow intermontana Humahuaca Basin, a N–S elongated basin bounded by inverse faults, with the Sierra Alta Range in the west, and the Tilcara and Hornocal ranges in the east (Pingel et al., 2013; Streit et al., 2015). The ranges bordering the basin are characterized by basement rocks from the Puncoviscan Formation (Proterozoic-Early Cambria), and by the Méson (Cambrian) and Santa Victoria (Cambro-Ordovician) groups (Aceñolaza, 2003), which were tectonically stacked with Cretaceous-Paleogene rift-related units from the Salta Group (Cónsole-Gonella et al., 2017; Sánchez & Marquillas, 2010). This complex set of thrust faults and folds is characteristic of the Alta Ranges, uplifted at ca. 14 Ma (Deeken et al., 2006).

The Salta Gr was deposited in an intra-continental rift environment, associated with the extension of the South American crust during the formation of the Atlantic Ocean (cf. Ramos, 1999). More recently, an evolution in a synorogenic foreland rift environment was proposed by Gianni et al. (2015), correlating this basin with the Atacama Basin, in the Western Cordillera (Gianni et al., 2015; Mpodozis et al. 2005; Reutter et al., 2006). The Salta Gp occurs in seven sub-basins, of which the Tres Cruces basin occurs in the Purmamarca and Humahuaca quebradas (Cónsole-Gonella et al., 2017 and references therein). It comprises the Subgroups Pirgua, Balbuena and Santa Bárbara, which mark different stages in the evolution of the rift (Carrera et al., 2006; Marquillas et al., 2004). The sin-rift period is marked by alluvial and fluvial sediments, composed by conglomerates and red sandstones (subgroup Pirgua), and alkaline volcanic rocks (110 to 100 Ma) interspersed. The post-rift stage is registered by the sub-groups Balbuena and Santa Bárbara. The Yacoraite Fm (Sub-group Balbuena) is formed by yellow stromatolitic limestones, calcareous sandstones and greenish shales, deposited in a shallow epeiric system, during Maastrichtian–Danian (Marquillas et al., 2004).

The Humahuaca Basin is filled with the Maimará (Miocene–Pliocene), Tilcara (Plio-Pleistocene) and Uquia (Plio-Pleistocene) Formations and with Quaternary

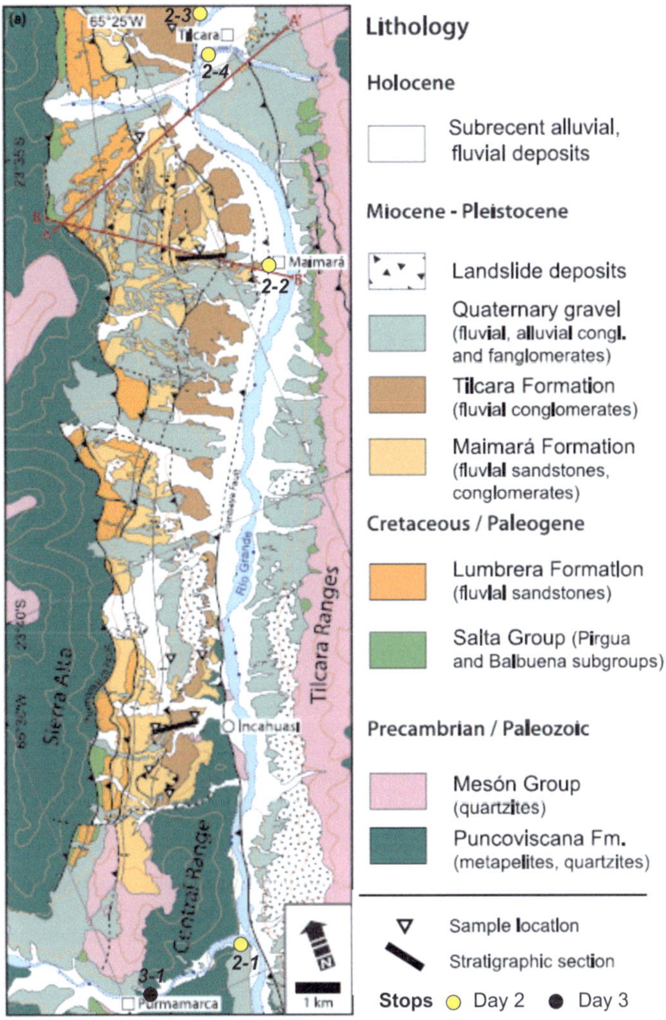

Fig. 7 Geological map of southern Humauaca Basin, with location of most of the stops from day 2 (excluded stop 2-6) and first stop of day 3 (modified from Pingel et al., 2013)

gravel deposits, all of which are deformed by a younger generation of thrusts faults (8.5 Ma to present day) (Pingel et al., 2013; Streit et al., 2015). The distribution of sediments in the Humahuaca Basin, variations on the paleocurrents and the young thrust faults suggest uplift of the Tilcara Range during the Neogene–Quaternary, with continued shortening in the basin and accompanied by the transition of a partially segmented foreland basin (Maimará Formation) to a fault-bounded intermontane basin (Pingel et al., 2013). A simplified map from the south Humahuaca Basin, from Pingel et al. (2013), is presented in Fig. 6.

Second Day—Comprises the route from Purmamarca Village towards Tilcara (Fig. 4). This day comprises stops along the main road (2-1, 2-2, 2-3 and 2-6) and also significant hiking (2-4 and 2-5), so it is imperative to be prepared for strong heat and Sun, and bring extra drinking water. The first stop (2-1), near Purmamarca, should preferably be done early in the day, aiming to take advantage of the morning light angle of incidence in the outcrops. To access stops 2-4 and 2-5, if travelling by bus, it is recommended to stop the vehicle in the Tilcara town entrance, and then walk about 1200 m through the village, from the bridge to the viewpoint of Huasamayo river. Small vehicles can get directly to this point. From the viewpoint to the stop 2-5 (*Garganta del Diablo*) it is a walk of approximately one and half hours.

Stop 2-1

Feature: Puncoviscana Fm.
Location: 20 K, 248,129, 7,371,900, 2235 m.
Description: In this location, on Ruta 52 south side near Purmamarca Village, it is possible to observe many features of the Proterozoic-Early Cambrian metasedimentary Puncoviscana Formation (Aceñolaza, 2003). This is one of the most expressive basement units cropping out in the Eastern Cordillera, in North Argentina, and can be easily distinguished from the surrounding ranges by its dark grayish green color. In this outcrop (Fig. 8). The metamorphic grade of the formation is considered low, but the metamorphism attains high grade in occurrences further south, in Tucuman and Catamarca provincies (see Aceñolaza, 2003; Escayola et al., 2011; Toselli et al., 2012). A well developed schistosity is observed parallel or at low angle with the S0, and tight to isoclinal folds affect both planar structures in outcrop scale. Deformation and metamorphism of the Puncovisana Fm is related to Pampean Orogenic Cycle, final stages of West Gondwana collage, in the early Cambrian, which closed the predominantly shallow marine Puncoviscana Basin (see Aceñolaza, 2003; Escayola et al., 2011; Toselli et al., 2012). Sedimentary features indicative of shallow marine environment, like wavy marks, marks of polyps and microbial mats described by Acenollaza (2003) can be observed in the sequence of green metasandstone-shales-slates of this outcrop, preferentially in the morning light (Fig. 7).

Stop 2-2

Feature: Maimará Fm and Tilcara Fm, Quebrada de Humahuaca.
Location: 20 K, 253,864, 7,385,290, 2428 m.
Directions: Ruta 9, Maimara town.
Description: Viewpoint on a hill near the Rio Grande River. In this location crops out units from the Maimará Fm, with characteristics of orogenic sedimentation. This unit is restricted to the Quebrada de Humahuaca and records the uplift and development of the Tilcara Range.

According to Pingel et al. (2013), the Maimará Fm. comprises ochre to yellowish beds of fluvial arkosic sandstones, interbedded with conglomerates with dominantly Paleozoic pebbles. The paleocurrents towards the east, perpendicular to the mountain

Fig. 8 Features of Puncoviscana Fm at locality 2-1, near Purmamarca. **a** General view of the metasedimentary package, which is tightly folded. The outcrop is ca. 10 m high and 20 m wide. **b** Detail of wavy marks (10 cm red pencil scale in the bottom right) and **c** microbial mats (20 cm wide)

strike, and the Miocene to Early Pliocene ages (5.92 ± 0.12 Ma and 4.18 ± 0.11 Ma—U–Pb zircon of volcanic ash layers) are characteristic of this formation. The change of paleocurrents towards the south, then parallel to the mountain strike, and Late-Pliocene ages (3.66 ± 0.20 Ma to 2.50 ± 0.10 Ma—U–Pb zircon of volcanic ash layers) are characteristic of Maimará Fm. The change in paleocurrent between the Maimará and Tilcara formations is interpreted as due to an evolution from foreland to intermontane basin configuration (Pingel et al., 2013).

Stop 2-3

Feature: Tilcara Fm.
Location: 20 K, 254,771, 7,390,551, 2478 m.
Directions: Ruta 9, Tilcara town.
Description: Road cut along Ruta 9, around 40 m high (Fig. 9a), where conglomerates of the Tilcara Fm are observed. The Tilcara Fm contrasts with the Maimará Fm due to the poorly consolidated strata, with dominant conglomerate and fanglomerate (and minor sandstones) composed by well-rounded, imbricated pebble to boulder-sized clasts, sourced mainly from the Proterozoic basement (Puncoviscana Fm). In this outcrop, the deposits comprise fluvial sandstones and poorly consolidated conglomerates, mainly with well-rounded quartzite clasts (in agreement with descriptions by Pingel et al. (2013). Layers show lateral discontinuity and are granulometry fining upwards (Fig. 9b). Clasts often present placoid shapes in which the major axis can reach up to 50 cm. Spatial arrangement of sediments suggests paleocurrents opposite to the layers dipping direction, that are tilted to NW with attitude (10/322), (dip/dip direction).

Stop 2-4

Feature: Maimará Fm. and Quaternary alluvial deposits affected by the Tilcara thrust fault.
Location: 20 K, 255,725, 7,389,593, 2547 m.
Directions: Av Huasamayo, Tilcara town.
Description: Cross-section along the Huasamayo River, where its alluvial sedimentation can be observed. In this location, the stratigraphy corresponds to five depositional sequences of Quaternary alluvial fans that overlies the Maimará Fm. These sequences are affected by a spectacular thrust fault with ramp-flat morphology and vergence to the east (Fig. 10a). The fault is responsible for thrusting the Maimará Fm red sandstone over Quaternary unconsolidated gravel (Pingel et al., 2013; Sancho et al., 2008). Looking in detail, it is possible to observe drag folds both in the hanging wall, affecting the sandstone, and in the footwall, affecting alluvial sediments, which confirms the top-to-the-E kinematics (Fig. 10b). Displacement estimates are between 15 and 20 m of vertical offset and ~40 m along the horizontal (Sancho et al., 2008). Geochronological studies indicate an age of 84.5 $\pm$ 7 ka for the Quaternary deposits, which suggests that compressive tectonics were active during the upper Pleistocene of the Eastern Cordillera (Sancho et al., 2008).

Stop 2-5

Feature: Puncoviscana Fm, Meson Gr, Tilcarian Unconformity.
Location: 20 K, 257,056/7,388,680, 2719 m.
Description: In this outcrop, at the entrance to the Garganta del Diablo, the Cambrian quartzites of the Gr Mesón occur in unconformity with the folded Puncoviscana Fm (Fig. 11). This surface is known as the Tilcarian Unconformity, which is associated with the Early Cambrian Tilcarian-Pampian orogenic

Fig. 9 Tilcara Fm conglomerate and sandstone outcrop on Ruta 9. **a** General view of the outcrop, which presents a fining upwards pattern on a broad scale. The yellow rectangle indicates the location of the detailed photo; **b** Close-up on a sandstone layer with pebble concentration in the bottom

cycle (Aceñolaza, 2003; Escayola et al., 2011; Toselli et al., 2012). The Meson Group comprises thousands of meters of thick massive sandstones and quartzites, with occasional conglomerate lenses (Aceñolaza, 2003). According to this author, the sequence is rich in fossiliferous occurrences, highlighting the presence of trilobites.

Stop 2-6

Feature: Purmamarca thrust fault.

Fig. 10 **a** Tilcara thrust along the Huasamayo river, SE–NW cross-section view; **b** detail of the thrust fault plane, with drag folds that confirms the apparent kinematics. Sketches are presented as insets

Fig. 11 Quartzites from the Meson Group overlay the Puncoviscana Fm, with the latter displaying folds under the Tilcarian Unconformity

Location: 20 K, 259,795, 7,409,257, 2705 m.
Directions: Ruta 9, after Huacalera town.
Description: This point allows a panoramic view of the Purmamarca Thrust Fault (Fig. 12).At distance, it is possible to observe thrusts and asymmetric folds on red pelites (Piriguá Fm) and yellow carbonates (Yacoraite Fm) from Salta Gp, thrusted over brown sediments (Uquia/Tilcara Fm). The greenish and gray rocks of the Puncoviscana Fm is seen in the background, thrusting over all these younger units. The main structure verges to the east.

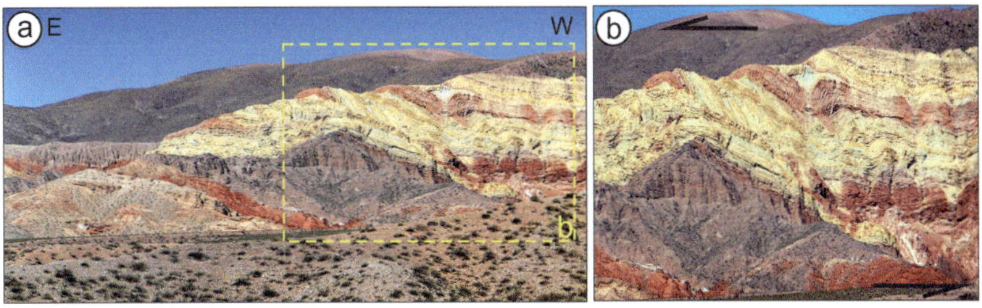

Fig. 12 **a** Purmamarca Thrust with vergence towards the E; **b** Colored rocks pertaining to Salta Group show asymmetric folds and thrust over brown-colored Uquia/Tilcara Formation. The folded red layer is at 3050 m, ca. 350 m high from the ground. The Proterozoic Puncoviscava Formation, in the background, thrusts over all these younger units

7 Complementary Visit Suggestion

In Tilcara one can visit the archaeological site *Pucará de Tilcara* (part of the Quebrada de Humahuaca World Heritage Site), located near the stop 2-4 and with easy access by foot.

7.1 Pucará de Tilcara

Pukara in Aymara and Quechua languages means "fortress", which stand out for their location in the landscape, situated in elevated areas next to rivers, clearly visible from nearby sites. The Pucará of Tilcara is the most important archaeological site of Quebrada de Humauaca (UNESCO), with records of occupation for > 10.000 years. It has features of a fortification built by the Omaguaca tribe (twelfth century), later occupied by the Incas (1470 AD), who incorporated their technology of working with rocks to fix dry stone walls, aiming for greater stability in case of seismic activity. In 1598, this site was conquered by the Spanish (Zaburlín, 2009). The location also has an Altitude Botanical Garden, where cactus (*cardones*) patches of ancient times are preserved.

7.2 Puna Orogenic Plateau

The Puna-Altiplano plateau is a major morphotectonic feature of the Central Andes, corresponding to the second largest and highest orogenic plateau in the world (elevation over 3600 m, around 700 $\times$ 200 km^2 in planar dimensions; Kay et al., 2008), behind only of the Tibetean Plateau. The Puna-Altiplano plateau shows differences from north to south, being generally divided into two internal sectors: Altiplano and

Puna (Allmendiger et al., 1997; Kay & Coira, 2009). While the Altiplano has a flatter topography, with elevations of 3600 m, the Puna has a more rugged topography and elevations above 4000 m (see Henríquez et al., 2020). Along the Puna, differences between Northern and Southern Puna are also noticed, mainly in the topography (higher average elevation in Southern Puna), in the thickness of lithosphere (thinner lithosphere in Southern Puna), and in structural, magmatic and sedimentological dynamics (see Kay & Coira, 2009; Kay et al., 2008).

The Northern Puna (Fig. 13) is the orogenic plateau segment between 21° S and 24° S (Kay & Coira, 2009) and is crossed during the itinerary proposed in this field guide. The high topography of Northern Puna is marked by the alternation of N–S trending mountain ranges with elongated salt flat basins in a "contractional basins and ranges" style (Fig. 4) (Allmendinger et al., 1997). This relief was originated by migration of the orogenic front from west, where it caused uplift and strong crustal shortening mainly in late Eocene (ca. 40–38 Ma), to the east, causing rapid exhumation from the Domeiko Cordillera to the Puna Plateau region in Oligocene, ca. 25 Ma (Carrapa & DeCelles, 2015; Henríquez et al., 2020), reactivating extensional structures and forming basement ranges (e.g. Del Cobre, Susques, Sierra ranges), that shoulders the sedimentary basin (e.g. Pastos Chico and Salinas Grandes basins; Steinmetz & Galli, 2015).

A distinctive feature of the Puna-Altiplano Plateau in relation to the Tibet, is the presence of a large volume of ignimbrites (greater than 8000 km^3), generated by a

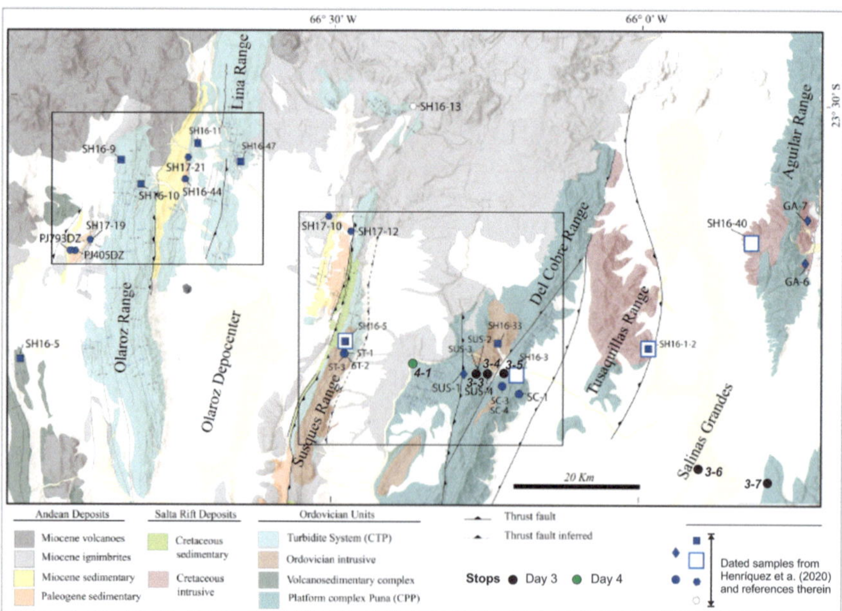

Fig. 13 Geologic map of the Northern Puna Plateau with most stops of day 3 and first stop of day 4 (modified from Henríquez et al., 2020)

set of large-scale caldera, between 23 and 1 Ma (de Silva & Kay, 2018; Kay & Coira, 2009; Kay et al., 2008). This magmatic flare-up event occurred after a period of arc magmatic inactivity, caused by flat slab subduction in the Central Volcanic Zone, in the early Miocene, followed by a slab roll-back, around 16 Ma, associated with lithospheric delamination and subduction of the Juan Fernández Ridge on the Nazca Plate (de Silva & Kay, 2018; Kay & Coira, 2009). During this period, a faster drift of South America Plate over Nazca Plate produced over 300 km of shortening on the Central Andes (Oncken et al., 2006), favoring eclogitization of the lower thickened crust (~50–70 km; de Silva & Kay, 2018).

The recurrence of variations on the subduction angle, with flat-slab periods, and the interaction of the resulting processes (crustal thickening, eclogitization of the lower crustal base, delamination, melting of the crust) are proposed as the causes of the Andean Orogenic Cycle (e.g. DeCelles et al., 2009; Ramos, 2009).

7.3 *Quebrada De Purmamarca—Del Cobre Range*

This route includes a section between the Eastern Cordillera and the Northern Puna. The itinerary starts in the Quebrada de Purmamarca, which extends from the Quebrada de Humahuaca to the west, cutting the Alta Range in a NW–SE direction. To the west of Alta Range, the direction of drainage reverses towards the interior of the Puna, and the high topography of the Eastern Cordillera gives way to the lower one of the Northern Puna. The segment of the route from Serra Aguillar to the west of Sierra Cobre, passing through Salinas Grandes, allows to observe typical features of the orogenic plateau. Along the way will be observed outcrops of the Santa Vitória and Salta groups, of the Cobres Plutonic Complex, as well as the Vizcachera Formation and several geomorphological and structural features of the Northern Puna.

Third Day—The route from Purmamarca towards Susques along the RN52 advances into increasing altitudes (Fig. 4), from ca 2300 m in stops 3-1 and 3-2 to 3800 m in Eastern Cordillera, and 3400 m in the Salinas Grandes Salar, therefore attention must be paid to the altitude sickness, temperature variations and high luminosity of the whiteness of the salar. The field stops can be done in sequence from east to west, finishing the day and staying overnight in Susques. Nevertheless, it is suggested to do the first stops (3-1 and 3-2) near Purmamarca, then go straight to the farthest point and return doing the sequence of stops on the way back to Purmamarca for the night. This is due to the high altitude (3600 m), low temperatures and the restricted lodging capacity for large groups in the village of Susques.

Stop 3-1

Feature: Puncoviscana Fm and Salta Gp thrust fault, Eastern Cordillera.
Location: 20 K, 245,732, 7,371,522, 2320 m.
Directions: Ruta 52, entrance of Purmamarca town.
Description: Viewpoint of thrust fault that places the Salta Gp (Cerro de Las Siete Colores) directly over the Puncoviscana Fm (Fig. 14). From this point towards

Fig. 14 Photomosaic panoramic view of a thrust fault juxtaposing the light colored Salta Gr rocks in the hanging wall with the greenish gray Puncoviscana Fm in the footwall, at Purmamarca entrance. The Cerro de Las Siete Colores standing out in the background

west, the Route 52 cuts through the magnificent Eastern Cordillera, presenting a sinuous and hilly trajectory with roadcuts and viewpoints up to the Puna Plateau. The diverse and outstanding geomorphology features of this Andes sector, which supports a continuous regional uplift until the present day, are well described by May (2008).

Stop 3-2

Feature: Yacorite Fm/Salta Gr.

Location: 20 K, 241,629, 7,375,990, 2614 m.

Directions: Ruta 52.

Description: At this stop it is possible to observe sedimentary features of Yacorite Fm, folded by Eastern Cordillera deformation (Fig. 15). In addition to stromatolites, palynomorphs and invertebrate fossils, the Yacorite Fm is well known for the presence of dinosaur tracks, which, together with sedimentary structures like desiccation cracks, are indicative of carbonate deposition in a coastal lagoon (Cónsole-Gonella et al., 2017 and references therein). In the visited outcrop, stromatolites and sedimentary features such as wavy marks and desiccation cracks can be observed.

Stop 3-3

Feature: Vizcachera Fm.

Location: 19 K, 777,061, 7,406,213, 3791 m.

Directions: Ruta 52.

Description: The Upper Vizcachera Fm and its Miocene sediments are observed in this point (Seggiaro et al., 2015), located at the Pastos Chicos–Coranzulí basin, west of the Del Cobres Range. This unit is unconformably covered to the north by ignimbrites of the Caldera Coranzulí. In this outcrop, a meter-scale trough cross-bedding sandstone layer occurs slightly tilted to the west (Fig. 16). The Vizcachera Fm represents a change from foreland basin to development of intramontane basins, with inversion of normal faults (Seggiaro et al., 2015).

Stop 3-4

Feature: Cobres Fm.

Fig. 15 **a** Yacoraite Fm, affected by open fold with high angle hinge, and the sedimentary structures seen in the outcrop (ca. 50 m wide, view to Azimuth 160°). **b** Desiccation cracks (10 cm pencil scale); **c** wavy marks, **d** stromatolites (40 cm hammer scale; photo minor axis has ca. 30 cm)

Location: 19 K, 780,583, 7,405,045, 3645 m.

Directions: Ruta 52.

Description: Following to the east, we start to cut the inverted/uplifted topography of Cobres Range. The Cobres Range is delimited by west-vergent (west flank) and two east-vergent reverse faults (east flank), uplifted by propagation of the Andean thrust front into the Del Cobre Range in late Eocene to early Oligocene time (ca. 37–31 Ma; Henríquez et al., 2020). The double vergent reverse faults that shoulder the range do not appear in surface, but seismic reflection profiles show that they are at accommodated horizontal shortening at the edges of the intermontane basins, such as the Salinas Grandes (Coutand et al., 2001; Steinmetz & Galli,

Fig. 16 General view of Vizcachera Fm sandstone in stop 3-3 with meter-scale trough cross-bedding, covered by Quaternary alluvial sediments (ca. 30 m wide, view to Azimuth 330°)

2015). Inside the Cobre Range, Ordovician quartzite and shale sequence of the Cobres Group, part of Complejo de Plataforma de la Puna (Henríquez et al., 2020; Seggiaro et al., 2015), also proposed as being of the Santa Victoria Group (Steinmetz & Galli, 2015), are folded, with wavelength depending on the quartzite/shale proportion (Henríquez et al., 2020). Along the path of the 52 Route the sequence with quartzites dominant over shales (Fig. 17) is open folded in km-scale.

Fig. 17 Quartzite sequence of Cobre Range with upright axial plane cleavage viewed in a NW–SE section. Geologist for scale is approximately 1.75 m

Stop 3-5

Feature: Syntectonic Plutonic Complex.
Location: 19 K, 783,637, 7,406,527, 3557 m.
Directions: Ruta 52.
Description: This point is located on the east flank of the Del Cobre Range and presents Paleozoic plutonic rocks which, along with metasediments, represent the basement of North Puna. Nearby, west from the outcrop, occurs one of the two east vergent thrust faults that delimit the east side of the range. About 10 km to east, the second hidden east vergent thrust fault limits the Del Cobres Range from the Salina Grande basin (Henríquez et al., 2020). According to Kay et al (2008) these rocks are part of the Syntectonic Plutonic Complex of Ordovician age (Kirschbaum et al., 2006), composed by plutons of granodiorite and monzogranite formed during a tectonomagmatic event 476 Ma ago (Haschke et al., 2006; Lork and Bahlburg, 1993). Such event represents the collision of the Precordillera terrane (Cuyania), originated from Laurentia, with the Gondwana Supercontinent (Seggiaro et al., 2015). The outcrop presents the Quepente Granodiorite (Seggiaro et al., 2015) intensely fractured by N–S upright fractures, with east and west dip (Fig. 18a). This granodiorite occurs as small N–S elongated bodies, intrusive in the Cobres Group during an extensional tectonic event in the Early Ordovician, with a ductile fabric produced during the compressive Ocloyic Phase of Famatinian orogeny, that caused regional metamorphism and deformation (Fig. 18b; Seggiaro et al., 2015).

Stop 3-6

Feature: Salina Grande.
Location: 20 K, 206,154, 7,387,163, 3412 m.
Directions: Ruta 52.

Fig. 18 Cobres Range east flank **a** Quepente Granodiorite intensely fractured (ca. 15 m wide, view to Azimuth 200°); **b** ductile oriented fabric Quepente Granodiorite marked by orientation of matrix biotite, elongated quartz and feldspar (10 cm pencil scale)

Description: On the northern part of the salar, this point allows a 360 degrees view of the Salinas Grande salt flat (Fig. 19a), which is the top of the southern part of the Guayatayoc–Salinas Grandes Basin. This is an immature clastic salar (Houston et al., 2011) with important deposits of borates, ulexite and borax (Seggiaro et al., 2015), from which evaporites, halite and borates, as well as potassium- and lithium-bearing brines are mined (Steinmetz, 2017). The basin is delimited by basement ranges of the Eastern Cordillera, to the east, and the Del Cobres Range, to the west, corresponding to a N–S oriented depocenter in which salt (mostly halite) is precipitated (Kay et al., 2008; Steinmetz, 2017; Steinmetz & Galli, 2015). It is one of many salars (salt flats) that are typical of Central Andes, which are a striking evidence of the strong aridity established in the internally drained plateau after the East Cordillera uplift. Contraction cracks with diameter up to 2 m are common in the salt (Fig. 19b), while artificial trenches of blue waters are made for salt precipitation, forming halite stalagmites (Fig. 19a). Seismic models indicate that inverted reactivation of high-dip normal faults systems, bounding the basin, controlled the synorogenic sedimentary filling (Coutand et al., 2001; Steinmetz & Galli, 2015). The boundary faults are covered by the latest sediments of the basin and quaternary alluvial fans. The endorheic and aridity condition of the basin, together with weathering and geotherm springs in the boundary ranges, are responsible for the boron–lithium system present in Guayatayoc–Salinas Grandes (Steinmetz, 2017). The framework of this salar is characterized by Strecker et al. (2012) as typical of orogenic plateaus, with various isolated depocenters, separated by basement highs.

Stop 3-7

Feature: Tucsa Range, Eastern Cordillera.
Location: 20 K, 222,151, 7,376,811, 3788 m.
Directions: Ruta 52.
Description: Eastwards from the previous point, occurs the limit between the Guayatayoc–Salinas Grandes basin and the Tucsa Range. The limit between these sectors is characterized by reverse and thrust faults, 1 km to the west of this point, in a similar way as observed in the other flank of the salar (Stop 3-5).Asymmetrical folds affected the S0 of Santa Victoria Group (Fig. 20), in agreement with thrust vergence sense to SW (Gonzalez et al., 2003).

8 Calderas in North Puna

Large ignimbrite calderas with tens of km in diameter are a striking feature of Northern Puna (Figs. 21 and 22). A synthesis of the geology and magmatic evolution of the Northern Puna calderas is presented by Kay and Coira (2009). The calderas are the result of an abundant generation of magma between ca. 16 and 0.5 Ma, which occurred after a period of absence of magmatism between ca. 30 and 16 Ma, associated with a Nazca Plate flat-slab subduction phase. The onset of the magmatic

Fig. 19 Typical aspects of Salar Grande. **a** Looking towards SSE view of a trench made for mining salt, within the remarkable flat landscape of Salar Grande. At the background, the Eastern Cordillera, with its highest peak Nevado de Chañi (5.896 m) (50 cm long hammer scale). **b** Contractional cracks filled by salt (10 cm pencil scale)

Fig. 20 Asymmetrical folds affecting S0 of the Santa Victória Group with a vergence towards the SW observed in a NE-SW section. Geologist for scale is approximately 1.8 m

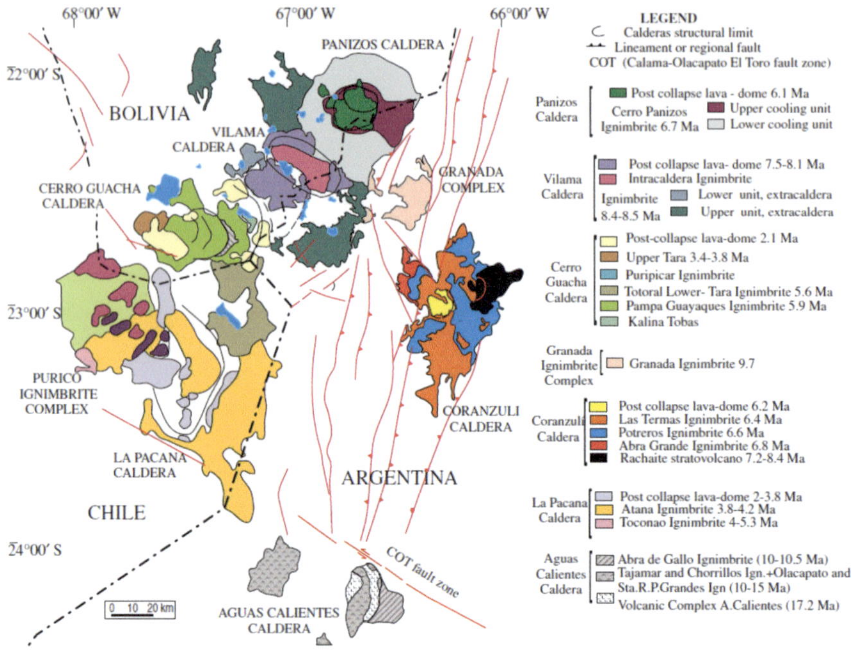

Fig. 21 Simplified map of Northern Puna Calderas (from Kay et al., 2008)

flare-up that originated the calderas occurred after the roll-back of the Nazca Plate, followed by delamination of the crust base in four stages: initially with formation of small centers with wide distribution (17–10 Ma) and later (9–5.5 Ma; 5.6–3 Ma; > 3 Ma) with progressive concentration for the modern arc region during this evolution (Kay & Coira, 2009).

Fourth Day—This day comprises travelling from Purmamarca, Argentina, to San Pedro de Atacama, Chile, the hosting site for the next two nights. We leave Purmamarca straight away to the first stop (4-1) near the village of Susque and the last field stop in Argentina. The Argentina–Chile border is crossed in Paso de Jama, near the small settlement of Jama (AR), ca. 4100 m altitude, where there are snack shops and gas station. In the border control there is joint exit/entry processing by Argentine and Chilean authorities. It is recommended to arrive early, because customs procedures can be time consuming. Remember that fresh food and seeds are not allowed to enter Chile.

On the route of the RN52, near San Pedro de Atacama, the road goes through the Los Flamencos National Reserve, created in 1990 to protect the rich diversity of flora and fauna of the Atacama Desert. San Pedro de Atacama, a very touristic town and has a large and varied infrastructure for lodging and groceries. Located at the same altitude as the Quebrada de Humahuaca, but on the western side of the Eastern Cordillera, this old town with dusty streets and adobe houses is an example of the desert culture of the dry north Chile.

Fig. 22 Oblique Google Earth image showing central-western main domains of Central Andes and the observation points of this guide (colored circles). Blue line marks the international border, yellow lines represent the main roads and red dashed lines the approximated limits of each domain. Note the N–S magmatic arc of the Western Cordillera (Central Volcanic Zone). Vertical exaggeration 3x

Stop 4-1

Feature: Ignimbrites of Conranzuli Caldera.
Location: UTM 19 K 770,032/7,409,086; 3697 m.
Directions: Ruta 52, near to Susques.
Description: This stop is located in the interior of the Pastos Chicos Basin, which is limited by the Del Cobres Range, to the east, and by the Susques Range, to the west, both bordered by inverse faults. About 45 km to the north of this Miocene basin, the Caldera Coranzulí formed four units of dacitic ignimbrite flows at ca. 6 to 7 Ma, which covered the sediments and basement of the basin (see Seggiaro et al., 2019). This caldera is an example of the second stage of Neogene magmatic evolution of Northern Puna, which is characterized by development of calderas with intermediate to giant size in arc-backarc position (Kay & Coira, 2009). N-, NE- and NW-trending faults of the Coranzulí caldera basement controlled its evolution and collapse (Seggiaro et al., 2019). In the outcrop, two layers of the second ignimbrite unit (see Guzmán et al., 2020 and Seggiaro et al., 2019) of the Coranzulí Volcanic Complex are deposited unconformably over the Cobres Group meta-sediments (Fig. 23). The lower and upper layers show lithoclast of ignimbrite, and the upper presents metasedimentary lithoclasts.

Stop 4-2

Feature: Ignimbrites of Purico Caldera.
Location: 19 K, 626,816, 7,465,227, 4749 m.
Directions: Ruta 27, road stop.

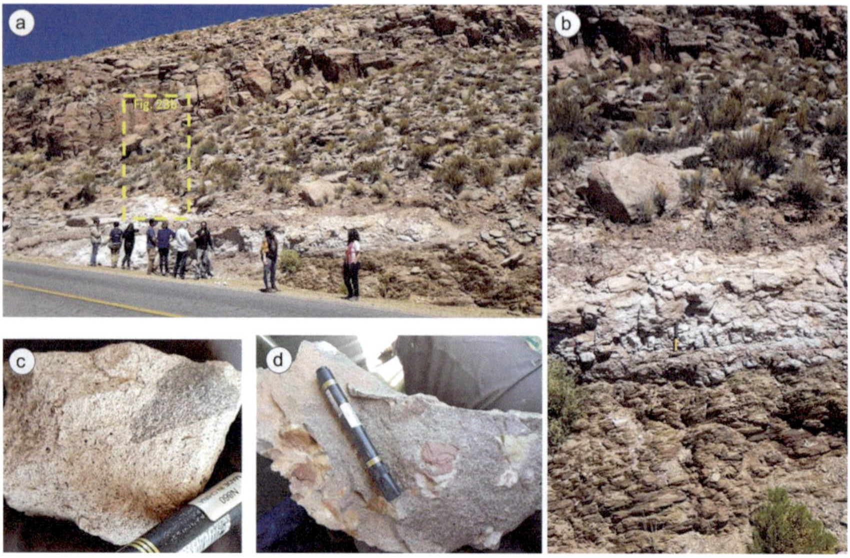

Fig. 23 Outcrop of ignimbrites of the Coranzulí Volcanic Complex (ca. 12 m wide, view to Azimuth 35^{5o}). **a** General appearance of the outcrop where two layers of ignimbrite are observed on the Cobres Group metasediments. It is observed that the whole package is slightly tilted to the west; **b** outcrop detail, showing the discordant contact of the lower ignimbrite layer on metasediments; **c**, **d** samples of the upper (right) layers, with lithoclasts of the metasedimentary rocks, and lower (left), with ignimbrite lithoclasts

Description: This is the first stop in Chile. From the Argentina border until this stop, we cross from E to W, three ignimbrite calderas (Figs. 21 and 22): the Cerro Guacha (5.9–2.1 Ma), La Pacana (5.3–2 Ma) and Purico (1.3 Ma). While Cerro Guacha and La Pacana calderas are from the third magmatic evolution stage of Northern Puna, the Purico Caldera is representative of the fourth stage described by Kay and Coira (2009). The three calderas, together with Coranzuli Caldera of the previous point, indicate progressively younger ages to W, suggesting magmatism migration from Northern Puna, since Miocene to Pliocene, towards the modern arc, at the Central Volcanic Zone (Kay & Coira, 2009; Kay et al., 2008). The Purico-Chascon Volcanic Complex ignimbrites erupted from ~1 Ma to 0.2 Ma and records the transition from ignimbrite flare-up (Purico Ignimbrite) to steady state arc volcanism (Cerro Chascon dacitic lavas) in Northern Puna (Burns et al. 2015). Three layers of the Purico ignimbrite form an approximately circular apron, over which dacitic domes ("D", El Cerillo, El Chasco), occur near the vent of ignimbrite eruptions (Schmitt et al., 2001) (Fig. 24). At this point, on the roadside, older Atana ignimbrites of La Pacana Caldera are covered by layers of Purico ignimbrite in the apron of the caldera ca. 200 m to the south (Fig. 24a). From a higher position at the Purico apron it is possible to have a better view of the huge Cerro Guacha Caldera, to the north. This point also marks the transition to Western Cordillera Domain, which corresponds to the active volcanic arc. The first arc volcanoes that can be seen from this point are the Juriques and Licancabur volcanoes (Fig. 24c), to the NW.

Fig. 24 Purico Caldera **a** layer of Purico ignimbrite over Atana Ignimbrite (ca. 60 m wide, view to Azimuth 164°); **b** Folded Ignimbrite (40 cm long hammer scale); **c** View to Azimuth 307°of stratovolcano chain of the CVZ magmatic arc—Western Cordillera) with Juriques and Licancabur volcanoes in front plane; **d** Google Earth image with view of south of D Dome and Caldera Purico

9 Western Cordillera

The Western Cordillera is the set of N–S high ranges that delimit the western edge of the Puna–Altiplano Orogenic Plateau (Fig. 22). This mountain range hosts the current magmatic arc of the Central Volcanic Zone, a strip of stratovolcanoes under the age of 3 Ma, with volcanoes which height that can be over 2000 m, defining summits elevations of more than 6000 m in the Western Cordillera (Wörner et al., 2018a, 2018b). The Western Cordillera began its elevation during the Incaic phase (50–30 Ma) by westward thrusting (Armijo et al., 2015). Its western limit is given by the West Andean Thrust, which, like other Incaic structures, is hidden by ignimbrite and sedimentary deposits (Armijo et al., 2015). Deformation during the Late Pliocene–Quaternary is indicated by coeval extension and contractional structures, that affect ignimbrites and lavas flows from the magmatic arc (Tibaldi & Bonali, 2018). According to these authors, compressional structures and crustal thickening is registered at lower altitudes (average of 3900), while crustal extension, with normal failures induced by orogenic collapse, is observed at higher altitudes (average of 4500 m).

The generation of arc magmatism by melting the asthenospheric wedge is caused by the release of fluids due the progressive metamorphism of the subduction plate, so that the distance of the arc in relation to the trench is defined by the subducting plate dip angle. The position of the modern CVZ magmatic arc, therefore, reflects the subduction angle of the Nazca Plate since the Pliocene, after the Nazca Plate flat-slab subduction period, when the absence of asthenospheric wedge over the subducted plate prevented the production of arc magmatism (Altiplano flat slab at 40–32 Ma and 27–18 Ma, and Puna flat slab at 18–12 Ma, according to Ramos & Folguera, 2009). The long history of subduction under the Andean margin is indicated by the occurrence of older magmatic arcs, of Jurassic (Coastal Cordillera) and Ordovician (Cordillera Domeiko) ages, whose positions, displaced in relation to the modern arc, indicate a change in the angle of subduction through time (e.g. Ramos, 2009). The tectonics during the development of each arc were different, varying from an extensional regime to neutral and, currently, compressional. The modern Pliocene–Quaternary arc, currently active in the Western Cordillera, developed after significant crustal thickening and marks the change from the ignimbrite flare-up magmatic regime, associated with the extreme melting promoted by the lithospheric delamination, to a steady-state regime (Kay & Coira, 2009; de Silva et al., 2015, 2018; Wormer et al., 2018b). While the flare-up regime is characterized by ignimbrites of acidic composition and associated with large calderas, the steady-state regime is typified by magmatic andesitic lavas and dacitic compositions associated with stratovolcanoes (Wormer et al., 2018b).

Stop 4-3

Feature: Licancabur and Juriques volcanoes.
Location: 19 K, 615,576, 7,464,271, 4093 m.
Directions: Ruta 27, road stop.

Description: In this stop we have a nice panoramic view, to the North, of Licancabur (west) and Juriques (east) stratovolcanoes, related to the modern volcanic arc (Fig. 25), that rest on the Purico ignimbrites. Licancabur has the most preserved volcanic cone and 3 packages of andesitic–dacitic lava flows, formed mainly after the last glaciation, that are easily identifiable (Figueroa et al., 2009). In its summit crater (6017 m) occurs the highest lake on Earth, that presents zoo-planktons (Tsyganov et al., 2017). The Juriques volcano is relatively older, with the top of the crater at 5704 m. It is noteworthy that Licancabur ("people's mountain" in kunza language, from Atacama) was an important ceremonial site for the Incas, as it preserves a complex of archeological sites between its summit and the Laguna Verde (Green Lake), in the Bolivian side of the border (Aros & Galeno-Ibaceta, 2017).

Fifth Day—This day is dedicated to the path of volcanoes through the RN28 (Figs. 22 and 26a). It is suggested to go directly to the furthest stop to the south, and then continue the sequence from there back to San Pedro de Atacama. The stops are near the road with short walks. In these high altitudes, temperatures are low, winds can be very strong and residual snow in the volcanoes flanks can be present.

Stop 5-1

Feature: Miñiques Volcano.
Location: 19 K, 626,277, 7,358,335, 4183 m.
Directions: Ruta 23, road stop.
Description: Along the route (Fig. 26a) it is possible to observe the aligned chain of stratovolcanoes of the modern volcanic arc (CVZ). This stop is in a lava lobe

Fig. 25 Panoramic view of Licancabur (6017 m) and Juriques (5740 m) volcanoes resting on Purico ignimbrites (view to North). The volcanoes are 5–6 km away from this point

Fig. 26 **a** Simplified map from Miñiques Volcano e deformational structures (Tibaldi & Bonali, 2018) with field stops 5.1, 5.2, 5.3 location; **b** panoramic view of Miñiques Volcano to the azimuth 330. The lava flow visited in point 5-1 can be partially seen in the left side of the volcano; **c** blocky lava lobes at the south flank of the volcano; **d** aspects of the lava flow at stop 5-1

located at the south flank of the Miñiques Volcano (Fig. 26b), a large basaltic-andesite to dacitic stratovolcano considered to have been active from the Pliocene to the Holocene (González-Ferrán, 1995). This Quaternary basaltic–andesite lava flow (Tibaldi & Bonali, 2018) forms a jumble of irregular centimetric to metric blocks of lava, with smooth, planar or slightly curved, and angular surfaces (Fig. 26c). The rock has plagioclase phenocrysts in a fine-grained matrix and small and sparse vesicles filled by epidote (Fig. 26d).

Stop 5-2

Feature: Miñiques lava flow.
Location: 19 K, 616,606, 7,368,039, 3816 m.
Directions: Ruta 23, road stop.
Description: This stop is located west of the N–S Miscanti Ridge, which is visible from this point, as well as the west face of the Miñiques stratovolcan (Fig. 27a). The Miscanti Ridge is one of the conspicuous N–S trending compressional ridges and reverse fault, that deforms the Pliocene–Pleistocene ignimbritic and volcanic flows (González et al., 2009). According to these authors, the Pliocene basaltic-andesite lava flow of this stop has spatial and temporal relationship with the development of the Miscanti Ridge, and present evidence of coeval compressional deformation and volcanism in this region. The rocks are exposed as morphologic highs (3–5 m high) of coherent and blocky lava flows (Fig. 27b, c). Levels of massive jointed cores and centimetric toes of pahoehoe lava are observed (Fig. 27d). The rock is porphyritic with plagioclase phenocrysts within a very fine-grained matrix, and a variable content of vesicles. Looking from this stop towards 290°, the hill on the opposite side of the road (Ruta 23) represents a thrust front verging east (Fig. 27e).

Stop 5-3

Feature: Ignimbrite Flows.
Location: 19 K, 615,108, 7,388,004, 3494 m.
Directions: Ruta 23, road stop.
Description: This stop at the east side of the road is a large outcrop (Fig. 28a) of whitish ignimbrite rock, composed predominantly by fragments of highly vesiculated rock, individual crystals of plagioclase, pyroxene, biotite, and lithic fragments of andesite and pumice, with the occurrence of native copper (Fig. 28b). The Pliocene–Pleistocene ignimbrite flows are the base for the Pliocene to Holocene andesitic to dacitic stratovolcanoes and the Pleistocene andesitic monogenetic volcanoes. According to González et al. (2009), the rocks in this area correspond to the 3.1–3.2 Ma Tucucaro-Patao Ignimbrite.

10 CVZ Forearc

To the west of the Western Cordillera—volcanic arc (over 5000 m elevation), the topography becomes progressively lower, until reaching sea level on the Pacific

Fig. 27 Miscanti Ridge and Miñiques lava flow: **a** view of the Miscanti Ridge in front of the Miñiques Volcano in point 5-2; Geologist in frontal plane is approximately 1.75 m; **b** general blocky aspect of the lava flow (ca. 60 m wide, view to Azimuth 104°); **c** rounded fronts of more coherent rock; **d** thin vesiculated flow with centimetric pahoehoe lava crust (yellow rectangle in b); blocks are 70 cm long; **e** view from the stop 5.2 to the azimuth 290° of a reverse fault front (100 m high) with tectonic vergence to east. The red bus (ca. 4 m high) is parked in Ruta 23

Fig. 28 **a** General view of ignimbrite exposition at stop 5-3; **b** aspects of the whitish vesiculated rock

Ocean coast, ca. 250 km distant. This domain represents the CVZ forearc (Armijo et al., 2015) and is marked by the alternation of depressions (Salar de Atacama and Central Depression) and mountain ranges (Domeiko Cordillera and Coastal Cordillera). During the Andean Orogeny, the structural and morphotectonic evolution of sectors of this domain had a strong influence of the structures inherited from the Mesozoic period, both extensional and transcurrent (e.g. Allmendinger & Gonzáles, 2010; Amilibia et al., 2008; Armijo et al., 2015; Lopez et al., 2019, 2020; Mpodozis et al., 2005; Reutter et al., 2006).

11 Salar de Atacama and Cordillera de la Sal

From the Western Cordillera to the west occurs the Atacama Salt Flat (Figs. 22 and 29), which marks the Salares Depression (or Preandean Depression), delimited from the Cordillera Domeiko by a narrow region with compressional features, represented by the Cordillera de la Sal, Llano de La Paciencia, Barros Arana Sinclyne and El Bordo Escarpment (Mpodzis et al., 2005; Arriagada et al., 2006; Lopez et al., 2020, among others). These structures exposed most of the lower part of the Salar de Atacama Basin infill during uplift periods of the Peruvian and Incaic phases (65–50 Ma and 50–28 Ma), due to migration of orogenic front to east of Domeiko Cordillera (Henríquez et al. 2019), with compressional tectonics continuing during the Pliocene–Quaternary (Arriagada et al., 2006; López et al., 2020).

The Salar de Atacama Basin contains up to 8 km of sediment (Arriagada et al., 2006), accumulated initially in a Mesozoic rift basin that evolved into a foreland basin, after tectonic inversion processes caused by the migration of the orogenic front (López et al., 2020; Mpodozis et al., 2005). The basin stratigraphy, composed of Upper Cretaceous–Oligocene sediments, is described by Mpodozis et al. (2005). According to these authors, the Purilactis Group (Tonel, Purilactis, Barros Arana and Totola formations) from the Upper Cretaceous to Paleocene, is covered in unconformity by the Orange Fm during the Paleocene–Eocene. The Loma Amarilla Fm, formed by thick blanket of syntectonic gravels from the Eocene–Oligocene and deposited in angular unconformity on the Fm Orange, is indicative of the uplift during the Incaic phase (Mpodzis et al., 2005). From the late Oligocene to early Miocene occurred the deposition of the Tambores formations on the uplifted areas of the basin, while the San Pedro Formation (synextensional evaporites) marks an extensional episode from ~28 Ma to the Middle Miocene (Henríquez et al., 2019; Rubilar et al., 2017).

Sixth Day—comprises the displacement from San Pedro de Atacama to Tocopilla towns, reaching the Pacific coast, the hosting site for the last 2 nights.

Tocopilla is a small port city that had its greatest development when saltpeter exportation in Chile was at its highest point. Currently, Tocopilla is an outflow port of ore from the Chuquicamata Mine and of saltpeter from El Toco. The city has a limited capacity for hosting. Some coastal inlets form pleasant artificial beaches, such as Balneario Covadonga and Balneario Caleta Boy.

Stop 6-1

Feature: San Pedro Fm., Cordillera De la Sal. Folds and faults.
Location: 19 K, 577,874, 7,465,492, 2592 m.
Directions: Ruta 23, Vale de la Luna panoramic viewpoint.
Description: From this stop begins the Cordillera de La Sal. To the E it is possible to observe the relatively low and flattened topography from the Atacama Salar and the Salar Depression. A double plunging asymmetrical mega antiformal fold can be observed, which is a major structure associated with the previous thrusting event (Fig. 30a). According to López et al. (2020) these mega folds involve basement

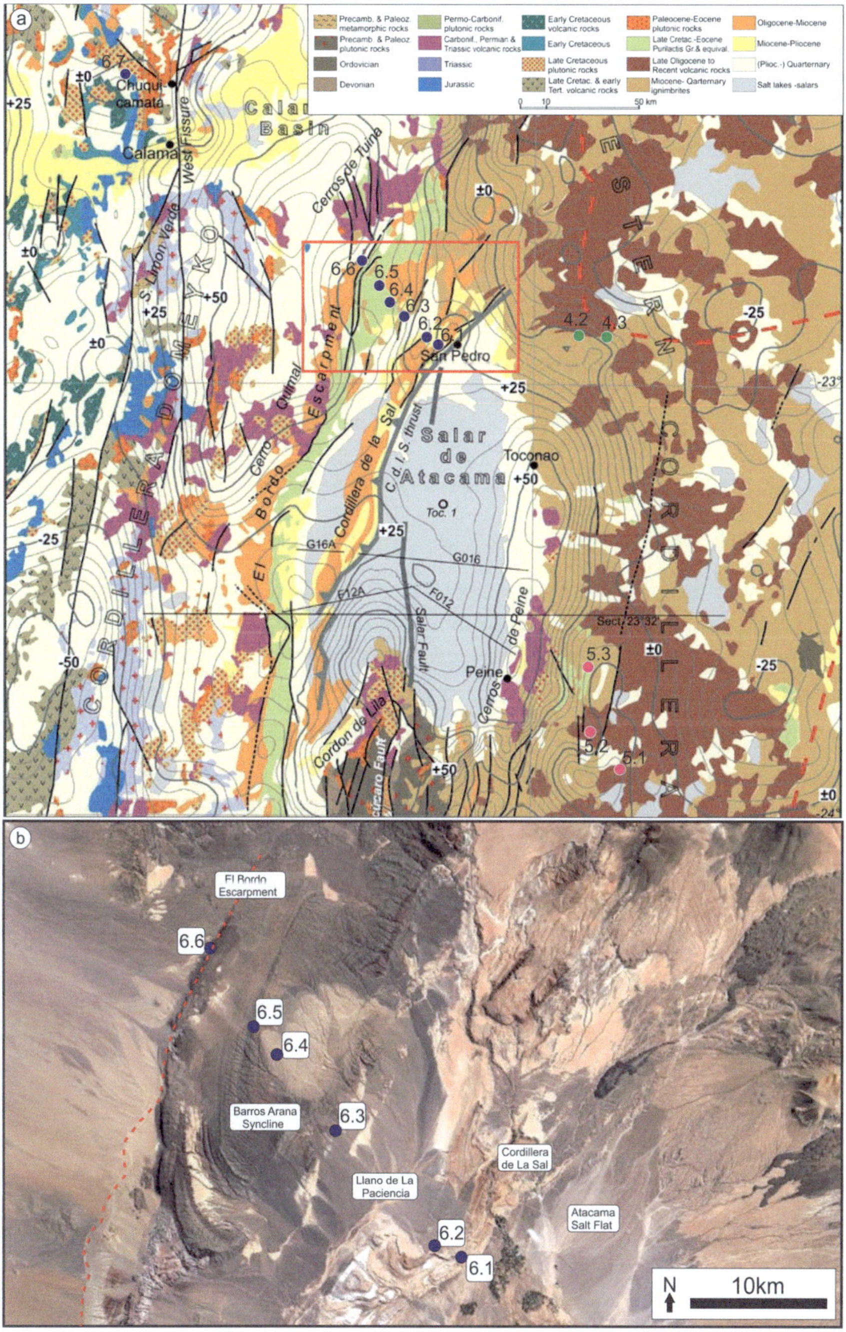

Fig. 29 **a** Geological map of the Salar de Atacama, Cordillera Domeiko and Western Cordillera (extracted from Reutter et al., 2006); **b** aerial view from Barros Arana Syncline (left side) and Cordillera de la Sal (right side), with Llano de La Paciencia narrow intermontane basin between both structures. To the right of stop 6.1 is Salar de Atacama salt pan north end (Google Earth)

slices (thick-skinned) and show top to SE vergence, while another group of folds affects only the shallow sediments (thin-skinned) and show top to NW vergence. In agreement, ~600 m to the E along the road, occurs a synformal fold with sub horizontal axis, as well as a disrupted asymmetric antiform which shows top-to-NW kinematics (Fig. 30b, c). A subvertical spaced fracture cleavage is also observed.

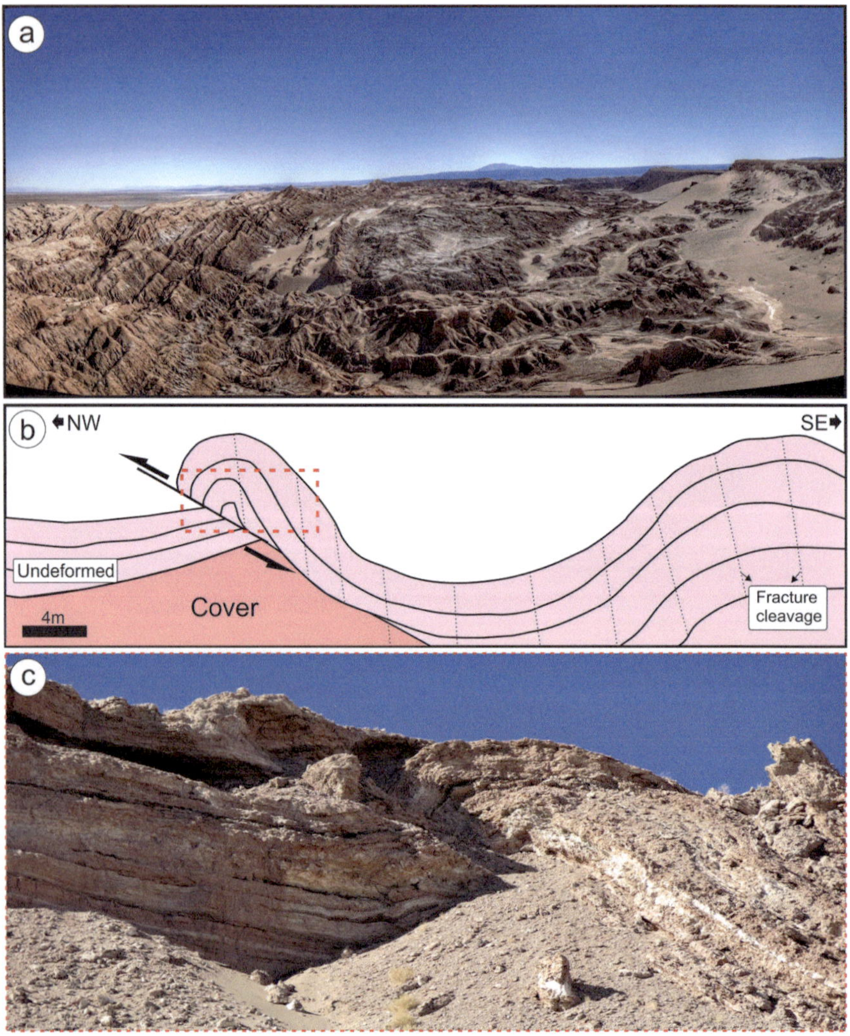

Fig. 30 Folds and faults in Cordillera de La Sal **a** Panoramic view of the mega antiformal fold (Looking to S); **b** Sketch from the roadcut in stop 6.1, showing an asymmetric disrupted anticlinal fold, with top-to-NW vergence; **c** detail of the disrupted fold

Stop 6-2

Feature: San Pedro Fm.
Location: UTM 19 K/576,835/7,466,209; 2617 m.
Directions: Ruta 23, road cut.
Description: In this stop it is possible to observe in detail the San Pedro Formation, which is the main lithology of Cordillera de La Sal. The stop shows a 4 m high road cut (Fig. 31a), with fine- to medium-grained sandstones with trough cross-bedding (Fig. 31b), intercalate with conglomeratic layers. Mm- to cm-calcrete layers occur within the sandstones. Locally, features of syn-sedimentary deformation are observed (Fig. 31c).
S0: (10/340).

Stop 6-3

Feature: Barros Arana Syncline/Purilactis Formation and Barros Arana Formation.
Location: UTM 19 K/567,932/7,475,528; 2788 m.
Directions: Ruta 23.
Description: East limb from the Barros Arana Syncline fold. According to López et al. (2020), it is a NNE-striking and asymmetrical east-verging syncline that deformed Upper Cretaceous-Lower Paleocene volcanic and sedimentary rocks of the Purilactis Formation and Barros Arana Formation. The Barros Arana Syncline contains a west-dipping axial plane. Its western limb maintains a dip of 60º–30º to E and the eastern limb of 50º–40º to W.

Fig. 31 Road cut outcrop from San Pedro Formation. **a** Panoramic view of the outcrop (Looking to N); **b** detail from the rock granulometry; **c** evidence of syn-sedimentary deformation

Outcrops display conglomerates, sandstones, and siltstones from the Purilactis Formation.

Stop 6-4

Feature: Barros Arana Syncline/Tambores Formation.
Location: UTM 19 K 563,534/7,481,661; 3280 m.
Directions: Ruta 23.
Description: This stop is located in the core of the Barros Arana Syncline fold. The outcrop exhibits fine intercalation of conglomeratic sandstones and conglomerates with a clay-rich matrix, with incipient cross-bedding, and syn-sedimentary tectonic features (Fig. 32). The rock association is interpreted as a fluvial system (López et al., 2020).
S0 (25/150).

Stop 6-5

Feature: Barros Arana Syncline/Purilactis Fm.
Location: UTM 19 K/562,096/7,483,046; 3380 m.
Directions: Ruta 23.
Description: West limb of Barros Arana Syncline fold, with road cut outcrops of volcano-clastic rocks from Purilactis Fm (Fig. 33a). S0: (37/111).

Fig. 32 General aspect of the rock, conglomeratic sandstones interleaved with conglomerate layers (35 cm long hammer for scale)

Fig. 33 **a** Purilactis Fm road cut outcrop; **b** panoramic view of the Central Volcanic Zone (looking to E from stop 6.5)

At this topographic high, it is possible to observe the regional syncline structure, thrusted with top-to-NW vergence, as well as a great view of the Central Volcanic Zone, to the east (Fig. 33b).

Stop 6-6

Feature: Cordón Barros Arana Fault/Barros Arana, Tambores and Purilactes Fm.
Location: UTM 19 K/558,736/7,488,848/3372 m.
Directions: Ruta 23.
Description: Topographic high, with panoramic view of Barros Arana Syncline west limb (Fig. 34), limited by Cordón Barros Arana Fault from the Domeiko Cordillera. This stop is located exactly in the Cordón Barros Arana Fault (López et al., 2020). This structure defines a limit that separates Triassic rocks from upper Cretaceous and Paleogene folded and thrust-faulted successions. The folds exposed on its eastern side mainly consist of narrow, upright and disharmonic anticlines and synclines commonly affected by subsidiary and shallow thrust faults. According to Arriagada et al. (2006) and Martínez et al. (2018) these structures are the morphological expression of a large blind east-verging reverse fault, which place basement and Mesozoic rocks over Upper Cretaceous and Cenozoic deposits. In this stop it is possible to see the truncation of the west limb of Barros Aranas Syncline, marked by the position of S0 of Cretaceous Puralicts Fm (Mpodozis et al., 2005).

Fig. 34 Cordón Barros Aranas Fault **a** Purilactis Fm at Flanco W of the Barros Arana Sinclinal, near Cordón Barros Aranas Fault S0 42°–162°; **b** panoramic view of Cordilheira Domeyko (Cerro Tuinas) to the azimuth 350°

12 Cordillera Domeiko–Precordillera

The Cordillera Domeiko, also known as Precordillera, is a well-defined geomorphological element, formed by elongated (30 to > 100 km) and wide (ca. 10 km) basement ridges, with NNE-SSW orientation, and that reaches average altitudes around 2800 m and maximum up to 4278 m (López et al., 2020). It is separated from the Coastal Cordillera by the Central Depression and from the Western Cordillera by the Salares Depression. It consists of major anticline structures, limited by inverse to high-angle oblique faults, developed during the Inca phase (Megard, 1984, and others). The N–S strike-slip fault system, called West Fault System, of the Domeiko Cordillera controlled the emplacement of Eocene–Oligocene intrusive complexes and associated porphyry copper deposits, being the main one located in northern Chile (Amilibia et al., 2008; Ossandón et al., 2001; Reutter et al., 1996).

The geology of the Cordillera Domeiko corresponds mainly to Paleozoic–Early Triassic basement, Late Triassic–Early Cretaceous synextensional sediments of the marine Domeyko-Tarapaca basin, and syn-orogenic Late Cretaceous up to present sediments (Amilibia et al., 2008). The presence of intrusive and volcanic arc rocks indicate the migration of the magmatic arc through the Cordillera Domeiko from Late Cretaceous to Paleogene ages (Henríquez et al. 2018). At late stages of arc position in this cordillera, during its last stage of shortening in the Middle Eocene–Oligocene, occurred the emplacement of acid porphyries associated to the giant porphyry copper ore deposits (e.g. Chuquicamata, La Escondida; Amilibia et al., 2008).

Stop 6-7

Feature: Granodiorite/Fortuna Intrusive Complex.
Location: UTM 19 K/499,128/7,531,935; 3329 m.
Directions: Ruta 24.
Description: The Fortuna Complex is a long-lived, composite magmatic system that contains intrusive phases emplaced during different stages of the Eocene to early Oligocene (45–33 Ma) Incaic phase. These porphyry copper-related intrusions were syntectonically emplaced along the West Fault System (Amilibia et al., 2008; Dilles et al., 1997). This stop is a road cut with a hornblende granodiorite (Fig. 35a) where the intense hydrothermal alteration forms a propylitic assemblage with chlorite + epidote and sulphides as chalcopyrite cutted by hematite and carbonate veinlets. The rock is transformed by intense supergene alteration to sericite + quartz, kaolinitic clay and residual K feldspar, accompanied by pyrite, chalcocite/bornite and also Cu-carbonate, as malachite (Fig. 35b). It has been explored in a small handmade exploration pit.

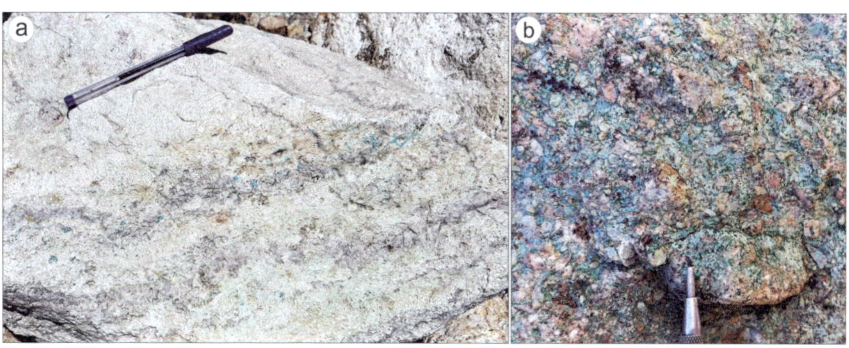

Fig. 35 **a** Highly altered Fortuna Granodiorite at stop 6-7; **b** detail of pervasive and fracture filling hydrothermal alteration and supergene malachite (Photo: Prof. Maria José Mesquita)

Complementary Visit Suggestion

Chuquicamata Mine.

Location: 19 K/509,956/753,072/*ca.* 3000 m.

From the city of Calama, it is possible to take a tour to Chuquicamata mine on weekday afternoons (Fig. 37). It is necessary to schedule in advance with Codelco, the state-owned company that runs the mine, which provides the transportation from Calama to the mine and a guided tour. It is suggested to be at the Codelco Norte visitor office at noon for the procedures of entrance.

The Cu(Mo–Au) Chuquicamata (or simply Chuqui) Mine, is the largest open pit mine (4.3 km long, 3 km wide, and 850 m deep) and one of the largest Cu producers on the planet (Fig. 36a, b), which mines some 600,000 tonnes of copper annually (Wörner et al., 2018b). Lying in the N–S orogen-parallel "porphyry copper belt" (Sillitoe, 1972), the Chuqui Porphyry Complex was developed from initial intrusions (36–33 Ma), through at least two major hydrothermal-mineralization events (~31 Ma), to final post-mineral brecciation and offset by the Domeyko fault system (Ossandón et al., 2001). The magmatic processes occur in a compressive environment related to the flattening of the subducting plate, and built up large magma reservoirs in the middle to lower crust. These feed the shallower porphyry system through recurrent, short-lived episodes of magmatic–hydrothermal activity (Fontboté, 2018, and references therein). The supergene oxidation and enrichment have occurred since ~40 Ma, controlled mainly by periods of surface uplift caused by contraction deformation. Also, lateral transport of copper from supergene profiles originated exotic gravel deposits such as the Mina Sur, that have been a key process in grade development of Chuquicamata porphyry copper system, the largest supergene enriched orebody in the world (Sillitoe, 2012).

Fig. 36 **a** Chuquicamata open pit; **b** hydrothermal and supergene Cu ore (Photo: Prof. Maria José Mesquita—UNICAMP)

Fig. 37 Oblique Google Earth image showing the main western most domains of Central Andes and the observation points of this guide (colored circles). Yellow lines represent the main roads and red dashed lines the approximated limits of each domain. Note the Loa River pathway to the sea, carved through the Coastal Cordillera. Vertical exaggeration 3×

13 Coastal Cordillera

The Coastal Cordillera presents scarps of 800–2000 m of elevation, being located 75 km to the east of the trench and 35 km above the subducting plate (Fig. 37). The main rock types are volcanic, subvolcanic and plutonic rocks of Jurassic age that represent early arc magmatism and Paleozoic/Mesozoic sedimentary rocks. Geochemical studies correlated volcanic and plutonic rocks in terms of a uniform subarc mantelic source through time (Lucassen et al., 2006). Mafic volcanic rocks from La Negra Formation comprise a sequence of (basaltic-) andesite lava flows with up to 2 m-thick volcaniclastic rocks (Németh et al., 2004), and presents crystallization ages of *ca.* 196 and 177 Ma (López et al., 2018). The plutonic rocks of the Coastal Batholith are intrusive in the La Negra Fm. (Scheuber & Gonzales, 1999) and presents crystallization ages that range between 170 and 150 Ma (López et al., 2018). Scheuber and Gonzales (1999) points the importance of normal-faulting during plutonic rocks emplacement at relatively shallow depth (*ca.* 6–9 km).

High obliquity of convergence is attributed as a cause to extensional/transtensional tectonics along the arc region, during periods of relatively low degree of coupling between the colliding plates (Scheuber & Gonzales, 1999). On the other hand, the hyperarid climate is responsible for low erosion rates and low supply of sediments to the subduction zone, which then caused high degree of coupling between subducted and overriding plates during the Late-Cenozoic, and high seismic activity as a consequence (Allmendinger & Gonzales, 2010). A good preservation of geological features and structures is also characteristic of such climate conditions.

Seventh Day—The stops are along the RN1 towards the north (Fig. 37). It is suggested to go through 230 km towards the entrance of Iquique town, enjoying the beautiful landscape of the narrow pacific coast, and to make the field stops on the return to Tocopilla.

Stop 7-1

Feature: Los Pacos thrust fault.

Location: UTM 19 K, 382,059/7,755,501, 20 m.

Directions: Coastal outcrop at the southern entrance to Iquique.

Description: The sea cliff in this area is 10 m high (Fig. 38a). It exhibits a E–W striking thrust fault which dips 30° towards the north. The Los Pacos thrust fault juxtaposed altered mesozoic volcanic rocks of La Negra Formation over a footwall of Pleistocene coarse-grained beach deposits (Fig. 38b, c). As the fault plane advances from the basement rocks in the footwall, it splays upwards into a triangular zone crosscutting the cover sequence, resembling the geometry of structures formed by trishear kinematics related to propagating thrust faults (Allmendinger & Gonzales, 2010). Cataclastic zones are marked by intense hydrothermal alteration, with illite + carbonate growth (Fig. 38).

According to Allmendinger and Gonzales (2010) margin-perpendicular reverse faults in the Coastal Cordillera are restricted to the latitudes between 19° S and

Fig. 38 **a** Overview to the North of the sea cliff and location of the outcrop in stop 7-1 seen from the RN1 near the southern entrance of Iquique; **b** Los Pacos thrust fault (ca. 10 m wide, view to Azimuth 90°); **c** detail of Pleistocene deposit (largest block has 40 cm diameter)

21° 40′ S, and symmetrically distributed on both sides of the axis of Bolivian Orocline. Shortening accommodated on these faults is related to the margin curvature (orocline) and also due to oblique subduction.

Stop 7-2

Feature: Andean Batholith.
Location: UTM 19 K, 380,417/7,670,639, 10 m.
Directions: Coastal outcrop in the Chomache Bay, southern of Caleta Lobos/Punta Lobos.
Description: In this area there are exposed plutonic rocks of the andean batholith, presenting magma mingling features. The outcrop corresponds to a biotite granodiorite with medium-grained, equigranular texture, and rare cm-sized microgranular mafic enclaves. The rock is massive and the enclaves are mostly rounded, without preferred orientation. Syn-magmatic mafic dikes (75/102) with lobate contacts cross-cut the granodiorite (Fig. 39). Fracture and cataclastic zones are observed with attitudes (72/005) and (77/029). Plutonic arc emplacement controlled by normal faults and successive diking is attributed by Scheuber and Gonzales (1999) to a period of slab-roll back and decoupling of the subducted plate, with the overriding plate in an overall trantensional system.

Stop 7-3

Feature: Loa river.
Location: UTM 19 K, 390,246/7,630,091, 5 m.
Directions: Loa River's mouth, on its north bank.
The Loa River mouth is commonly dammed by a gravel bank (Fig. 40a), where it can be found diverse gneiss blocks that display a discontinuous banding, mostly

Fig. 39 Andean Batholith main structures along the Coastal Cordillera. **a** General aspect of the coastal outcrop, with synplutonic mafic dike (75/102) crosscutting biotite granodiorite; **b** synplutonic mafic dike with irregular boundaries; **c** detail of the irregular contact between biotite granodiorite and mafic dike; **d** fault (72/005) affecting the granodiorite, with damage zone (right) and breccia (center) development

Fig. 40 Loa River aspects at the coast. **a** Loa's rivermouth; **b** panoramic view looking towards the Coastal Cordillera

tonalitic and more rarely mafic. In the riverbed there is a cm-thick mud layer. The water presents a high concentration of arsenium (Peel et al., 2013). Looking towards the Coastal Cordillera one can observe how the river valley carved its way to the sea, cutting ca. 1000 m of rocks, and affected by normal faulting (Fig. 40b). It is noteworthy that the Loa River's mouth is an important ecological spot for diverse species of the region, since it is the only perennial river that crosses the Atacama Desert and one of the very few places with fresh water reservoirs along ca. 400 km of shoreline (Peel et al., 2013; Zárate et al., 2020). It is also an important site for marine biodiversity, due to a coastal circulation upwelling (Palma et al., 2006).

Stop 7-4

Feature: La Negra Formation.
Location: UTM 19 K, 374,617/7,553,751, 6 m.
Directions: Caleta Boy beach, south of Tocopilla.
At this place a large coastal outcrop is accessible, with more than 200 m of N–S extension, just south of Caleta Boy. The volcanoclastic layering is tilted (Fig. 41a)

with orientations (strike/dip) as (024/58) and (024/43). At this outcrop it is possible to recognize two distinct layers (Fig. 41b, c).
The lower bed is interpreted by Németh et al. (2004) as a hialoclastite, being a lapilli-tuff rich in plagioclase fragments (30–40%) and presents sparse, cm-sized felsic and mafic rock fragments (lapilli), partially replaced by epidote with radial habit (Fig. 41d). The upper layer is andesitic and presents a minimum of 20 m of thickness and is characterized by 10–15% of plagioclase phenocrysts. It presents, in its lower contact, a very fine-grained, cm-thick layer, above an epidotized band, which is interpreted as a chilled margin (Fig. 41e). The andesite above the chilled margin is well foliated. Pervasive alteration (epidotization) in the lower bed, when compared to the upper one, is in agreement with interpretations of Németh et al. (2004).

After this last stop of the tour, you can return to San Salvador de Jujuy, using the same roads used to get here, or take a plane directly at one of the regional airports (Iquique, Antofagasta, Calama).

14 Final Remarks

An active mountain range is a unique observatory of the relationships between diverse endogenous and exogenous processes that act in its evolution. Aiming to offer the opportunity to know the Andes Cordillera to undergraduate and graduate geology students from the Institute of Geosciences of the Universidade Federal do Rio Grande do Sul (IGEO/UFRGS), and other universities in Brazil, it has been, since the 2016 (see Albers et al., 2017), taught a course known as *Geology of the Central Andes*, taking 155 participants to the Central Andes (2016 to 2019). Although it was not possible to keep going with the course during the pandemic, we intend to re-establish this field course in 2022.

This is a field based open course integrated with a regular course for MSc and PhD students through the Postgraduate Program in Geosciences (PPGGEO) of UFRGS. The open course is an extension course offered by IGEO/UFRGS, mainly attended by geology students of this university, but that also welcomes students from other universities and other undergraduate courses (e.g. geography, biology, physics, geophysics), as well as lecturers of geology courses and professionals of geology and of related areas. No previous background is demanded to participate. Certificates to participants of the open course are granted by UFRGS. The open course *Geology of the Central Andes* is divided into two parts: the first part includes 12 h of pre-field activities developed in three to four reunions to arrange the logistics of the tour, planning field activities, division of tasks, revision lectures and seminars on the geology of the Central Andes, aiming to create a framework of knowledge about the geology that will be visited. This course part deeply involves the students, many of which would be on their first international trip, and turn them protagonists of activities such as searching for lodging and booking a visit to the Chuquicamata Mine, the

Fig. 41 La Negra Fm aspects south of Tocopilla. **a** General aspect of the coastal outcrop. Note the tilted beds, gently dipping towards the east, also marked in the Cordillera Costera scarp at background (ca. 1000 m high); **b** general view of the outcrop looking from above; **c** contact zone between two volcanic layers; **d** Lapilli-tuff (hialoclastite) of La Negra Fm, with a lithoclast partially replaced by radial epidote; **e** detail of contact zone between two volcanic layers

largest open pit mine and a major Cu producer of the world. The second part of the course is the 112 h of field trip, departing by bus from Porto Alegre (Brazil), near the Atlantic coast, and crossing the South American continent during a 11-day field trip (15 days including commuting). During the tour, the participants of the course, gathered in small teams, are responsible for organizing accommodation (mostly in campsites), as well as groceries and food preparation for the big group. In our experience, this format of geology course, with early theoretical preparation and significant involvement of participants in logistics activities since the beginning, added important learning experiences to all the course participants.

The M.Sc. and Ph.D. regular post-graduation course is the GEB0139A—GEOLOGIA DOS ANDES CENTRAIS of PPGGEO/UFRGS (https://www.ufrgs.br/ppggeo/ppggeo/wp-content/uploads/2019/07/GEOLOGIA-DOS-ANDES-CENTRAIS.pdf). It is inserted in the open course and has a strong link with it. The GEB139A has 5 credits, being 1 credit (15 h) of theoretical classes and 4 credits (60 h) of field based classes. The course can be attended by external students as special students (graduated with no link to post-graduations programs) or as post-graduation students from other universities from Brazil and other countries. Aiming at the internationalization of this course, classes can be given in English. The theoretical classes are divided in 5 meetings of 3 h each, in which the geology and tectonic evolution of the visited sectors are reviewed, including: Central Andes Orogeny, Central Andes Magmatism, Evolution and Geology of Main Central Andes Domains (Coastal, Domeiko, Western, Eastern Cordilleras, Santa Bárbara System and N Puna), Central Andes Mineral resources, Central Andes Topographic and Climatic evolution. In the field work, the main exercise involves a multiscale description and analysis of outcrops representative of the mountain section (i.e. from Purmamarca, Argentina to Tocopilla, Chile) and the spatial and temporal integration of these records in a field report, using this cross referenced information (field-based—bibliography) that support a proposal to the evolution of Central Andes Mountain Belt. Students are also faced with main themes of Andean geology through the intense collaboration of researchers, namely Professors Guillermo Acenollaza, Fernando Sardi, José G. Viramontes and Alfonso Sola from Argentina and Rodrigo Gonsalez from Chile, which join the field work in different sectors and give lectures during visits to universities and research centers (e.g. INSUGEO—Tucumãn; LA.TE. Andes—Salta; Universidad de Antofagasta). This collaborative network has proven to be an opportunity for new contacts and a fruitful exchange of scientific experiences for all participants. During the field trip, the post-graduation students are expected to act as teaching assistants especially to the undergraduate students.

Future goals of this course are to create a permanent and virtual exhibition of the sample collection (already gathered in the previews field works), photos, videos and other materials produced in this field trip, and also a didactic collection (samples and thin sections) for use in geology classes. Since its first edition, we have created a field trip guide (longer and more detailed than the one presented here), which has been edited and improved continuously. This support material was fundamental to link the various outcrops visited during field work, and to contextualize them in relation to the Central Andes tectonic evolution.

In our quest to conduct a field trip, we faced the difficulty of finding references with precise location and description of outcrops along main roads, that would allow the observation of a complete and logical sequence of the cross-section of the Central Andes, intended to be done with a large bus. The field stops presented here were gathered from several publications, and also recommended by fellow geologists from Argentina and Chile, whom we were fortunate to have as collaborators of this tour. This field guide aims to share our experience and to facilitate a field trip across the Central Andes Cordillera for every geologist and nature enthusiastic.

Acknowledgements The authors acknowledge the support from the Institute of Geosciences of the Federal University of Rio Grande do Sul, Brazil, particularly to the former director Prof. André S. Mexias, and from the Postgraduate Program in Geosciences of the same university. We are grateful to Profs Guillermo Acenollaza, Fernando Sardi, José G. Viramontes and Alfonso Sola (Argentina), and Rodrigo Gonsalez (Chile) for their valuable support during fieldwork. We are also thankful for the partnership of all participants of the excursions (2016–2019), and specially to Max Albers and Gabriel Drago by their initiative and help organizing the first edition of the field trip. Soumyajit Mukherjee (IIT Bombay) invited, edited and reviewed this article. Mukherjee (2023) summarized this chapter.

Supplementary Data

See Table 1.

References

Aceñolaza, G. F. (2003). The Cambrian system in Northwestern Argentina: Stratigraphical and palaeontological framework. *Geologica Acta, 1*(1), 23–39.

Albers, M., Drago, G., De Toni, G. B., Porcher, C. C., Gomes, M. B., Mônaco, G., Martini, A., Angonese, B., Betella, C.M., Trevisan, C., Canarim, D., Leyen, E. B., Abreu, E. P., Silva, F. D., Link, G., Junqueira, I. S., Nedel, I. L., Hoffmann, I. B., Oliveira, J. B., Bilhar, J. D., Machado, J. P., Ferronatto, J. P., Sobiesiak, J. S., Silveira, L. M., Radmann, L., Boffil, L., Pezat, L., Silva, L. A., Loureiro, P., Rossoni, R. B., Quillfeldt, S., Cruz, V. G., Fornari, V., & Capelari, V. (2017). Projeto de extensão Geologia dos Andes Centrais: ensino, pesquisa e extensão em um laboratório orogênico ativo. In *X Simpósio Sul-Brasileiro de Geologia, 10, 2017, Curitiba—PR.* http://ssbg2017anais.siteoficial.ws/

Allmendinger, R. W., & Gonzales, G. (2010). Invited review paper: Neogene to Quaternary tectonics of the coastal Cordillera, northern Chile. *Tectonophysics, 495*, 93–110.

Allmendinger, R. W., Jordan, T. E., Kay, S. M., & Isacks, B. L. (1997). The Evolution of the Altiplano-Puna Plateau of the Central Andes. *Annual Review of Earth and Planetary Sciences, 25*(1), 139–174.

Alonso, R. N., Bookhagen, B., Carrapa, B., Coutand, I., Haschke, M., Hilley, G. E., Schoenbohm, L., Sobel, E. R., Strecker, M. R., Trauth, M. H., & Villanueva, A., et al. (2006). Tectonics, Climate, and Landscape Evolution of the Southern Central Andes: The Argentine Puna Plateau and adjacent regions between 22 and 30° S. In O. Oncken (Ed.), *The Andes* (pp. 265–283). Springer.

Amilibia, A., Sàbat, F., McClay, K. R., Muñoz, J. A., Roca, E., & Chong, G. (2008). The role of inherited tectono-sedimentary architecture in the development of the central Andean mountain belt: Insights from the Cordillera de Domeyko. *Journal of Structural Geology, 30*(12), 1520–1539.

Armijo, R., Lacassin, R., Coudurier-Curveur, A., & Carrizo, D. (2015). Coupled tectonic evolution of Andean orogeny and global climate. *Earth-Science Reviews, 143*, 1–35.

Aros, F., & Gakeno-Ibaceta, C. A. (2017). Alguien llamado Licancabur. Cuadernos De Arquitectura. *Habitar El Norte, 4*, 30–35.

Arriagada, C., Cobbold, P. R., & Roperch, P. (2006). Salar de Atacama basin: A record of compressional tectonics in the central Andes since the mid-Cretaceous. *Tectonics, 25*(1), TC001770.

Burns, D. H., de Silva, S. L., Tepley, F., III., Schmitt, A. K., & Loewen, M. W. (2015). Recording the transition from flare-up to steady-state arc magmatism at the Purico-Chascon volcanic complex, northern Chile. *Earth and Planetary Science Letters, 422*, 75–86.

Carrapa, B., & DeCelles, P. G. (2015). Regional exhumation and kinematic history of the central Andes in response to cyclical orogenic processes. In: P. G. DeCelles, M. N. Ducea, B. Carrapa, & P. A. Kapp (Eds.), *Geodynamics of a Cordilleran Orogenic system: The Central Andes of Argentina and Northern Chile* (p. 212). Geological Society of America Memoir.

Carrapa, B., Huntington, K. W., Clementz, M., Quade, J., Bywater-Reyes, S., Schoenbohm, L. M., & Canavan, R. R. (2014). Uplift of the Central Andes of NW Argentina associated with upper crustal shortening, revealed by multiproxy isotopic analyses. *Tectonics, 33*. https://doi.org/10.1002/2013TC003461

Carrera, N., Muñoz, J. A., Sabat, F., Mon, R., & Roca, E. (2006). The role of inversion tectonics in the structure of the Cordillera Oriental (NW Argentinean Andes). *Journal of Structural Geology, 28*, 1921–1932.

Cónsole-Gonella, C., Valais, S., Marquillas, R. A., & Sánchez, M. C. (2017). The Maastrichtian-Danian Maimará tracksite (Yacoraite Formation, Salta Group), Quebrada de Humahuaca, Argentina: Environments and ichnofacies implications. *Palaeogeography, Palaeoclimatology, Palaeoecology, 468*, 327–350.

Coutand, I., Cobbold, P. R., de Urreiztieta, M., Gautier, P., Chauvin, A., Gapais, D., Rossello, E. A., & López-Gamundí, O. (2001). Style and history of Andean deformation, Puna plateau, northwestern Argentina. *Tectonics, 20*(2), 210–234. https://doi.org/10.1029/2000tc900031

de Silva, S. L., & Kay, S. M. (2018). Turning up the heat: High-flux magmatism in the Central Andes. *Elements, 14*(4), 245–250.

de Silva, S. L., Riggs, N. R., & Barth, A. P. (2015). Quickening the pulse: Fractal tempos in continental arc magmatism. *Elements, 11*(2), 113–118.

DeCelles, P. G., Carrapa, B., Horton, B. K., & Gehrels, G. E. (2011). Cenozoic foreland basin system in the central Andes of northwestern Argentina: Implications for Andean geodynamics and modes of deformation. *Tectonics, 30*(TC6013). https://doi.org/10.1029/2011TC002948

DeCelles, P. G., Ducea, M. N., Zandt, G., & Kapp, P. (2009). Cyclicity in Cordilleran orogenic systems. *Nature Geoscience, 2*, 251–257. https://doi.org/10.1038/ngeo469

Deeken, A., Sobel, E. R., Coutand, I., Haschke, M., Riller, U., & Strecker, M. R. (2006). Development of the southern Eastern Cordillera, NW Argentina, constrained by apatite fission track thermochronology: From early Cretaceous extension to middle Miocene shortening. *Tectonics, 25*(6), TC6003. https://doi.org/10.1029/2005tc001894

Dilles, J. H., Tomlinson, A. J., Martin, M., Blanco, N. (1997). El Abra and Fortuna complexes: A porphyry copper batholith sinistrally displaced by the Falla Oeste. In: *VIII Congresso Geológico Chileno, Antofagasta*, vol. 3, pp. 1883–1887

Echeverría, J., & Muñoz, C. (1988). Maiz, regalo de los dioses (197 p). Instituto Otavaleno de Antropologia (IOA).

Escayola, M. P., van Staal, C. R., & Davis, W. J. (2011). The age and tectonic setting of the Puncoviscana Formation in northwestern Argentina: An accretionary complex related to Early Cambrian closure of the Puncoviscana Ocean and accretion of the Arequipa-Antofalla block. *Journal of South American Earth Sciences, 32*, 437–458.

Figueroa, O., Déruelle, B., & Demaiffe, D. (2009). Genesis of adakite-like lavas of Licancabur volcano (Chile—Bolivia, Central Andes). *Comptes Rendus Geoscience, 341*(4), 310–318.
Fontboté, L. (2018). Ore deposits of the Central Andes. *Elements, 14*(4), 257–261.
Gansser, A. (1973). Facts and theories on the Andes. *Journal of the Geological Society, 129*(2), 93–131.
Gianni, G. M., Navarrete, C. G., & Folguera, A. (2015). Synorogenic foreland rifts and transtensional basins: A review of Andean imprints on the evolution of the San Jorge Gulf, Salta Group and Taubaté Basins. *Journal of South American Earth Sciences, 64*, 288–306.
González, G., Cembrano, J., Aron, F., Veloso, E. E., & Shyu, J. B. H. (2009). Coeval compressional deformation and volcanism in the central Andes, case studies from northern Chile (23° S–24° S). *Tectonics, 28*(6), TC002538.
González, M. A., Pereyra, F., Ramallo, E., & Tchilinguirian, P. (2003). Hoja Geológica 2366-IV, Ciudad de Libertador General San Martín, provincias de Jujuy y Salta. Instituto de Geología y Recursos Minerales, Servicio Geológico Minero Argentino. Boletín 274, 109 p. Buenos Aires.
González-Ferrán, O. (1995). Volcanes de Chile. Instituto Geográfico Militar.
Guzmán, S., Doronzo, D. M., Martí, J., & Seggiaro, R. (2020). Characteristics and emplacement mechanisms of the Coranzulí ignimbrites (Central Andes). *Sedimentary Geology, 405*, 105699.
Haschke, M., Günther, A., Melnick, D., Echtler, H., Reutter, K.-J., Scheuber, E., & Oncken, O. (2006). Central and Southern Andean tectonic evolution inferred from arc magmatism. In O. Oncken, G. Chong, G. Franz, P. Giese, H.-J. Götze, V. A. Ramos, M. R. Strecker, & P. Wigger (Eds.), *The Andes: Active subduction orogeny. Frontiers in earth sciences*. Springer-Verlag.
Henríquez, S., DeCelles, P. G., & Carrapa, B. (2019). Cretaceous tomiddle Cenozoic exhumation history of the Cordillera de Domeyko and Salar de Atacama basin, northern Chile. *Tectonics, 38*, 395–416. https://doi.org/10.1029/2018TC005203
Henríquez, S., DeCelles, P. G., Carrapa, B., Hughes, A. N., Davis, G. H., & Alvarado, P. (2020). Deformation history of the Puna plateau, Central Andes of northwestern Argentina. *Journal of Structural Geology, 140*, 104133.
Horton, B. K. (2018). Sedimentary record of Andean mountain building. *Earth-Science Reviews, 178*, 279–309.
Houston, J., Butcher, A., Ehren, P., Evans, K., & Godfrey, L. (2011). The evaluation of brine prospects and the requirement for modifications to filing standards. *Economic Geology, 106*(7), 1225–1239.
Isacks, B., & Molnar, P. (1971). Distribution of stresses in the descending lithosphere from a global survey of focal-mechanism solutions of mantle earthquakes. *Reviews of Geophysics, 9*(1), 103.
James, D. E. (1971). Plate tectonic model for the evolution of the central Andes. *Geological Society of America Bulletin, 82*, 3325–3346.
Kay, S. M., Coira, B., & Mpodozis, C. (2008). Field trip guide: Neogene evolution of the central Andean Puna plateau and southern Central Volcanic Zone. In: S. M. Kay, & V. Ramos (Eds.), *GSA Field Guide 13: Field trip guides to the backbone of the Americas in the Southern and Central Andes: Ridge Collision, Shallow Subduction, and Plateau Uplift* (pp. 117–181).
Kay, S. M., & Coira, B. L. (2009). Shallowing and steepening subduction zones, continental lithospheric loss, magmatism, and crustal flow under the Central Andean Altiplano-Puna Plateau. In S. M. Kay, V. A. Ramos, & W. R. Dickinson (Eds.), *Backbone of the Americas: Shallow subduction, plateau uplift, and ridge and terrane collision* (pp. 229–259). Geological Society of America Memoir.
Kearey, P., Klepeis, K. A., & Vine, F. J. (2009). *Global tectonics* (3rd ed., p. 482). Wiley-BlackWell.
Kirschbaum, A., Hongn, F., & Menegatti, N. (2006). The Cobres plutonic complex, eastern Puna (NW Argentina): Petrological and structural constraints for Lower Paleozoic magmatism. *Journal of South American Earth Sciences, 21*, 252–266.
López, C., Martinez, F., Maksymowicz, A., Giambiagi, L., & Riquelme, R. (2019). What is the structure of the forearc region in the Central Andes of northern Chile? An approach from field data and 2-D reflection seismic data. *Tectonophysics, 769*, 228187.

López, C., Martinez, F., Ventisette, C., Bonini, M., Montanari, D., Muñoz, B., & Riquelme, R. (2020). East-vergent thrusts and inversion structures: An updated tectonic model to understand the Domeyko Cordillera and the Salar de Atacama Basin transition in the western Central Andes. *Journal of South American Earth Sciences, 103*, 102741.

López, C., Riquelme, R., Martínez, F., Sanchez, C., & Mestre, A. (2018). Zircon U-Pb geochronology of the mesozoic to lower Cenozoic rocks of the Coastal Cordillera in the Antofagasta region (22°30′–23°00 ′ S): Insights to the Andean tectono-magmatic evolution. *Journal of South American Earth Sciences, 87*, 113–138.

Lork, A., & Bahlburg, H. (1993). Precise U-Pb ages of monazites from the Faja Eruptiva de la Puna oriental and the Cordillera oriental, NW Argentina. In *XII Congreso Geológico Argentino y II Congreso de Exploración de Hidrocarburos, Actas IV*, pp. 1–6.

Lucassen, F., Kramer, W., Bartsch, V., Wilke, H.-G., Franz, G., Romer, R. L., & Dulski, P. (2006). Nd, Pb, and Sr isotope composition of juvenile magmatism in the Mesozoic large magmatic province of northern Chile (18–27° S). Indications for a uniform subarc mantle. *Contributions to Mineralogy and Petrology, 152*, 571–589.

Marcato, G., Pasuto, A., & Rivelli, F. R. (2009). Mass movements in the Rio Grande Valley (Quebrada de Humahuaca, Northwestern Argentina): A methodological approach to reduce the risk. *Advanced Geosciences, 22*, 59–65.

Marquillas, R. A., del Papa, C., & Sabino, I. F. (2004). Sedimentary aspects and paleoenvironmental evolution of a rift basin: Salta Group (Cretaceous–Paleogene), northwestern Argentina. *International Journal of Earth Sciences, 94*(1), 94–113.

Martínez, F., López, C., Bascuñan, S., & Arriagada, C. (2018). Tectonic interaction between Mesozoic to Cenozoic extensional and contractional structures in the Preandean Depression (23°–25°S): Geologic implications for the Central Andes. *Tectonophysics, 744*, 333–349.

May, J.-H. (2008). A geomorphological map of the Quebrada de Purmamarca, Jujuy, NW Argentina. *Journal of Maps*, 211–224.

Mégard, F. (1984). The Andean orogenic period and its major structures in central and northern Peru. *Journal of the Geological Society, 141*(5), 893–900. https://doi.org/10.1144/gsjgs.141.5.0893

Mpodozis, C., Arriagada, C., Basso, M., Roperch, P., Cobbold, P., & Reich, M. (2005). Late Mesozoic to Paleogene stratigraphy of the Salar de Atacama Basin, Antofagasta, Northern Chile: Implications for the tectonic evolution of the Central Andes. *Tectonophysics, 399*(1–4), 125–154.

Mukherjee, S. (2023). Introduction to structural geology and tectonics field guidebook—Volume 2. In S. Mukherjee (ed.), *Structural geology and tectonics field guidebook—Volume 2* (pp. xi–xiv). Springer Nature Switzerland AG. ISBN: 978-3-031-19575-4.

Németh, K., Breitkreuz, C., & Wilke, H.-G. (2004). Volcano-sedimentary successions within an intra-arc related Jurassic Large Igneous Province (LIP): La Negra Formation, Northern Chile (a preliminary scientific report on the Br 997/22–1 DFG Pilot Project). In *A Magyar Állami Földtani Intézet Évi Jelentése 2002 (2002 annual report of the Geological Institute of Hungary), Budapest*, pp 233–256

Oncken, O., Hindle, D., Kley, J., Elger, K., Victor, P., & Schemmann, K. (2006). Deformation of the Central Andean upper plate system—Facts, fiction, and constraints for plateau models. In O. Oncken, G. Chong, G. Franz, P. Giese, H.-J. Götze, V. A. Ramos, M. R. Strecker, & P. Wigger (Eds.), *The Andes: Active subduction orogeny. Frontiers in earth sciences* (pp. 3–28). Springer-Verlag.

Ossandón, C. G., Fréraut, C. R., Gustafson, L. B., Lindsay, D. D., & Zentilli, M. (2001). Geology of the Chuquicamata mine: A progress report. *Economic Geology, 96*(2), 249–270.

Palma, W., Escribano, R., & Rosales, S. A. (2006). Modeling study of seasonal and inter-annual variability of circulation in the coastal upwelling site of the El Loa River off northern Chile. *Estuarine, Coastal and Shelf Science, 67*, 93–107.

Peel, A., Márquez, A., López-Sánchez, J. F., Rubio, R., Barbero, M., Stegen, S., Queirolo, F., & Díaz-Palma, P. (2013). Occurrence of arsenic species in algae and freshwater plants of an extreme arid region in northern Chile, the Loa River Basin. *Chemosphere, 90*, 556–564.

Pingel, H., Strecker, M. R., Alonso, R. N., & Schmitt, A. K. (2013). Neotectonic basin and landscape evolution in the Eastern Cordillera of NW Argentina, Humahuaca Basin (~24°S). *Basin Research, 25*(5), 554–573.

Ramos, V. A. (1999). Plate tectonic setting of the Andean Cordillera. *Episodes, 22*(3), 183–190.

Ramos, V. A. (2009). Anatomy and global context of the Andes: Main geologic features and the Andean orogenic cycle. In: S. M. Kay, V. A. Ramos, & W. R. Dickinson (Eds.), *Backbone of the Americas: Shallow subduction, plateau uplift, and ridge and terrane collision* (vol. 204, pp. 31–65). Geological Society of America Memoir. https://doi.org/10.1130/2009.1204(02)

Ramos, V. A., & Folguera, A. (2009). Andean flat-slab subduction through time. *Geological Society, London, Special Publications, 327*(1), 31–54.

Reutter, K. J., Charrier, R., Götze, H. J., Schurr, B., Wigger, P., Scheuber, E., Giese, P., Reuther, C.-D., Schmidt, S., Rietbrock, A., Chong, G., & Belmonte-Pool, A. (2006). The Salar de Atacama Basin: a subsiding block within the western edge of the Altiplano-Puna Plateau. In: O. Oncken, G. Chong, G. Franz, P. Giese, H.J. Götze, V. A. Ramos, M. R. Strecker, & P. Wigger (Eds.), *The Andes: Active subduction orogeny. Frontiers in earth sciences* (pp. 303–325). Springer-Verlag.

Reutter, K. J., Scheuber, E., & Chong, G. (1996). The Precordilleran fault system of Chuquicamata, northern Chile: evidence for reversals along arc-parallel strike-slip faults. *Tectonophysics, 259*(1–3), 213–228.

Rojas Vera, E. A., Giampaoli, P., Gobbo, E., Rocha, E., Olivieri, G., & Figueroa, D. (2019). Structure and tectonic evolution of the Interandean and Subandean Zones of the central Andean fold-thrust belt of Bolivia. In: B. K. Horton, A. Folguera (Eds.), *Andean tectonics* (pp. 399–427).

Rubilar, J., Martínez, F., Arriagada, C., Becerra, J., & Bascuñán, S. (2017). Structure of the Cordillera de la Sal: A key tectonic element for the Oligocene-Neogene evolution of the Salar de Atacama basin, Central Andes, northern Chile. *Journal of South American Earth Sciences, 87*, 200–210.

Sánchez, M. C., & Marquillas, R. A. (2010). Facies y ambientes del Grupo Salta (Cretácico-Paleógeno) en Tumbaya, Quebrada de Humahuaca, Porvincia de Jujuy. *Revista De La Asociación Geológica Argentina, 67*(3), 383–391.

Sancho, C., Penã, J. L., Rivelli, F., Rhodes, E., & Muñoz, A. (2008). Geomorphological evolution of the Tilcara alluvial fan (Jujuy Province, NW Argentina): Tectonic implications and palaeoenvironmental considerations. *Journal of South American Earth Sciences, 26*, 68–77.

Scheuber, E., & Gonzales, G. (1999). Tectonics of the Jurassic-early cretaceous magmatic arc of the north Chilean Coastal Cordillera (22°–26°S): A story of crustal deformation along a convergent plate boundary. *Tectonics, 18*(5), 895–910.

Scheuber, E., Mertmann, D., Ege, H., Silva-González, P., Heubeck, C., Reutter, K.-J., & Jacobshagen, V., et al. (2006). Exhumation and basin development related to formation of the Central Andean Plateau, 21° S. In O. Oncken (Ed.), *The Andes* (pp. 285–301). Springer.

Schildgen, T. F., & Hoke, G. D. (2018). The topographic evolution of the Central Andes. *Elements, 14*(4), 231–236.

Schmitt, A., de Silva, S., Trumbull, R., & Emmermann, R. (2001). Magma evolution in the Purico ignimbrite complex, northern Chile: Evidence for zoning of a dacitic magma by injection of rhyolitic melts following mafic recharge. *Contributions to Mineralogy and Petrology, 140*, 680–700.

Seggiaro, R. E., Becchio, R., Bercheñi, V., & Ramallo, L. (2015). Hoja Geológica 2366-III Susques, provincias de Jujuy y Salta. Instituto de Geología y Recursos Minerales, Servicio Geológico Minero Argentino, Boletín N°414, 103 pp. Buenos Aires.

Seggiaro, R. E., Guzmán, S. R., & Martí, J. (2019). Dynamics of caldera collapse during the Coranzulí eruption (6.6 Ma) (Central Andes, Argentina). *Journal of Volcanology and Geothermal Research, 374*, 1–12.

Sillitoe, R. (1972). A plate tectonic model for the origin of porphyry copper deposits. *Economic Geology, 67*(2), 184–197.

Sillitoe, R. H. (2012). Copper provinces. *Society of Economic Geologists Special Publication, 16*, 1–18.

Steinmetz, R. L. (2017). Lithium-and boron-bearing brines in the Central Andes: Exploring hydrofacies on the eastern Puna plateau between 23 and 23 30′ S. *Mineralium Deposita, 52*(1), 35–50.

Steinmetz, R. L. L., & Galli, C. (2015). Basin development at the eastern border of the Northern Puna and its relationship with the plateau evolution. *Journal of South American Earth Sciences, 63*, 244–259.

Strecker, M. R., Hilley, G. E., Bookhagen, B., & Sobel, E. R. (2012). Structural, geomorphic, and depositional characteristics of contiguous and broken foreland basins: Examples from the eastern flanks of the Central Andes in Bolivia and NW Argentina. In: C. Busby & A. A. Pérez (Eds.), *Tectonics of sedimentary basins* (664p). Wiley-Blackwell.

Streit, R. L., Burbank, D. W., Strecker, M. R., Alonso, R. N., Cottle, J. M., & Kylander-Clark, A. R. C. (2015). Controls on intermontane basin filling, isolation and incision on the margin of the Puna Plateau, NW Argentina (~23°S). *Basin Research, 29*, 131–155.

Tibaldi, A., & Bonali, F. L. (2018). Contemporary recent extension and compression in the central Andes. *Journal of Structural Geology, 107*, 73–92.

Toselli, A. J., Aceñolaza, G. F., Miller, H., Adams, C., Aceñolaza, F. G., & Rossi, J. N. (2012). Basin evolution of the margin of Gondwana at the Neoproterozoic/Cambrian transition: The Puncoviscana Formation of Northwest Argentina. *Neues Jahrbuch Für Geologie Und Paläontologie—Abhandlungen, 265*(1), 79–95. https://doi.org/10.1127/0077-7749/2012/0246

Tsyganov, A. N., Shatilovich, A. V., Esaulov, A. S., Chernyshov, V. A., Mazei, N. G., Malysheva, E. A., & Mazei, Y. A. (2017). Morphology and phylogeny of the testate amoebae Euglypha bryophila Brown, 1911 and Euglypha cristata Leidy, 1874 (Rhizaria, Euglyphida). *European Journal of Protistology, 61*, 76–84.

Walk, J., Stauch, G., Reyers, M., Vásquez, P., Sepúlveda, F. A., Bartz, M., Hoffmeister, D., Brückner, H., & Lehmkuhl, F. (2020). Gradients in climate, geology, and topography affecting coastal alluvial fan morphodynamics in hyperarid regions–The Atacama perspective. *Global and Planetary Change, 185*, 102994.

Wörner, G., Mamani, M., & Blum-Oeste, M. (2018a). Magmatism in the Central Andes. *Elements, 14*(4), 237–287.

Wörner, G., Schildgen, T. F., & Reich, M. (2018b). The central Andes: Elements of an extreme land. *Elements, 14*(4), 225–230. https://doi.org/10.2138/gselements.14.4.225

Zaburlín, M. A. (2009). Historia de ocupación del Pucará de Tilcara (Jujuy, Argentina). *Intersecciones En Antropología, 10*(1), 89–103.

Zárate, A., Dorador, C., Araya, R., Guajardo, M., Florez, Z. J., Icaza, G., Cornejo, D., & Valdés, J. (2020). Connectivity of bacterial assemblages along the Loa River in the Atacama Desert, Chile. *PeerJ, 8*, e9927. https://doi.org/10.7717/peerj.9927

Structures Associated with the Dynamics of Granitic Rock Emplacement (NW Portugal)

Jorge Pamplona, Benedito C. Rodrigues, Mark Peternell, Alex Lorenz, Alex Schimdt, Melissa Mengert, Thomas Altmeyer, and Jonas Köpping

Abstract This chapter proposes a fieldtrip on NW of Portugal (Oporto metropolitan area), which is a sequence of thematic stops. The stops began on deep crustal level at a magmatic feeder zone—Leça da Palmeira Metamorphic Complex, and follows to a middle-upper crustal level with magmatic features associated with magmatic chamber dynamics of post-orogenic biotite granite (Bt-granites)—the Lavadores granite. Each theme is methodological divided into several sections: short introductory text about each feature highlighted on the outcrops; stop description supported with sketches and photos illustrating what can be really seen at the outcrops; field and/or drawing activity; discussion and conclusion section, focused on new approaches and interpretations of the geological results. On Leça Palmeira Metamorphic Complex the major issue is a granite-tonalite relationship on a deep shear zone, where a magmatic feeder zone was evolved as a gneissic complex. The Lavadores granite shows all a sequence of mesostructures related with the interaction between mafic microgranular enclaves and the host granite; a morphological potash feldspars (Kfs) classification is used; hybridization mechanism involving Kfs and enclaves motion and an Enclave Disruption Mechanism (EDM) is proposed; feldsphatic plume structure recorded the feeding process into the magmatic chamber

Keywords Granitic rocks · Metamorphic complex · Magmatic structures · Magmatic chamber dynamics · Mafic micogranular enclaves

J. Pamplona (✉)
ICT—Instituto de Ciências da Terra, Pólo da Universidade Do Minho, Campus de Gualtar, 4710-057 Braga, Portugal
e-mail: jopamp@dct.uminho.pt

B. C. Rodrigues
CCT—Centro de Ciências da Terra, Universidade Do Minho, Campus de Gualtar, 4710-057 Braga, Portugal

M. Peternell
Institutionen För Geovetenskaper, Göteborgs Universitet, Guldhedsgatan 5C, 41320 Göteborg, Sweden

A. Lorenz · A. Schimdt · M. Mengert · T. Altmeyer · J. Köpping
Tectonophysics, Institute of Geoscience, University of Mainz, 55128 Mainz, Germany

S. Mukherjee (ed.), *Structural Geology and Tectonics Field Guidebook–Volume 2*, Springer Geology, https://doi.org/10.1007/978-3-031-19576-1_2

Abbreviations

A	Area of the grain
AMS	Anisotropy of magnetic susceptibility
BLG	Bulging recrystallization
Bt	Biotite
C	Circularity
C/C'-S	C, shear plane surface; C', secondary shear plane surface; S, foliation surface
CIZ	Central Iberian Zone
CMF	Critical melt fraction
cm-sized	Centimetric sized
CRSS	Critical resolved shear stress
CSD	Crystal size distribution
CZ	Cantabrian Zone
D_1	First Variscan deformation phase
D_2	Second Variscan deformation phase
D_3	Third Variscan deformation phase
DBDSZ	Dúrico-Beirã Ductile Shear Zone
dm-sized	Decametric sized
EAD	Equal area diameter
EDM	Enclave disruption mechanism
FD	Foz do Douro metamorphic complex
f_{O2}	Oxygen fugacity
GBM	Grain boundary migration
GN Bt	Biotitic gneiss
GN Bt-f	Fine grain biotite gneiss
GN Bt-rest	Biotitic gneiss with lenses of restitic material
GN Kfs	K-feldspar gneiss
GN Ton	Tonalitic gneiss
GN-2 m	Two-micas gneiss
GTMZ	Galicia Trás-os-Montes Zone
h1, h2, h3, h4	Stages enclave hybridization mechanism
HP	High pression
HREE	Heavy rare earth elements
HT	High temperature
IAA	Ibero-Armorican Arc
IL	Inner layer
Kfs	Potash feldspar
L	Length of grain longest diameter
L, R	Left and right side 1st order shear zone
L', R'	Left and right side 2nd order shear zone
L_{mag}	Magmatic lineation
LPMC or LP	Leça da Palmeira Metamorphic Complex

LPO	Lattice preferred orientation
LREE	Light rare earth elements
Lx	Stretching lineation
Mig	Pelitic migmatites
MLDSZ	Malpica-Lamego Ductile Shear Zone
MME	Mafic microgranular enclaves
Mnz	Monazite
Ø	Crystal fraction
OL	Outer layer
OMZ	Ossa-Morena Zone
P	"Primary/stable" microgranular enclave
Pg	Late granitic pegmatites dykes and sills
Pl	Plagioclase
PT1	First percolation threshold
PT2	Second percolation threshold
PTDSZ	Porto-Tomar Ductile Shear Zone
Qtz	Quartz
R	Grain geometry roundeness
REE	Rare earth elements
R_i	Initial ellipse ratio
Rf/ϕ	Finite strain method
R_s	Strain ellipse ratio
RT1	First rheological threshold
RT2	Second rheological threshold
Rb/Sr	Radiogenic method based on rubidium and strontium
SGR	Sub-grain rotation
S_{mag}	Magmatic foliation
SHRIMP	Sensitive high-resolution ion microprobe
Sill	Sillimanite
Sn	Foliation
SPO	Shape preferred orientation
SPZ	South Portuguese Zone
t	"Transient" microgranular enclave
t1, t2, t3	Transient stages on enclave disruption mechanism
U/Pb	Radiogenic method based on uranium and plumb
VVT	Vila Verde Thrust
WALZ	West Asturian-Leonese Zone
WR	Whole rock
YX-plane	Foliation plane
Zr	Zircon

Highlights

(1) Fieldtrip on NW of Portugal (Oporto metropolitan area) with a thematic stops approach.
(2) Magmatic features associated with magmatic chamber dynamics of a post-orogenic biotite granite and its mafic micogranular enclaves—proposal of new approaches.
(3) Magmatic feeder zone evolved as strike-slip shear zone on a metamorphic complex.

1 Introduction

1.1 A Thematic Field Book: The Concept

In this chapter, we propose a fieldtrip, which is thematically structured. During a more traditional field trip, geographic stops are chosen to show a sequence of outcrops, landscapes, etc., and based on geological points of interest.

Here, the field trip follows a sequence of thematic stops at approximately the same geographic locality. The thematic stops are related with deep crustal level magmatic feeder zone—Leça da Palmeira Metamorphic Complex and to magmatic features associated with a middle-upper crustal level magmatic chamber dynamics of post-orogenic biotite granite (Bt-granites)—the Lavadores granite (Porto, NW Portugal).

Within the Leça da Palmeira Metamorphic Complex the five thematic stops a one-half day field tripare (Sect. 2):

(1) Deformation microstructures on granitic gneisses (Sect. 2.2.1)
(2) Mesoscale shear zones (Sect. 2.2.2)
(3) Others structures (Sect. 2.2.3)
(4) Transvers to a mesoscale shear zone (Sect. 2.2.4)
(5) Magmatic feeder zone (Sect. 2.3).

Within the Lavadores granite, we choose five themes for a one-half day field trip (Sect. 3):

(1) Magmatic flow and lineation structures (Sect. 3.2)
(2) Schlieren (Sect. 3.3)
(3) Microgranulares enclaves (Sect. 3.4)
(4) Interaction between more felsic and basic magmas (Sect. 3.5)
(5) Feldspathic plume (Sect. 3.6).

Each theme is divided into:

(a) A short introductory text about each feature highlighted on the stop;
(b) A stop description supported with sketches and photos illustrating what can be really seen at the outcrops;

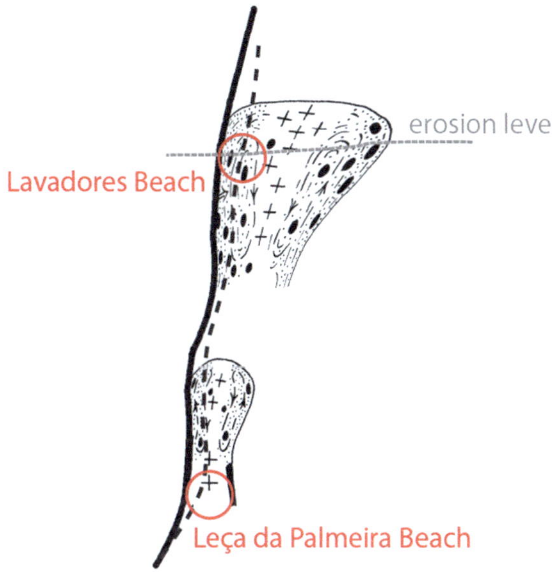

Fig. 1 Spatial location sketch of Leça da Palmeira and Lavadores outcrops on a magmatic arising chamber (base sketch adapted from Hutton, 1988)

(c) A field and/or hand drawing activity such as measuring long/short axes of feldspar phenocrysts or measuring angular values of C/C'-S structures;
(d) A discussion and conclusion section, were put a focus on new approaches and interpretations of the geological results.

A conceptual geometric link between Leça da Palmeira and Lavadores outcrops related with the arising and the emplacement process of a magmatic body is shown on Fig. 1.

1.2 NW Portugal Geological Framework

The Variscan Belt—Palaeozoic mountain system is ~1000 km broad and 8000 km long. It extends from the Caucasus to the Appalachian and Ouachita mountains of northern America (e.g. Matte, 2000). It is the result of the Late Paleozoic collision between Gondwana, Laurentia and Baltica with small and intermediate continental plates—microplates—and correspondent oceans between them (Franke, 1989; Hatcher, 1989; Martínez-Catalán, 2011; Martínez-Catalán et al., 1997; Matte, 2000) (Fig. 2).

The European Variscan Belt has a classical architecture including a metamorphic inner complex with mafic and ultramaphic rocks, multiply deformed plutons—pre-, syn- and post-orogenic, high angle shear zones, crystalline nappes, ophiolites, marginal slate and foreland fold-and-thrust belts, and fore deep basins (Fig. 2).

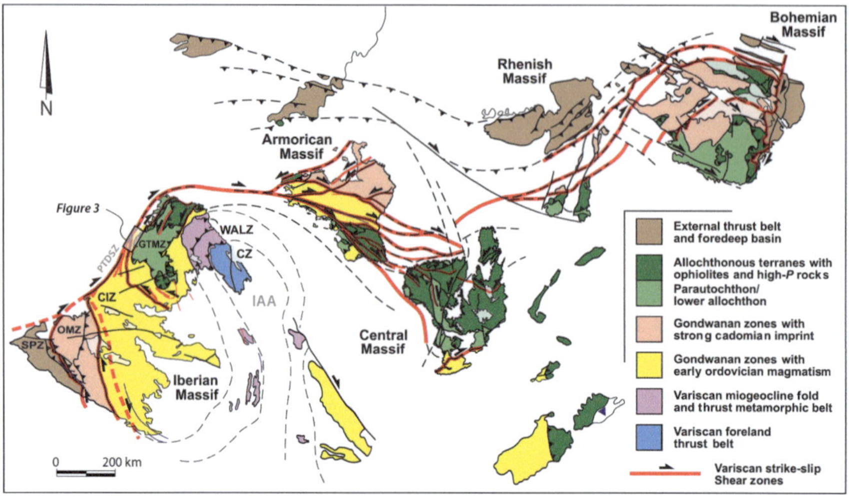

Fig. 2 Simplified sketch of European Variscan Belt. *IAA* Ibero-Armorican, *Arc* Iberian Massif Zones: *CIZ* Central Iberian Zone; *CZ* Cantabrian Zone; *GTMZ* Galicia-Trás-os-Montes Zone; *OMZ* Ossa-Morena Zone; *SPZ* South Portuguese Zone; *WALZ* West Asturian-Leonese Zone; *PTDSZ* Porto-Tomar Ductile Shear Zone (adapted from Martínez-Catalán (2011). The location of Fig. 3 is shown

As a relative autochthonous unit of Gondwana, the European Variscan Belt is organized around an internal zone, which extends from Iberian Massif (Central Iberian Zone—CIZ) to the Bohemian Massif (Teplá-Barrandian Zone), to an external fold-and-thrust belt and fore deep basin that is well known on the Rhenish Massif and on the South Portuguese Zone—SPZ (Martínez-Catalán, 2011). Despite regional variations, this model increases its complexity with the emplacement of nappes of allochthonous terranes—mainly ophiolites and HP metamorphism units—, the development of strike-slip ductile shear zones, and later brittle deformation (Franke, 2000; Martínez-Catalán, 1990, 2011).

The European Variscan Belt can also be split in a Late Devonian-Carboniferous tectonic-metamorphic event that is dominant in the northern realm (Rhenish Massif and Bohemian Massif) and in a major late Carboniferous-Permian collisional event superimposed on previous events that are prevalent on the southern realm (Armorican Massif and Iberian Massif). This interpretation results from several major orogenic cycles, which were superimposed in space and time (Schulmann et al., 2014).

The architecture of the European Variscan Belt is variously interpreted with complementary and contrasting models (Faure et al., 1997, 2009; Franke, 2000; Martínez-Catalán et al., 2009; Matte, 1986; Schulmann et al., 2009). The general accepted plate tectonic scenario for the Variscan Belt is based on a polyphase accretionary evolution during a bulk transpressive continental collision (e.g. Martínez-Catalán et al., 2009). One of the more remarkable resultant structures is the genesis of the Ibero-Armorican Arc—IAA, which links the Armorican Massif to the northwest of the Iberian Massif. Schulz (1858, in Ballèvre et al., 2014) made the first reference to the existence of this arcuate structure in Oviedo province (Spain). This structure received full acceptance after the work of Bard et al. (1971).

The Ibero-Armorican Arc of the Iberian Massif presents one tectonic-stratigraphic zonation, originally proposed by Lotze (1945), which currently includes three main terranes (Ribeiro, 2013): (i) The Iberian Terrane (Cantabrian Zone—CZ; West Asturian-Leonese Zone—WALZ; Central Iberian Zone—CIZ; Ossa-Morena Zone—OMZ), (ii) the South Portuguese Terrane (South Portuguese Zone—SPZ), and (iii) exotic terranes. This tectonic-stratigraphic zonation is currently under discussion (e.g. Arenas et al., 2016).

Large stacked nappes overly the Central Iberian Zone, following the general sequence: parautochthonous thrust sheet, and allochthonous continental and ophiolitic nappe stack. These zones are sutured by major tectonic structures such as complex strike-slip shear zones (e.g. Porto-Tomar Ductile Shear Zone—PTDSZ; Burg et al., 1981; Dias et al., 2013; Ribeiro et al., 1980, 1995) and basal thrusts (e.g. Vila Verde Thrust—VVT; Ribeiro & Pereira, 1986; Ribeiro et al., 1990; Pamplona et al., 2006). These terrains were, then, affected by late Variscan crustal sub-vertical strike-slip ductile shear zones, with the Malpica-Lamego Ductile Shear Zone—MLDSZ (Ferreira et al., 1987; Llana-Fúnez & Marco, 1998; Pamplona et al., 2016) is one of the most important on the NW of Iberia (Fig. 3).

1.3 NW Iberian Granitic Rocks—Some General Notes

The section is mainly based on selected papers about NW Portugal Variscan belt granitic rocks: Dias et al. (1998), Ferreira et al. (1987), Neiva et al. (2009), Ribeiro et al. (2019), Valle Aguado et al. (2005).

During long time, the first magmatic event documented on the northwester part of the Central Iberian Zone was of Precambrian age (Cadomian Orogeny), recorded on the Miranda do Douro orthogneiss by Lancelot et al. (1985) and with an absolute age of 618 ± 9 Ma (U/Pb, Zr). This age was corrected to Early Ordovician by Bea et al. (2006) (483 ± 3 Ma—U/Pb, Zr) and Zeck et al. (2007) (496 ± 3 Ma—U/Pb, Zr). Taking this correction into account, the oldest Precambrian magmatism was related to the Foz do Douro biotitic orthogneiss dated by Andrade et al. (1983) (560 ± 10 Ma—Rb/Sr, WR) and confirmed by works of Noronha and Leterrier (1995) (575 ± 5 Ma—Rb/Sr, WR) and Leterrier and Noronha (1998) (567 ± 6 Ma—U/Pb, Zr). As a scientific curiosity, this age has again been questioned by Pinto et al. (1987) and Sousa et al. (2014). Using the reference isochrones, lead to a revised age of 604 Ma (Pinto et al., 1987), and U/Pb (SHRIMP) analysis on zircons revealed an emplacement age of 452–456 ± 7.8 Ma (Sousa et al., 2014).

In a few cases, the Cambrian tectonic framework would have allowed the genesis of granitic crustal magma batches. Possible examples for such an event could have been the Gandra orthogneiss (Ferreira et al., 1987) and the Figueiró dos Vinhos granite (Pereira & Macedo, 1983, 514 ± 9 Ma—Rb/Sr, Bt).

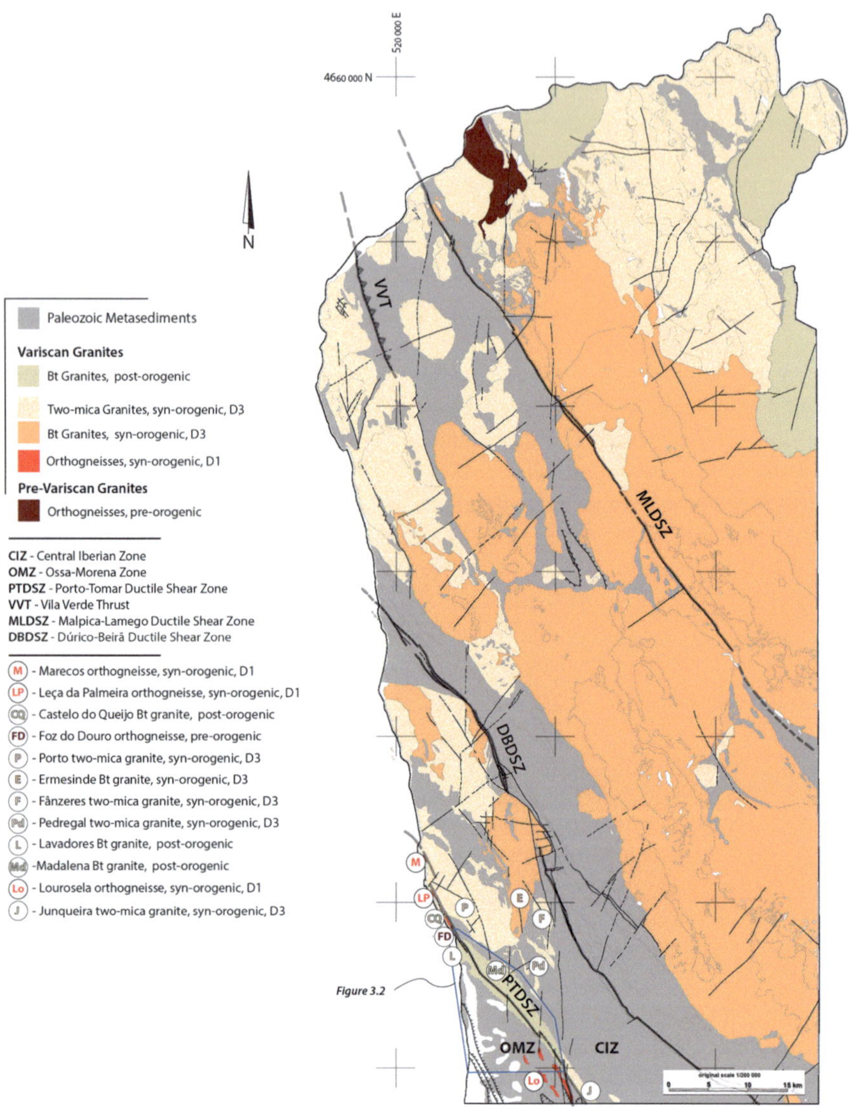

Fig. 3 NW of Portugal simplified geological map. There are represented the most significant igneous events: the pre-Variscan igneous rocks; the Variscan igneous rocks, splitting between Bt and two-mica granites; and post-Variscan granitic rocks, which include the Lavadores Granite [adapted from Fernandez et al. (2003) and Pereira (1989)]. The location of Figs. 5 and 28 are shown

During the Ordovician the post-variscan magmatic events result in magmatic rocks of granitic-granodioritic compositions (e.g. Oledo granite, Antunes et al., 2009, 481 ± 1 Ma—U/Pb, Zr). This event was followed up by a granitic magmatic and result e.g. in the Galicia orthogneisses dated by Priem (Pinto et al., 1987), 469 ± 8 Ma—Rb/Sr, WR for the Vigo orthogneiss).

With the intention to build a global framework for all variscan related granitic rocks at CIZ and Galicia Trás-os-Monte Zone (GTMZ), Ferreira et al. (1987) developed a model that links the generation of granitic magmatic rocks with the three mains variscan deformation phases (Noronha et al., 1979; Ribeiro, 1974). These pre-orogenic granites, synorogenic granites (syn-D_1, syn-D_2, syn-D_3 and late- to post-D_3), and late- to post-orogenic granites will be discussed in more detail in this chapter.

1.3.1 Pre-Orogenic Granites (Pre-Variscan Granites)

Pre-Variscan granites evidences of a Precambrianbasement, and are rare and only represented by the Foz do Douro biotitic orthogneiss (Fig. 2; Andrade et al., 1983; 560 ± 10 Ma—Rb/Sr, WR) and the Fontoura/Gandra ortogneiss (Ferreira et al., 1987).

1.3.2 Synorogenic Granites

The Devonian age emplacement of the Oliveira de Azeméis—Porto blastomylonitic belt syn-D_1 granitic rocks is controlled by the PTDSZ geometry and kinematics, with the occurrence of e.g. Lourosela orthogneisse (Fig. 2; Chaminé et al., 1998; 421 ± 4 Ma—U/Pb, Zr), Leça da Palmeira and Mareco orthogneisse (Ferreira et al., 1987).

Two-mica granite syn-D_2 intrusions are spatially linked with major thrust planes that are interpreted as top-to-the-East variscan allochthonous basal nappes. This episode marks the North Variscan ocean closure (Pamplona et al., 2006; Pereira, 1988).

On the CIZ and GTMZ northwestern tip, Bt and two-mica syn-orogenic granites (Variscan D_3) are the main units. This granitic association is typically generated during continental two-plate convergence and collision (e.g., Chappell & White, 1992; Lefort, 1981). After the subduction and/or obduction onset, the continental collision with crustal thickening started, which partially melted the continental crust (Barbarin, 1999). The typical continental collision associated with these granites is of oblique type (Pitcher, 1979; Ribeiro, 1984).

During ongoing intracontinental collision, two-mica granites melt is generated. Two-mica granites are related to the rework of former shear planes as wrench faults, and with thermal domes ascending on D_3 antiforms (Pereira, 1988).

A previous tectonic anisotropy (ductile shear zones) and a favorable regional metamorphic pathway are mandatory factors for generation and the up-rise of granitic magmas. This relationship explains the reason why the root zone of two-mica granites derives from major crustal ductile, transcurrent or thrusting-type shear zones (e.g. Barbarin, 1996).

The huge amount of granitic rocks—two-mica and Bt-granites—related with the D_3 variscan orogenic phase, immediately raises the classical space problem. According Pereira (1988) this magma volume only can ascent and emplace because of a tensional gap between continental convergence and collisional compressive regimes. In fact, on a regional scale, a distentional period between D_2 and D_3 has been recorded, and was related with the eastward movement of allochthonous nappes.

E.g. syn-orogenic two-mica granites are represented by the Cabeceiras de Basto Granite (Almeida et al., 1998, 311 ± 1 Ma—U/Pb, Zr Mnz) and by the classical Porto Granite (Martins et al., 2001, 318 ± 2 Ma—U/Pb, Mnz). The latter is another good example of multiple and discordant geochronological data produced over times: since eight different ages are reported, ranging from 354 ± 58 Ma (Mendes, 1968, Rb/Sr, WR) to 306 ± 7 Ma (Almeida et al., 2014, Rb/Sr, WR).

The other main group of syn-orogenic granitic rocks (syn-D_3, late-D_3 and late to post-D_3) are the biotitic granites with calcic plagioclase (Pereira, 1989). These bt-granites can have a deep crustalorigin, contain granulitic restites, and result in some cases from partial melting from gneisses and, pelitic and quartz-feldspar schists (Albuquerque, 1971).

The most deformed Bt-granites (syn-D_3)—the older granodiorites on the Capdevila et al. (1973) classification framework—are linked to four NW–SE major tectonic lineaments (Ferreira et al., 1987). They are named the occidental lineament (DBDSZ, Fig. 3), the intermediate lineament (MLDSZ, Fig. 2), with e.g. the Sameiro granite 316 ± 2 Ma (Dias et al., 1998, U/Pb, Zr Mnz), the eastern lineament and, external northeastern lineament.

According to Ferreira et al. (1987) the late-D_3 Bt granites occur along the MLDSZ (e.g. Celeirós granite 306 ± 2 Ma, Dias et al., 1998, U/Pb, Zr Mnz) and the DBDSZ marginal zones. The late- to post-D3 Bt granites are intrusive into the axial zone of previous late-D_3 granitic rocks (e.g. Ferreira de Aves granite 295 ± 11 Ma, Costa et al., 2006, K/Ar, Ms). Their emplacement is controlled by conjugate fault sets (NW–SE and NE–SW) and NNW–SSE fractures (Fig. 3).

1.3.3 Post-Orogenic Granites

The uprise and emplacement of late- to post-orogenic granites such as the Lavadores Granite (Fig. 3), occurred from Late Carboniferous to Permian ages. In this period, the regional tectonic extensional direction (σ_3) changed from N–S (Upper Pennsylvanian) to E–W (Permian), causing the fracturing of Variscan terrains and the development of late-Variscan NW–SE and NNE–SSW strike-slip faults, which improve the ascent and emplacement of this granitic batholiths (Pereira, 1988).

Table 1 shows a chronological sequence of granitic rocks correlated to the PTDSZ and organized into orthogneisses, and two-mica and biotite types, both are spatially

Table 1 Chronological sequence of the granitic rocks correlated to the PTDSZ and spatially related with the Lavadores Granite

			Orthogneisses	Two-mica Granites	Bt Granites
Permian Carboniferous (Upper Pennsylvanian)	*Late to Post-orogenic*				Castelo do Queijo[a] (292 ± 15 Ma) Lavadores[b] (298 ± 11 Ma)
Carboniferous (Middle-Upper Pennsylvanian)	*Syn-orogenic*	D3		Junqueira[c] (308 ± 1 Ma) Pedregal[d] (311 ± 5 Ma) Porto[e] (318 ± 2 Ma) Fânzeres[f] (332 ± 11 Ma)	Ermesinde[g]
Devonian		D1	Leça da Palmeira[h] Marecos[h] Lourosela[i] (421 ± 4 Ma)		
Precambrian (Proterozoic)	*Pre-orogenic*		Foz do Douro[j] (Bt gneisse) (560 ± 11 Ma)		

(a)—Mendes (1968), (b)—Martins et al. (2011), (c)—Valle Aguado et al. (2005), (d)—Ferreira et al. (2014), (e)—Martins et al. (2001), (f)—Pinto (1984), (g)—Carrington da Costa and Teixeira (1957), (h)—Ferreira et al. (1987), (i)—Chaminé et al. (1998), (j)—Andrade et al. (1983)

related with the Lavadores Granite. Related with Variscan orogeny is represented pre-orogenic to post-orogenic granitic rocks.

2 Leça Da Palmeira Beach

In order to get to Leça da Palmeira Beach takes the highway A28, north of Porto. Leave the highway approximately 4 km north of Porto at exit number 7, which is called Exponor/Sta Cruz do Bispo. Follow the signs to Exponor, immediately after the exit until you reach the "wave roundabout" near Tryp Porto Expo Hotel. Once there, follow the signs to "Praias" and cross two roundabouts in the direction of Leça da Palmeira lighthouse. Turn south in front of lighthouse. You reach Leça da Palmeira/Beijinhos Beach parking after approximately 750 m (Coordinates: 41.195614°N, −8.708933°W)—see details on Fig. 4.

Fig. 4 Road map and Leça da Palmeira beach outcrops aerial view (background image source: Google Earth Pro)

2.1 Brief Geological Setting and Outcrop General Map of Outcrop

The Leça da Palmeira Metamorphic Complex (LPMC or LP on Fig. 5) was pre- to syn-tectonically emplaced during the Variscan orogeny (Ferreira et al., 1987; Pereira, 1989). Taking the structural similarities with the Foz do Douro Metamorphic Complex (FD) into account—Fig. 5a, b—the LPMC could be older than the proposed Variscan age. Andrade et al. (1983) and Noronha and Leterrier (1995) indicated a late pre-Cambrian to early Cambrian age.

This metamorphic complex was controlled by the development of the Porto-Tomar Ductile Shear Zone two-micas gneiss (GN-2 m), flakes of pelitic migmatites (Mig) and late granitic pegmatites (Pg) and quartz (Qtz) dykes and sills.

Gneisses

Gneisses belong to regional metamorphic rocks, which were built under medium–low pressure and medium–low temperature conditions. The gneiss at Leça da Palmeira beach can be classified as orthogneiss of lower metamorphic conditions, probably at the transition from amphibolite to greenschist facies. Furthermore it can be subdivided into two varieties—K-feldspar gneiss (GN Kfs) and biotitic gneiss (GN Bt).

K-feldspar Gneiss—GN Kfs

K-feldspar gneiss occurs medium to coarse grained (Fig. 6). It contains a high amount of K-feldspar (45%), quartz (20%) and plagioclase (15%). In the outcrop, the K-feldspar gneiss shows orange to brownish color and feldspar grains occur in an irregular manner, according to the grain size. Some areas contain feldspar grains, with euhedral to subhedral shape, up to 3 cm in length and 1 cm in width. These grains partly show chemical zonation in macroscopic scale. Feldspar in thin section is partly sericitized and also shows chemical zonation. The K-feldspar gneiss shows moderate foliation, which is mainly built by biotite.

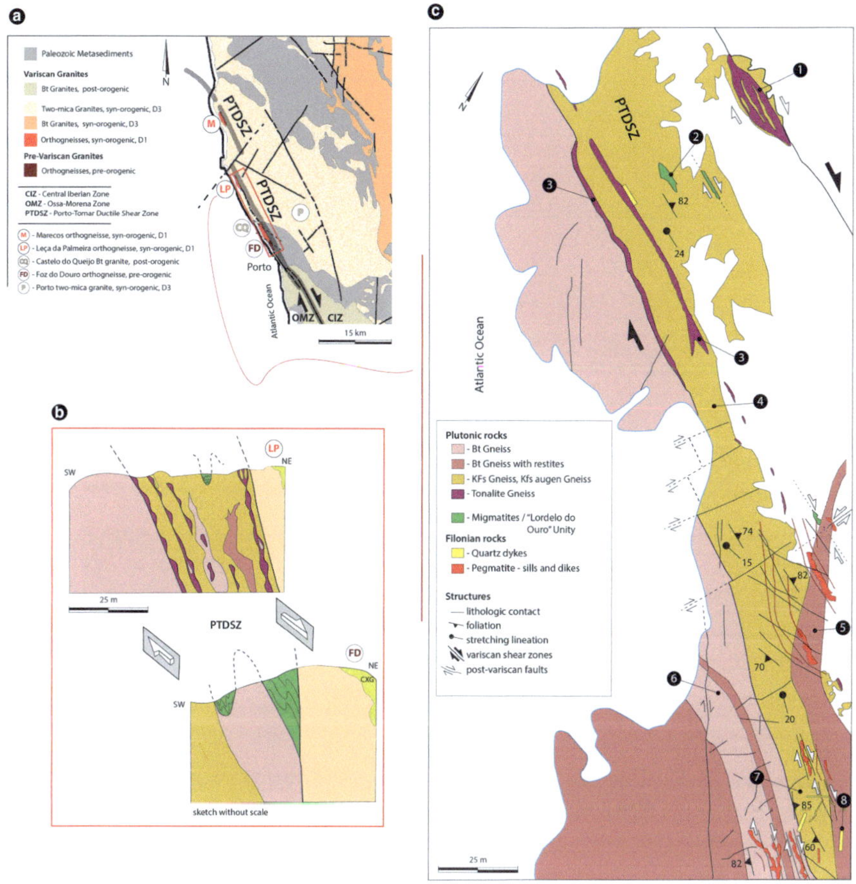

Fig. 5 Geological maps and cross-sections showing field relationships between lithological units of the LPMC and PTDSZ. **a** General geological sketch highlighting the position of Marecos, Leça da Palmeira and Foz do Douro along the PTDSZ. **b** Geological cross-sections along the Leça da Palmeira and Foz do Douro metamorphic complexes. **c** Geological map of the Leça da Palmeira beach, Beijinhos sector. Modified after Altmeyer et al. (2014) with fieldwork contributions of students from Minho and Porto Universities. Stops: 1—Tonalite, 2—Migmatite, 3—Magmatitic feeder zone, 4—Kfs Gneiss and Kfs Augen gneiss, 5—Bt Gneiss with restites, 6—Bt Gneiss, 7—Sinistral late shear zone, 8—Qtz dike

Biotitic Gneiss—GN Bt

The biotitic gneiss occurs mainly medium grained, but partly also contains feldspar grains with a size up to 1 cm in length and width (Fig. 7). In the outcrop, the biotitic gneiss appears in a slightly lighter color than the K-feldspar gneiss. It consists of ~35% K-feldspar, 20% quartz, 15% plagioclase and 5% muscovite. The foliationand the lineation are formed by biotite and muscovite.

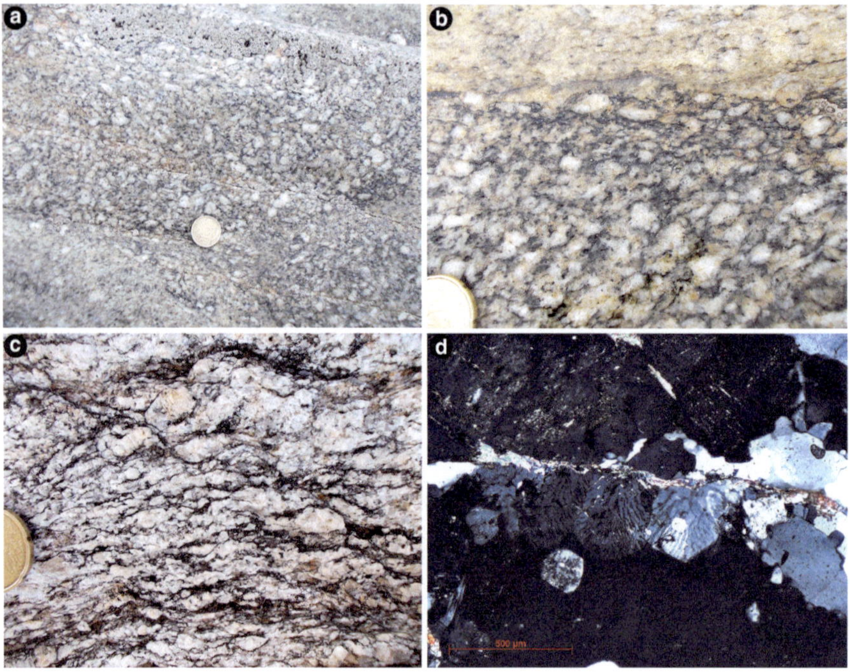

Fig. 6 Meso and microscope photos from Gneiss Kfs and Kfs augen gneiss. **a** GN Kfs-Aug on a random pattern of Kfs porphyroclats, showing a subtle—magmatic like—transition to the GN Kfs on the left bottom corner. **b** Transition between GN Kfs-augen (bottom) to GN-2 m. The metamorphic foliation (Sn) that is well marked as parallel planes on GN-2 m develops C^S foliations on GN-Kfs-augen. **c** Detail of GN-Kfs-augen with Kfs porphyroblasts sigma microstructures. **d** Myrmekitic development along Sn surface on Kfs crystal border (microphotography N+)

The biotitic gneiss can also develop a mesocratic, fine, equigranular grain texture—Fine grain biotite gneiss (**GN Bt-f**).

The biotitic gneiss with lenses of restitic material (GN Bt-rest) is a slight variation of the biotitic gneiss (Fig. 8). It occurs medium grained and also shows irregular appearance of larger K-feldspar minerals. It contains a large amount of biotite (~25–30%) and shows irregular appearance of large biotitic lenses that can be interpreted as restites. The lenses have a size up to 4 cm in length, partly with a sillimanitic core, and are highly eluviated. The mineral assemblage shows the same composition as biotitic gneiss, only with some variations in percentage. K-feldspar occur with an amount of 30%, quartz 25%, plagioclase 10%, muscovite 5% and chlorite < 5%. Just as in the biotitic gneiss form biotite and muscovite the foliation.

Tonalitic Gneiss—GN Ton

The field relationship of the Tonalitic Gneiss (GN Ton) at Leça da Palmeira beach with other gneisses looks like intrusive dikes (Fig. 9). The GN Ton occurs fine-grained with grey to brownish color and shows various flow structures marked by elongated

Fig. 7 Macro to microscope photos from Gneiss Bt and Bt-f. **a** GN Bt-f body installed on GN Bt medium so coarse grain with sin-magmatic relationship, exhibiting some interlayering. **b** Detail of contact GN Bt and GN Bt-f, highlighting the grain-to-grain, magmatic type, boundary. **c** Typical outcrop of GN Bt with medium to coarse grain with finer grain variations. **d** Chloritized biotite with a secondary Ms border around Pl grain (microphotography N+)

Bt flakes and ellipsoidal Qtz ocelli. The tonalitic gneiss is mainly equigranular, but contains some larger developed biotites. Biotite and hornblende occur with an amount of 25%, therefore it can be defined as melatonalite (> 40% mafic minerals). Biotite shows a moderate to distinctive parallel structure and occurs euhedral to subhedral, hornblende mainly subhedral. The felsic part of the GN ton is composed of quartz (25%) and plagioclase (25%), which is sometimes sericitized. Several plagioclase grains are almost euhedral and show polysynthetic twinning and chemical zonation. The tonalite contains further a small amount (< 3%) of opaque minerals, probably titanite or magnetite.

Metasediment rocks and migmatites

Migmatites—Mig

Migmatites originate byanatexis, the partial to complete melting of a metamorphic rock during regional metamorphism (e.g. Mukherjee & Koyi, 2010). It is described a medium to coarse-grained mingled rock with a content of two well-defined parts.

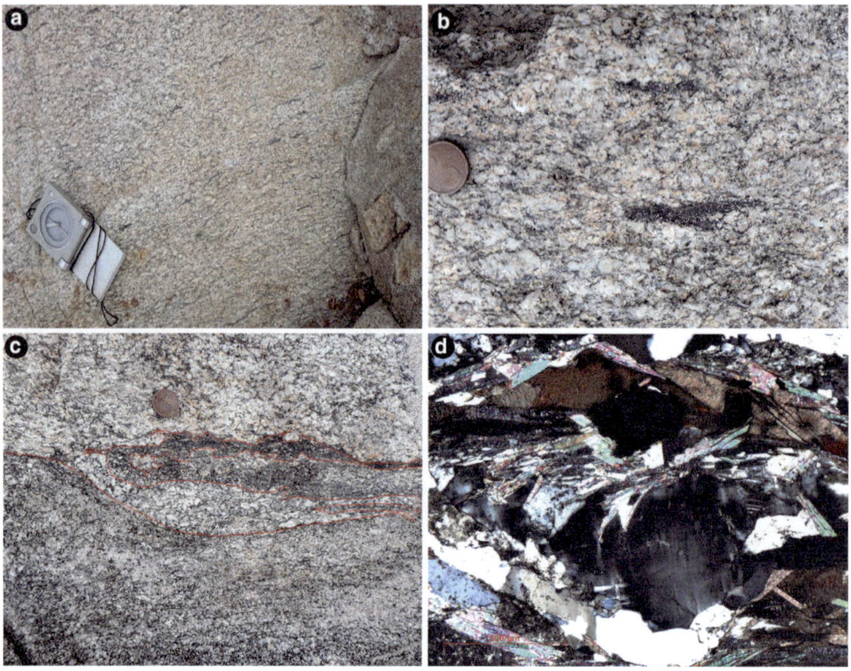

Fig. 8 Meso to microscope photos from Gneiss Bt with lenses of restitic material. **a** Outcrop of GN Bt-rest with parallel lenses of restitic material, enhancing the gneissic banding foliation. **b** Detail of the restites formed by biotite generally involving a sillimanitic core. **c** GN-Bt-rest (top) to GN-Bt (bottom) boundary with small apophysis (red lines) between them what testifies the igneous character of the relationship. **d** Biotitic lenses with secondary Ms parallel to gneissic banding (microphotography N+)

One part defines the largely unaltered metamorphic source rock, also called mesosome or palaeosome. The other part, the neosome, shows typical characteristics of igneous rocks and can be subdivided into two further parts—leucosome and melanosome (e.g. Menhert, 1968). The light colored (leucocratic) leucosome of Leça da Palmeira migmatites (Fig. 10) is rich in quartz and feldspar and the mineral grains show no preferred orientation. The fabric of the leucosome is medium to coarse grained and contains parts of larger grained pegmatites. The melanosome exhibits a previous texture (Sn) and is enriched with mafic minerals, which implies a darker (melanocratic) color (Fig. 10). It is mainly provided at the contact between mesosome and leucosome. The foliated melanosome indicates that during the diatexis, the dark restitesactivated and recrystallized. Vein-like structures with folded mesosome, leucosome and melanosome are common, as well as ptygmatic folding of leucosomes.

Fig. 9 Meso to microscope photos from Gneiss Tonalitic. **a** General outcrop of GN Ton dikes with small veins of GN Kfs catch inside. **b** 3-D image of a GN Ton dike emphasized by the differential weathering. **c** Boudin tip of GN Kfs inside a GN Ton matrix (detail on D). **d** Detail of the GN Ton/GN Kfs magmatic type boundary (red line) and of an ellipsoidal quartz ocelli inside GN Ton acting as a strain mark. **e** Microphotographs sequence of N//, N+ and Δ550 highlighting the gneissic deformation (dashed red line), Bt2 and new Pl2 + Qtz2 cross-cutting the magmatic feldspar (Pl) and quartz (Qtz) grains

Fig. 10 Meso to microscope photos from Migmatite. **a** Phlebitic structure; **b** Folded and boudinated neosomic veins on migmatite outcrop (Sn-1 migmatitic foliation) in contact with GN Bt (Sn gneissic banding). **c** Gradual and syngenetic contact between folded migmatite (bottom) and KFs gneiss (top). **d** Folding of Sn-1 on paleosome (bottom) with quartz feldspathic neosome (top) (microphotography N+)

Filonian rocks

Pegmatite lenses, dykes and sills—Pg

Granititic pegmatites (Pg) are coarse-grained rocks with variable fabric characteristics. The pegmatites at Leça da Palmeira beach are irregular shaped, as lenses, sills or dykes (Fig. 11a, b). Locally, the boundaries are diffuse, have a schlieren-type shape, or are sharp. The grain sizes vary from 0.5 to 5 cm and partly increase from the boundary to the center (comb structure). The pegmatite outer regions are neither structured nor zoned. The main minerals are K-feldspar and quartz. K-feldspar, mainly microcline, shows perthitic texture (exsolution lamellae). According to the rock composition, the Leça da Palmeira pegmatites correspond to granitic plutons.

Quartz lenses and dikes

Quartz lenses (Qtz) occur in late-open small fractures, crosscutting the main structures (Fig. 11c, d). They are of typical white to gray color and form boudin structures or small straight dikes. The majority of quartz dykes are not deformed, because of syntaxial growth of mineral fibers. This is shown on a one meter broad, and 15 m long and straight quartz dyke.

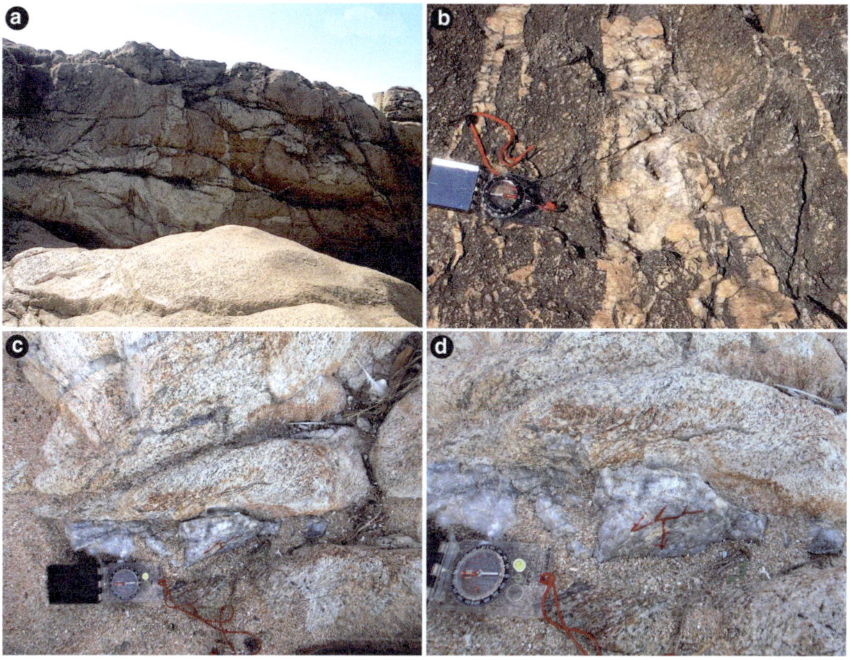

Fig. 11 Meso to microscope photos from filonian bodies. **a** Pegmatite horizontal sills in continuity with vertical dykes; **b** Comb structures in pegmatites. **c** Gray quartz infilling GN Kfs slightly boudinated **d** Detail of quartz boudin from picture C

2.2 *Regional Metamorphic Framework*

The regional metamorphic framework of Leça da Palmeira Metamorphic Complex is dominated by the rework of a major crustal ductile shear zone that it is proposed to be linked to the PTDSZ (see Fig. 5). This crustal ductile shear was active from the Cambrian (500 Ma) to Permian age (280 Ma), and indicates dextral kinematics during its whole life spam (Ribeiro et al., 2007).

The metamorphic event that affects the entire LPMC had a metamorphic peak at upper amphibolitic facies (zone of sillimanite—orthoclase) and at low to intermedium pressure conditions. The multiphase deformation events generated a NW–SE striking sub-vertical to the E plunging, penetrative foliation gneiss banding (Sn). The stretching lineation plunges at low-angle to SE.

2.2.1 STOP 1—Deformation Microstuctures

Regional deformation in association with a medium to high metamorphic grade crustal-scale shear zone, left an imprint on the related myloniticand gneissic rocks, indicated by several micro and mesostructures (e.g. Mukherjee & Mulchrone, 2015;

Passchier & Trouw, 2005). Theses microstructures allows the interpretation of shear zones: foliation zoning, orogenic stretching direction, deformation kinematics and vorticity, among others. The foliation related with the deformation is of the gneissic banding type—as we already describe for all the gneiss types outcropping at Leça da Palmeira beach (Sect. 2.1). They develop a clear C/C'-S foliation (Berthé et al., 1979a, 1979b; Mukherjee, 2013; Mukherjee et al., 2020) with dextral kinematics (Fig. 12a–c). In case of high strained regions (Fig. 12d), an ultramylonitic banding occurs parallel to C and S planes.

One of the most important structures is the stretching lineation (Lx) because it allows the identification of large-scale orogenic mass movement. In addition, the correct identification of Lx on foliation plane is the key to well understand all the other kinematic microstructures.

At Leça da Palmeira gneisses the Lx microstructure is identified by orientation of the sillimanite fibers inside restitic Bt-Sill lenses of the GN Bt-rest (Fig. 13a) or by the quartz stretched lenses on gneissic banding (Fig. 13b).

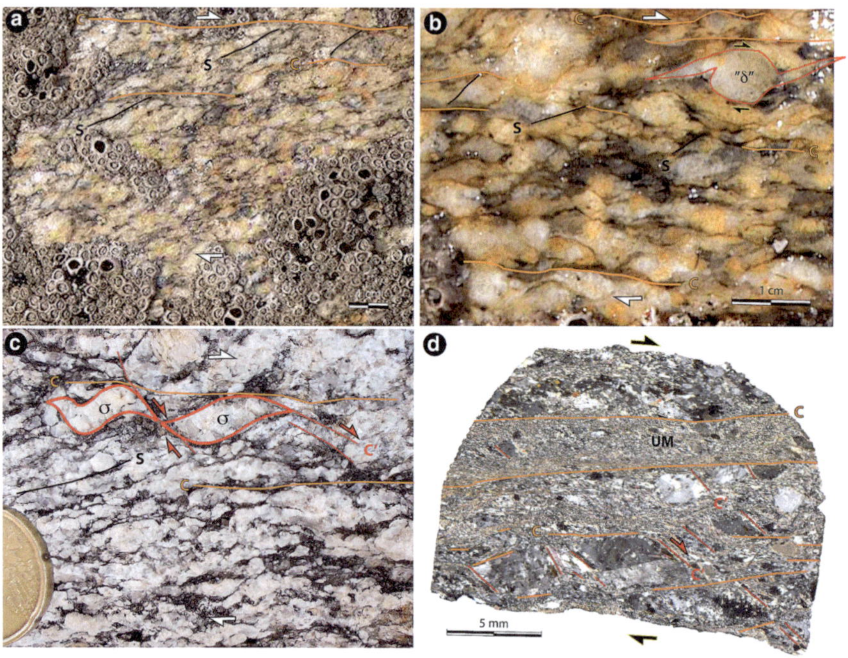

Fig. 12 C/C'-S microstructures. **a** C–S kinematic criteria on GN-Bt gneissic banding. **b** The same structure at a closer view with the C planes marked by recrystallized quartz bands. An initial stage of a delta porphyroclast development (δ) is seen at the right-up corner. **c** Sigma porphyroclasts (σ) highlighting the relationship between C, C' and S planes. **d** Microphotography of a GN-Bt revealing the development of ultramylonitic zones (UM) parallel to C surfaces (S is also almost parallel to C) and the C' surfaces plunging to the right giving a general dextral shear sense. (S—black lines; C—orange lines; C'—red lines)

Fig. 13 Gneissic stretching lineation. **a** Sillimanite fibers recording the stretching lineation inside Bt + Sill restitic lenses (GN Bt-rest). **b** Exposition of the Sn and shear planes of the GN Bt. The Lx direction on Sn plane can be seen. The shear plane shows a linear feature that corresponds to slickensides

In rare cases magmatic structures, relicts from the magmatic episode of the orthogneiss esorigin can appear. On the Fig. 14a a Kfs phenocryst "pincement" shows a paleo relative sinistral NW–SE magma flow. Otherwise, porphyroblasts on different stages of evolution can give a dextral kinematic criteria to the deformation along the shear zone: Ø—type object, σ—type object and δ—type object, respectively (Fig. 14b–d).

2.2.2 STOP 2—Meso-Scale Shear Zones

Mesoscales shear zones are quite frequent all around Leça da Palmeira gneiss outcrops. It is possible to distinguish meso-scalemedium—high grade (Scholz, 1988) dextral shear zones with strike NW–SE and steep dipping to East, parallel to PTDSZ, from that ones, with lesser importance and dimension, that belongs to latter deformation episodes.

Meso-shear zones transpose the gneissic banding on deformation corridors. On these corridors GN Bt, GN Fks, migmatites and metatonalites (GN Ton) are associated. Different types of meso-structures could be seen: boudin trains, isolated boudins, migmatitic folding and asymmetric folding in quartz veins (Fig. 15a). The presence of the paragenetic association of Sill + Qtz + Bt + Pl + Kfsandthe kind of related structures indicates conditions of deformation on Amphibolitic Facies.

Meso scale shear zones crosscut all the previous structures enhanced their later character. These structures can strike N110°, N090°or N045° with steep dipping. Two types were identified: low-grade type with only a very narrow fault zone and a remarkable gneissic banding drag (Fig. 15b, example of a sinistral strike-slip shear zone; Mukherjee (2014), for review on drag pattern in general); high-grade type with a centimetric ultramylonitic zone with abundant quartz and feldspar dynamic recrystallization embedding rounded feldspar grains (Fig. 15c, example of a dextral strike-slip ultramylonite shear zone). Symmetric structures (e.g. Mukherjee, 2017) are present

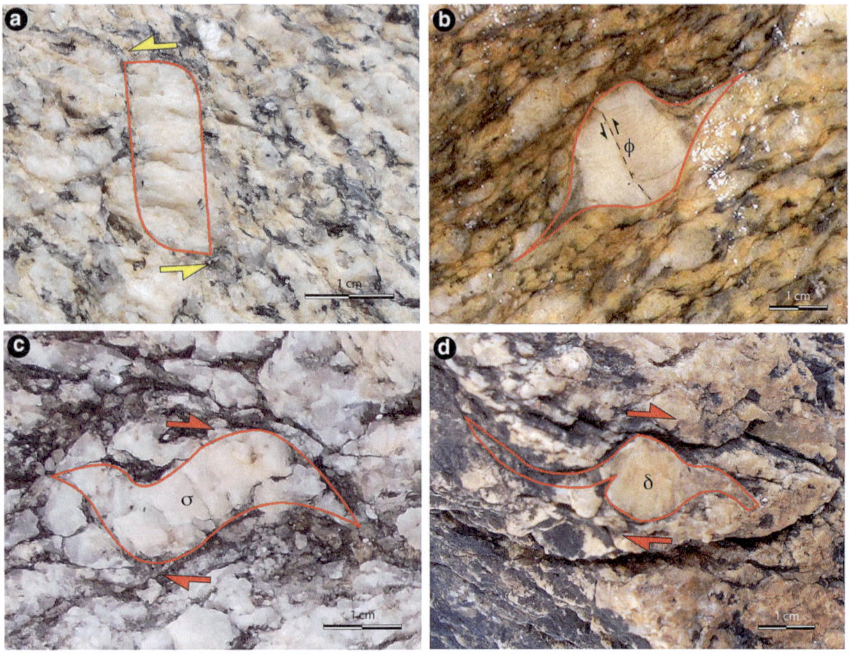

Fig. 14 Relict phenocrysts, and porphyroblasts giving dextral kinematic criteria. **a** "Pincement" of a relict Kfs phenocryst indicative of an igneous origin to the gneiss. **b** Porphyroblast Ø—type object with an internal antithetical domino movement making a dextral kinematic interpretation easier. **c** Porphyroblast of σ—type object. **d** Porphyroblast δ—type object

and are not useful in shear sense determination. Besides, a brittle latter fault with an infill structure (cryptocrystalline quartz?) remembering pseudotachylites can also occur (Fig. 15d).

2.2.3 STOP 3—Other Structures

Outcrops show other kind of structures and microstructures. Some of them are magmatic (Fig. 16a, magmatic oscillatory zoning in plagioclase), previous to the gneissic deformation, others are synchronous with the main deformation (Fig. 16b, isolated boudin) and, finally, some are later to the main deformation episode (Fig. 16c, syntaxial quartz vein).

Table 2 shows a summary of the different local deformation episodes with the respective orientations and associated structures.

Fig. 15 Macro- to meso-scale shear zones. **a** Macro scale shear zone parallel and into the PTDSZ. **b** Mesoscale shear zone with gneiss banding drag. **c** Mesoscale shear zone with ultramilonitization of gneissic matrix with quartz and feldspar dynamic recrystallization. **d** Mesoscale brittle shear zone with a pseudotachylite lookalike structure

2.2.4 STOP 4—Transvers to a Mesoscale Shear Zone (Profile 44)

Several later ductile shear zones crosscut the gneissic massif. It was performed two detailed profiles on N120° and on a N090° meso-scale shear zones. For this field guide it was chosen the N120° shear zone. On this meso-scale shear zone a transverse sampling was made with five samples collected at different distance to the shear zone center (Fig. 17): a—sample at 17.0 cm, it is the most distant and probably the one with the regional gneiss textural characteristics; b—sample at 11.0 cm; c—sample at 5.5 cm; d—sample at 2.0 cm; and e—sample on the center of the shear zone, it is most affected by the this deformation event.

The description of the results referring to the grain geometry and the grain texture of the samples A to E of profile 44 (Fig. 17) are graphically shown as line plots, histogram and rose diagrams. In this context, calculated values are plotted against their distance to the center of the shear zone—transversal profile on the Fig. 18. Set in relation to the parameters area, roundness, and circularity, the description is

Fig. 16 Magmatic to post-deformation structures. **a** Magmatic oscillatory zoning in plagioclase into a gneissic rock (GN Bt). **b** Asymmetric isolated quartz bound giving dextral shear kinematic criteria. **c** Syntaxial growth of a latter quartz vein

Table 2 Episodes of local deformation

Local deformation episodes	Orientation	Structures
Ante-D_1	Not determined	Migmatitic paleossomic foliation transposing an older metamorphic foliation
D_1	Sn = N115°/82°NE Lx = 20°/110°	Gneissic banding (Sn) with the development of C/C'-S with dextral kinematics; ultramylonites banding
D_2	N140°/subvertical	Meso-scale dextral shear zones
D_3	N110° to N045°	Meso-scale later sinistral ultramylonitic shear zones

divided into two parts, which attend to K-feldspar and quartz minerals. Furthermore, R_i (initial ellipse ratio) and R_s (strain ellipse ratio) values of quartz, calculated with the Rf/φ-method as well as its nucleation density are illustrated as line plots likewise. Other depicted parameters are the EAD of recrystallized K-feldspars (histogram) and the shape preferred orientation (long axial orientation) of K-feldspar and quartz in relation to their distance to the shear zone (rose diagrams).

Fig. 17 Drilled profile 44 from the outside (A) into the center of the shear zone (E). Shear zone with orientation N120°

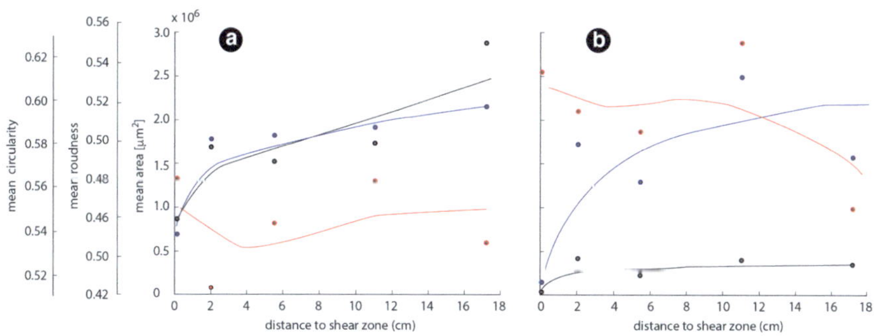

Fig. 18 Grain geometry parameters for K-feldspar (**a**) and Quartz (**b**) on a transversal profile to the shear zone: mean area (black line), mean roundness (blue line) and circularity (red line)

Grain Geometry

Area (A), Roundness (R) and Circularity (C)

To describe the grain geometry Area (A), Roundness (R) and Circularity (C) were used. All this parameters were obtained by digital means of a toolbox for quantitative textural analysis on Matlab and polyLX (Lexa, 2003). Area—A, Polyhedral object area of the grain; Roundness—$R = \frac{4A}{\pi L^2}$; Circularity—$C = \frac{4A}{PL}$; A: area of the grain, L: Length of the longest diameter of the grain and P: Perimeter of the grain.

The mean values of grain area are presented in square microns (Fig. 18). On Kfs, a reduction tendency in area is clear from about 2.8×10^6 µm² (17 cm distance to shear zone) to 0.9 × 10^6 µm^2 (shear zone center)—a 3 × grain means area reduction. Compared to the area of Kfs, measured values for Qtz minerals are much smaller and are located in a range of 0.35 × 10^6 µm^2 to 0.07 × 10^6 µm^2—a 5 × grain mean area reduction. Both minerals considered, a decreasing area is noticed from the outside to the center of the shear zone.

Relating to the parameter roundness, Kfs shows a hardly changing with a range from 0.517 (sample 44 A) to 0.453 (sample 44 E). Compared to the collected data ahead, this may look like a strong decay at first glance. Related to the range of roundness from 0 to 1, however, the difference of 11% is not so extreme. Similar to Kfs minerals, Qtz shows a constant tendency to reduction of roundness when getting closer to the shear zone center. Starting with a value of 0.487 (sample 44 A) its roundness rises to 0.427 inside the shear zone (sample 44 E). Noisy part ahead the roundness reduction of ~12% is almost the same of the Kfs.

The circularity of Kfs can be divided into 3 sections: outside the shear zone with a value of about 0.545 (samples 44 A, B, C), the border of the shear zone with a lower circularity of 0.513 (sample 44 D), and the center of the shear zone with the highest value measured of 0.564 (sample 44 E) even still increasing. Compared to the circularity of Kfs, values of Qtz minerals are located in a range of 0.549–0.623 and are therefore classified higher. Starting with value of 0.549 (sample 44 A) a value of 0.611 in the center of the shear zone can be observed (sample 44 E).

Taking in account all the three parameters together, it is possible conclude that general tendency of the deformation related to the shear zone is: reduction of area and roundness of Fks and Qtz recrystallized grains with a gentle increment of grain circularity.

Equal Area Diameter (EAD)

The Equal Area Diameter (EAD $= 2 \times \sqrt{\frac{A}{\pi}}$, A: area of the grain) measurements were made only on sample in the center of the shear zone (Sample 44 E) (Fig. 19). Recrystallized Kfs exhibit a mean EAD of 96.14 µm. With consideration of the standard deviation, values of 58.71–133.57 µm can be noticed. Otherwise, the EAD of recrystallized Qtz minerals gives a measured mean EAD of 49.5 µm plus minus the standard deviation of 11 µm.

Crystal Size Distribution (CSD)

The trend of nucleation density (n_0) of quartz minerals in relation to their distance to the shear zone is shown on Fig. 20. Overall, it is sensible to split it into two segments. The first contains all samples outside the shear with an average ofnucleation density (mm^{-4} = number of crystal per size per volume) of 0.21 * 10^{-10} mm^{-4}. Subsequently, the second part follows with a steep increase to 2.70 * 10^{-10} mm^{-4} and is located inside the shear zone.

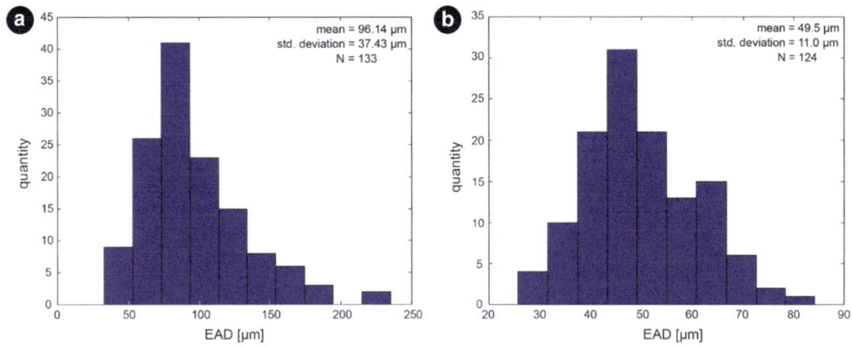

Fig. 19 EADs of recrystallized Kfs minerals (**a**) and recrystallized Qtz minerals (**b**) within the center of the shear zone (sample 44 E)

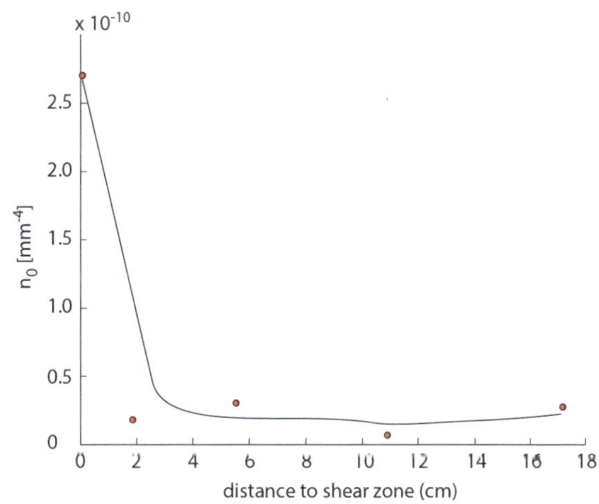

Fig. 20 Trend of nucleation density of quartz in relation to its distance to the shear zone

Shape and Lattice Preferred Orientation of Quartz

Shape Preferred Orientation (SPO)

Shape preferred orientation is concerned with the long axial orientation of Kfs and Qtz minerals (Fig. 21). In former case the long axial orientation of sample 44 A is aligned to NE with a mean direction of circa N036°. Sample 44 B exhibits a broad distribution for which reason no mean direction can be defined. Closer to the shear zone, sample 44 C, the preferred orientation is SE with a given mean direction of N117°. The following, sample 44 D, is placed 2 cm apart from the shear zone and falls out of alignment. No adjustment (E–W) is identifiable, but at least a mean direction of N021° becomes evident. Inside the shear zone, the long axial orientation is aligned to E–W and displays just a small diversification. Furthermore, the mean direction of N094° almost exactly constitutes the E–W direction.

Compared to the long axial orientation of Kfs, Qtz behaves in a similar way. The first two samples outside the shear zone, sample 44 A and B, are aligned to NE with a mean direction of N023°. Sample 44 C shows an alignment to E-W, nevertheless, with a major dispersion. As stated above to Kfs, sample 44 D falls out of alignment. This applies to Qtz as well. Similar to Kfs, Qtz minerals at that place are oriented to NE. The mean direction amounts to N046°. Within the shear zone a perfect E-W orientation is reached. There is just a minor distribution and a mean orientation of N092°.

Lattice Preferred Orientation (LPO)

While deformation and possible dynamic recrystallization of quartz grains occur, there is an influence to c-axis geometry and its lattice-preferred orientation. In the case of non-coaxial strain, some typical patterns of plotted c-axis pole figures are required (Passchier & Trouw, 2005). A non-coaxial deformation case is considered depending on prior acquired information (Sect. 2.2).

Figure 22 displays the results of plotted c-axis of profile 44. The labeled sinistral shear sense depends on our field observations and prior researches (Altmeyer et al., 2014).

Sample 44 A shows one clear partial vectored maximum in the lower pole area of X and a slight girdle along the periphery. This represents a major basal-<a> and rhomb-<a> slip activation. Further, a light prism-[c] slip could be interpreted. Between sample 44 A and 44 B a transition occurred. A part of the maximum migrated and shared now more regular to the Z poles whereby higher density is now located on the upper pole. A symmetry to foliation plane also occurred. Slip systems are still similar to sample 44 A. In sample 44 C a change could be established. Now, a clear concentration of density appears on the right side of pole figure. The distribution, more or less symmetric to shear plane, and consisting of one prominent maximum in X and two secondary maxima beneath the Z poles could be described as an inverse 'c'. This indicates a major activation of prism-[c] slip combined with lower basal-<a> and rhomb-<a> slip. Further a slight activation of prism-[a] could be initiated. This process is repeated in samples 44 D and 44 E combined with a light clockwise rotation. Maximum density of polar plots within profile 44 ranges from 2.01 at sample 44 B to 2.69 within sample 44 A.

Rf/ϕ-Method

There are a few methods, which can be used for strain analysis. The Rf/ϕ-Method first introduced by Ramsay in 1967 (Bose et al., 2018; Lisle, 1985; Ramsay & Huber, 1983) was used in this work.

In order to be able to employ this method it is crucial to have approximately circular or elliptical shapes on unstrained as well as on strained state. Furthermore, the grains have to display a significant display on the orientation. The evaluation of

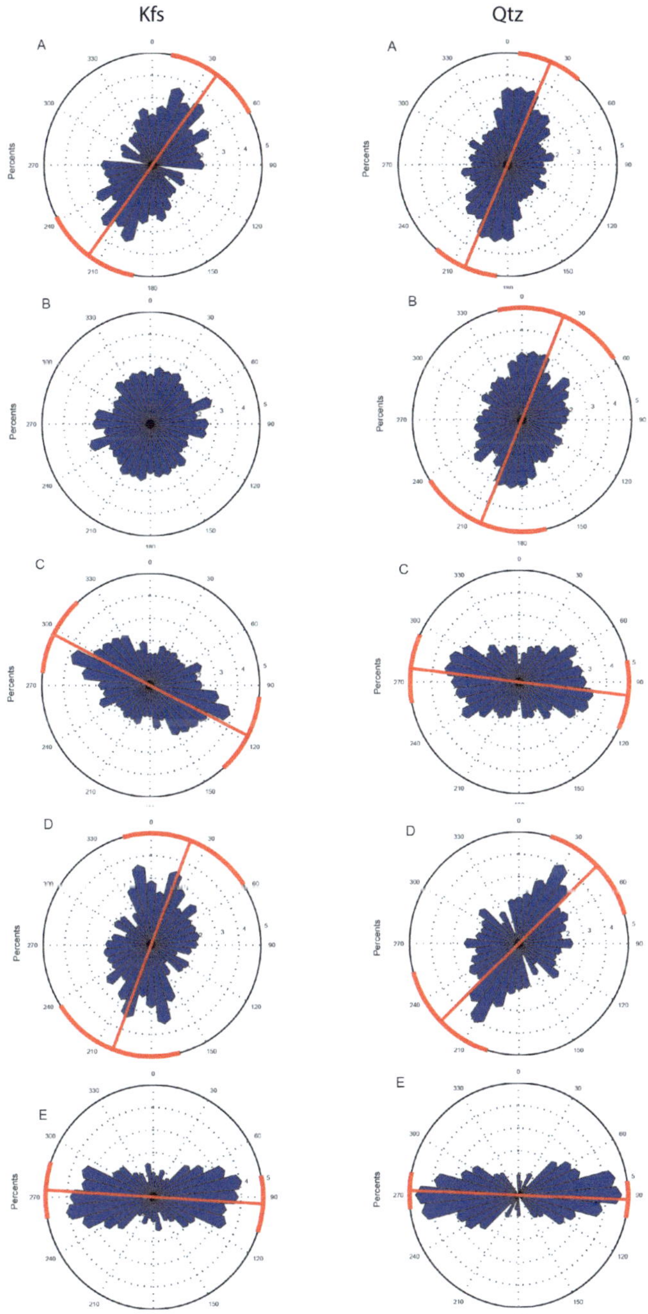

Fig. 21 Shape preferred orientation of Kfs and Qtz minerals. Labels A to E represent the position of the samples (Fig. 17)

Fig. 22 Pole figures of profile 44: horizontal line represents the foliation including lineation (black dots), oblique line describes shear plane. Plots are created with respect to the structural reference frame, Z normal to foliation, X parallel to lineation, Y perpendicular to X and Z. Equal area projection, lower hemisphere, multiples of uniform distribution (MUD), n gives number of measurements

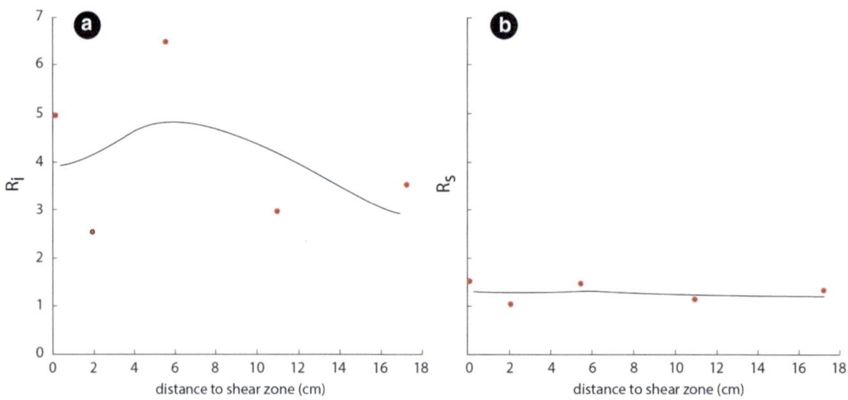

Fig. 23 R_i-values (**a**) and R_s-values (**b**) of quartz in relation to their distance to the shear zone

the Rf/ϕ-Method is only used for quartz minerals (Fig. 23), because on ortogneisses the quartzgrains (porphyroclasts) are the only that record the strain state and could be adequately measured. R_i as well as R_s values are selected from the given templates the Rf/ϕ-graph is plotted. These measurements can be split into three parts: samples 44 A, B and D show comparatively minor values of approximately $R_i = 3$. Samples 44 C and E show considerably high R_i-values by contrast: one outside, the other inside the shear zone.

While looking closely at the strain values it is striking that they are classified in a small range of R_s values between 1.1 and 1.5.

Discussion and Conclusions About Profile 44 Shear Zone

The low area of K-feldspar and quartz minerals inside the shear zone of profile 44 in comparison to the outside can be ascribed to a predominating high strain rate. Compared to general geological strain rates between $10^{-13}\ s^{-1}$ and $10^{-15}\ s^{-1}$ (Bose & Mukherjee, 2020; Carter & Tsenn, 1987; Paterson and Tobisch, 1992; Pfiffner & Ramsay, 1982) higher values of approximately $10^{-12}\ s^{-1}$ are to be accepted under the consideration of analyzing a high-strain zone. However, a declining strain rate to temperature ratio implicates a decrease in nucleation density to growth rate ratio with the result of developing coarser grain sizes. Moreover, this equation is comparably appropriate the other way round (Lexa, 2003). This case, taking all aspects into consideration, presumably eventuates in the profile of the shear zone 44. Except for the increasing strain rate from the outside to the center of the shear zone, latter one shows an additional increasing within the center of the shear zone, trending SE–NW direction. The resulting small grain sizes are also reflected in the steep increase of nucleation density of quartz minerals in profile 44.

Within the shear zone, minerals align their long axial orientation parallel to the X-axis of the finite strain ellipsoid and build the stretching lineation (Passchier & Trouw, 2005). On this understanding, the distinct orientation in E–W direction of the longaxial orientation of Kfs and Qtz minerals within the center of the shear zone 44 (sample 44 E) consequently shows the orientation of the finite strain ellipsoid. On the supposition that the X-axis of the strain ellipsoid is approximately perpendicular to the X-axis of the stress ellipsoid, the acted stress at that place shows an orientation in N–S direction.

Based on the change of the long axial orientation of minerals (SPO, Fig. 21) from the outside to the center of the shear zone 44, indications about the sense of shear can be obtained. Going from the outside to the center of the shear zone, an initial orientation in NE–SW direction is shown. Analyzing Fig. 21, an anti-clockwise rotation ending in E–W direction can be assumed. On this conjecture, a sinistral sense of shear may be accepted and can be verified with the help of the analysis of microstructures especially of foliation drag near shear zone.

The matter of fact of existing C/S fabrics within this shear zone enables to extrapolate to a medium metamorphic-graded shear zone and implicates temperature conditions of 450–600 °C (Passchier & Trouw, 2005).

Taking in account Equal Area Diameter of recrystallized Kfs minerals within the shear zone (sample 44 E), it is possible to obtain information about the peak temperature during the deformation. Measured values can be compared with a graph which is shown in Kruhl (2003) to read out the peak temperature that lead to the conclusion that temperatures between 515 and 545 °C were reached during the deformation inside this late shear zone.

The Passchier and Trouw (2005) values to metamorphic grade shear zone coincide with the measured temperature on EAD values of recrystallized Kfs.

Scattered measurements with Kfs EADs up to 235 μm indicate short-period peak temperatures of 570 °C. K-fs start to recrystallize at temperatures $\geq$ 500 °C (Voll, 1976, 1980). The small size of these minerals leads to the conclusion that there was just a minor time range with temperatures $\geq$ 500 °C.

The EAD of recrystallized Qtz minerals gives information about the differential stress during the deformation, paleopiezo meter of Twiss (1977). With the measured values of quartz EAD on sample 44 E, it results on a flow stress for quartz of approximately 40–60 MPa during this later deformation episode.

Concerning the examined sample 44 E no perfectly shaped mantle is identifiable. Instead, smaller areas with recrystallized Kfs show the beginning development of core-and-mantle structures and confirm the assumption of temperatures slightly > 500 °C affecting short-period of non-measurable areas.

As well known, there is a relation between the grain size of recrystallized Qtz minerals, the differential stress, which interacted during the deformation, and the temperature. Differential stress values between 40 and 60 MPa result from the measured mean grain size and fit to the foregone assertion of medium-graded conditions. Thus, temperatures between 515 and 570 °C can be affirmed.

All in all, the tectonic evolution of this area can be divided into different segments, which can be divided as follows. Starting with the magmatic development of the rock, a foliation defined by preferred orientation of platy minerals (e.g. biotite and muscovite) follows. Thereafter, the first episode of deformation, caused by a dextral shear movement, occurs and leads to a ductile deformation. During the shear movement, C/S fabrics come to existence where by shear bands transect the foliation. An uplift of the whole area up to shallow depth sentails changes in strain rate as well as in the acting differential stress, which are described above and causes the second episode of deformation, this time a ductile–brittle one.

The observations during microstructure analysis on profile 44 shows that bulging recrystallization (BLG) and sub-grain rotation (SGR) recrystallization occurred. The grain geometry additionally suggests a high temperature grain boundary migration (GBM) partly took place. That implies a relative increase in temperature. From the presence of lower temperature dynamic recrystallization appearances, a temperature from the transition stage between SGR and GBM is considered (Stipp et al., 2002).

The pole figures of profile 44 do not connect to the typical patterns that are given, for example in Passchier and Trouw (2005). Nevertheless a slight relation is established and will be discussed. At sample 44 A there is a maximum pole distribution that indicates the major activation of basal-<a> slip. Sample 44 B there is similar to sample 44-A, but with a planar symmetry around YX plane. Becoming closer to the shear zone, the pole maxima seem to be moving to Y direction. In addition to closer distance to shear zone this seems to be a change in major slip activities from prior basal-<a> to an activation of rhomb-<a> slip and prism-<c> slip systems.

In sample 44 D, amount of prism-[c] slip growths to symmetrical maxima next to X whereas basal-<a> slip loosing relevance. Sample 44 E, from the center of shear zone, shows a slight broader distribution and weaker significant maxima. Reason for this could be a later overprinting by later fluid circulation within the shear zone and as a consequence to this a slight influence to lattice preferred orientation.

The main factors determining dominant slip systems are temperature, strain rate and defect chemistry. Furthermore, it is assumed that temperature could be the most important factor to control the critical resolved shear stress (CRSS) on slip systems within quartz (Okudaira et al., 1995). With respect to the activated slip systems and the observed microstructures a peak temperature of 500–550 °C is considered (Stipp et al., 2002). Remarkable to this profile is the absence of prism-[a] slip activation. The main activation of prism-[a] slip activation occurs at a temperature of approximately 600 °C (Kurz et al., 2002).

2.3 STOP 5—Magmatic Feeder Zone

There are three models broadly accepted for explain the rising of magmas though a cold continental crust: pervasive migration, diking and the diapirism (Weinberg, 2003).

It is well known that felsic magmas can ascend through the cold continental crust via narrow channels associated with deep fractures (e.g. Shaw, 1980; Vigneresse, 1995; Weinberg, 2003). During this high-speed magma rising narrow brittle channels or ductile shear zones are used. They are the upper crust magmatic chamber feeder zones.

Plutonic complexes are sometimes related to processes occurring at depth, within deep-seated channels. However, the outcrops related with processes occurring in deep-seated magmatic conduits are not frequent. The eroded level of orogenic belts with ancient magmatic channels is rare, and in many cases these structures are not easily identified as magma conduits (Castro et al., 1995).

So, the opportunity to observe deep-seated magmatic feeder channels allows direct testimonies of dynamics and associated mixing and mingling phenomena that occurs repeatedly over time.

The field relationship of these feeder zones show composite bodies, typically with strong foliated granitic rocks include small-elongated more basic rocks (Castro et al., 1995).

At Leça da Palmeira beach outcrops a magmatic feeder zone composed by coarse-grained biotite granitic gneiss (GN-Bt) and feldspatic granitic gneiss (GN Fks), strongly foliated, which include small-elongated bodies (dykes) of tonalitic composition (GN Ton) (Fig. 24).

The emplacement of tonalitic rock (GN Ton) seems to be coeval with the crystallization of granitic rocks (GN Kfs and GN Bt) in an active shear zone tectonic regime. The GN Kfs and GN Bt rocks were emplaced controlled by the active shear on a process of syntectonic crystallization: a magmatic fabric followed by a tectonic fabric. Before the apex of this double process, the GN Ton dikes intruded the granitic

Fig. 24 Magmatic feeder zone. **a** General view of the outcrops—GN Ton on purple color (background image source: Google Earth Pro). The feeder zone strikes N130°. **b** Interlayer between tonalitic gneiss (GN Ton) and granitic gneiss (GN Kfs) with a GN KFs shearband boudin into a tonalitic "host rock". **c** 3D view of the relationship between GN Kfs e GN Ton. **d** Dextral shearband boudin (GN Kfs) and GN Kfs veins (to right) on a hybrid mixture with GN Ton **e** Detail of turbulent mixture between GN Kfs and Gn Ton

gneissic complex and a mixing and mingling process starts. The shear zone functioning goes on until the gneissic complex (GN Kfs, GN Bt and GN Ton) reach solid state.

2.4 *Field Activity/Geological Challenge*

On this stop, our challenge is to characterize qualitatively the strain of the gneissic rocks. It will be measure the angular values between C/C'-S structures.

It will be necessary use a contact goniometer (Fig. 25) and follow the instructions of field data identification work sheet (Fig. 26).

After chosen a suitable outcrop of gneissic rock, GN Fks or Gn Bt, the first step is to identify the C/C'-S structures (Figs. 12 and 14 will help on this identification). Once a good identification of C/C'-S fabrics, the outcrop is valid for this field activity. Then, use a tape to measure the distance to the shear zone center, record it as (L) if you are in the left side or (R) if you are in the right side. If there are a less important shear zone (2nd order shear zone), which influences the outcrop fabric, with any orientation, record the distance to this 2nd order shear zone as L′ or R′ using the same criteria. Using the contact goniometer—or the compass—to measure the dihedral angle between the two structures (C^S or C'^S). Alternately, that could be measured the attitudes of C and the S planes, and either stereographically or graphically, find out their angles. For a better understanding of this work project all

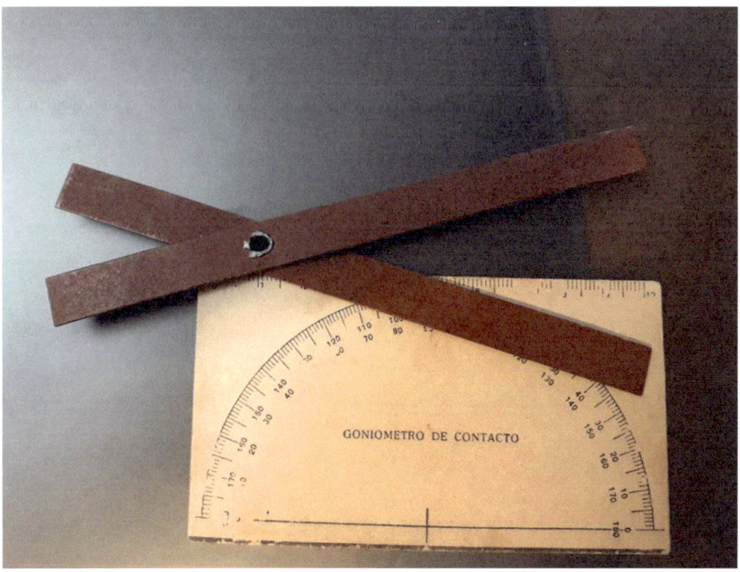

Fig. 25 Contact goniometer used on crystallography

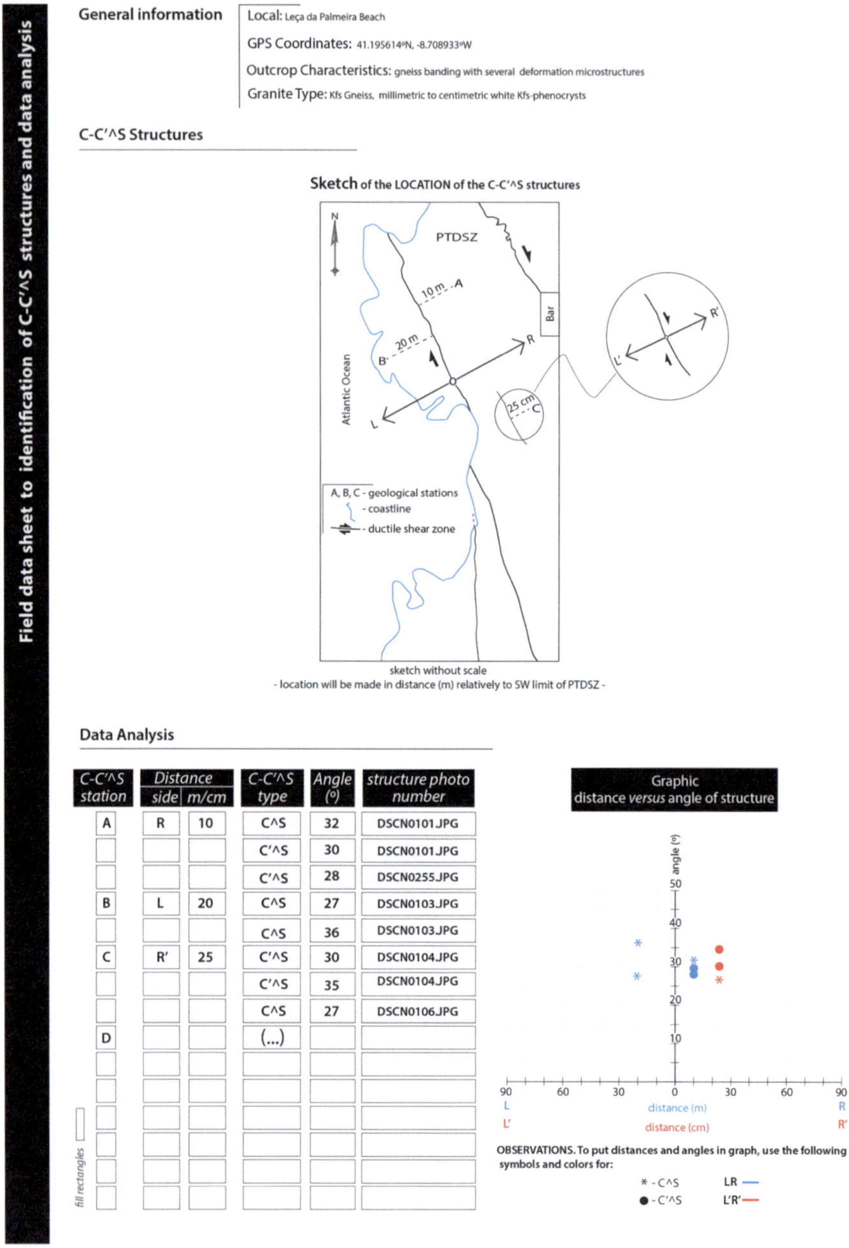

C-C'^S station	Distance side	Distance m/cm	C-C'^S type	Angle (º)	structure photo number
A	R	10	C^S	32	DSCN0101.JPG
			C'^S	30	DSCN0101.JPG
			C'^S	28	DSCN0255.JPG
B	L	20	C^S	27	DSCN0103.JPG
			C^S	36	DSCN0103.JPG
C	R'	25	C'^S	30	DSCN0104.JPG
			C'^S	35	DSCN0104.JPG
			C^S	27	DSCN0106.JPG
D			(...)		

Fig. 26 Field data sheet activity

your data on the graphic "distance versus angle of structure". At the end of the activity, all the data will be projected on the same graphic for global statistical analysis.

3 Lavadores Beach

In order to get to Lavadores Beach, take the A1 highway to Porto city. Near Porto city—2.2 km south of Arrábida Bridge exists the A1 at Canidelo (Devesas). Follow the direction to Madalena until you reach the coastal road (Rua 25 de Abril/Av. Da Beira-Mar). Then, turn right in north direction and continue 2.7 km until the Lavadores Beach. You can park the car near the Casa Branca Restaurant. In front of the Restaurant, follow the walkway along 35 m to the south and use the stairs downwards to the exposures (details on Fig. 27).

3.1 *Brief Geological Setting and Outcrop General Map*

The Lavadores is a Variscan, post-tectonic granite. The granite intruded at the western boundary of the CIZ at the transition to the OMZ, along the Oliveira de Azeméis—Porto Blastomilonitic Belt (Oliveira et al., 1991), which locally corresponds to one of the most important Variscan structures, the PTDSZ (Fig. 3). The fault system of subvertical strike-slip faults strikes approximately to the south and stretches out approximately 170 km from Porto to Tomar. The 2–8 km width dextral shear zone (Martins et al., 2011; Pereira, 2010) has a complex history, and was active since Devonian and Carboniferous, until the Quaternary (e.g., Ribeiro et al., 1980). It constitutes an important basement weakness zone that supports the emplacement of the syntectonic Lavadores Granite, which is aligned along the trace of the shear zone (Fig. 28).

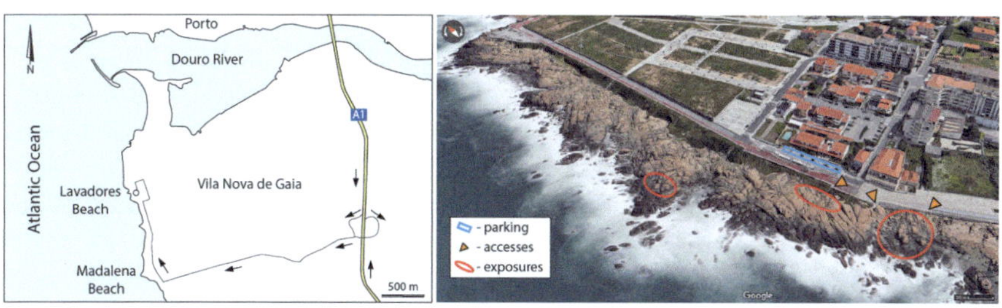

Fig. 27 Road map and Lavadores Beach outcrops aerial view (background image source: Google Earth Pro)

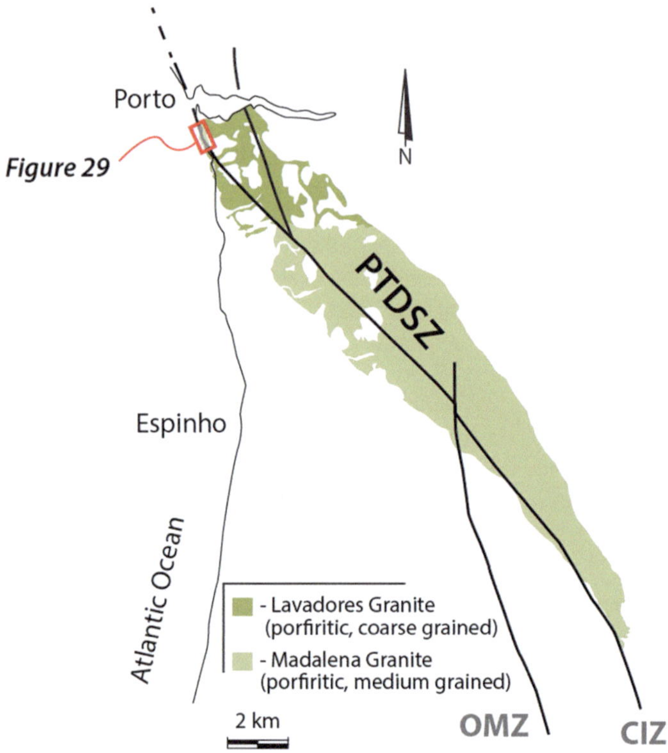

Fig. 28 Geological sketch with the field relationship between Lavadores Granite and PTDSZ. The Geological basic map information is from "Carta Geológica de Portugal", scale 1/50 000, sheets 9-C Porto (Carrington da Costa & Teixeira, 1957), and 13-A—Espinho (Teixeira et al., 1962). The location of Fig. 29 is shown

Several authors describe syntectonic pluton emplacement along shear zones (e.g. Hutton et al., 1990; Marre, 1986; Speer et al., 1994) and suggest that it is a relatively common process. The Lavadores intrusive body consists of a gradual transition between a porphyritic facies, the Lavadores Granite, and one equigranular facies, the Madalena Granite. It intruded predominantly into different metamorphic rocks including gneisses, amphibolites and folded quartz-pelitic metasediments (micaschists and quartz-micaschists) of Middle/Upper Proterozoic and Upper Paleozoic age. All these metamorphic rocks exhibits a penetrative shear caused by the third deformation phase—D_3 (Chaminé et al., 2003; Martins et al., 2011), which is in contrast to the Lavadores Granites, in which the primary magmatic structures preserve.

According with anisotropy of magnetic susceptibility (AMS) studies developed by Martins et al. (2011) and Sant'Ovaia et al. (2014), the Lavadores granite massif, and its non-porphyritic facies (Madalena Granite) have a steep magnetic foliation (W–E to WNW–ESE with mean dip 61–79°S). The Lavadores Granite also has a deep magnetic lineation interpreted as the result of the proximity to the feeder channel, probably the PTDSZ. Its emplacement was controlled by local extensional structures associated to this ductile shear zone (Martins et al., 2011). In turn, the sub-horizontal magnetic lineation of the Madalena Granite identifies a laminar spreading to the East (Sant'Ovaia et al., 2014).

The magnetic fabric pattern probably records the evolution of regional tectonics during or just after the complete crystallization of the magma, because only magmatic microstructures are present rather than low temperature solid-state microstructures (Sant'Ovaia et al., 2014).

The model for ascent and emplacement of the Lavadores Granite can be compared to the Hutton model 4 without a late ballooning imprint (Hutton, 1988). Taking into account the bulk geometry, the Lavadores Granite covers ~22 km × 4 km^2 as the exposed area, wedge-shaped pluton (Ameglio et al., 1997) with a probably single feeder channel with unknown vertical thickness.

Similar emplacement mechanism are reported from others places of the CIZ, e.g., the Vila Pouca de Aguilar Pluton related with the Penacova-Régua-Verin fault (Martins et al., 2009) and the Caria–Vila da Ponte Plutons (Gonçalves et al., 2019) probably related to the Huebra Shear Zone (Pereira et al., 2017), along the extension of the southern tip of the MLDSZ.

U–Pb geochronology of the Lavadores Granite reveals an age of 298 ± 11 Ma (Martins et al., 2011), which underwent late to post-orogenic emplacement described by others (Ribeiro et al., 2019 and references herein). This model is mainly based on the lack of intra-crystalline deformation and the absence of a shape-preferred orientation (SPO) of the main tectonically controlled minerals. The timescale for granite-magmatism partial melting, segregation, ascent and emplacement processes in the continental crust is estimated to be ~ 9–12 Ma (Petford et al., 2000). Therefore, the Lavadores Granite genesis has started with first-melting probably during D_3, which is in accordance to emplacement relationship of the Lavadores Granite with the Junqueira syn-D_3 granite (308 ± 1 Ma, Vale Aguado et al., 2005).

Silva (2010), described the Lavadores Granite as a porphyritic, biotitic, magnesian, and high-K granite-granodiorite. Based on the (87Sr/86Sr)-ratio (0.7045–0.7046) the author concluded that it is infra-crustal granite with a small upper crustal component. The temperature of crystallization was determined to be 750 ± 40 °C with a pressure of $PH_2O = 2.6 \pm 0.6$ kbar. The εNd-value is −1.04. Contrasting to these results the included microgranular enclaves show values of (87Sr/86Sr) = 0.7043–0.7046, 849–714 ± 40 °C, 4–2.3 ± 0.6 kbar and ε Nd = (−1.49) – (0.55). The enclaves are richer in REE and LREE/HREE, lower than in the Lavadores granite and implicating that these two phases originated from different sources. The calculated depth of emplacement is 7–11 km (Silva & Neiva, 2001).

This beach outcrop is an excellent locality to study magmatic structures formed during emplacement of the Lavadores Granite. To the sea, the granite is bordered by a vertical wall contact between the Lavadores Granite and migmatites (Fig. 29). Due to this contact the granite consist of K-feldspar accumulation (cumulates) zones, biotitic and K-feldspar curved-planar flux plans (schlieren), polygenetic enclave-rich accumulations intruding the Lavadores Granite (swarms enclaves) and, exotic plume structures. The later K-feldspar aplites to pegmatites crosscut all this structures.

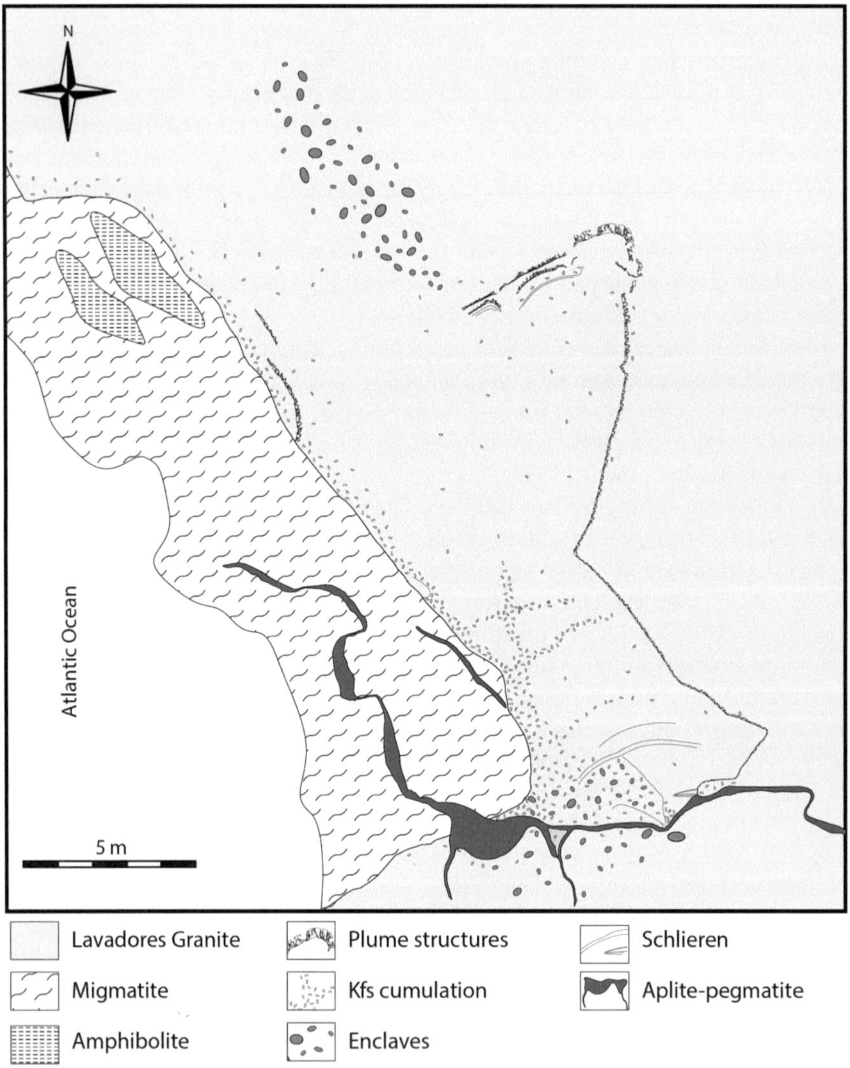

Fig. 29 Lavadores Beach geological sketch with magmatic key features (adapted from Lorenz, 2014)

3.2 STOP 1—Magmatic Flow and Lineation Structures

3.2.1 Ascent and Emplacement Mechanism and Rheological Behavior of the Magma

We will use a simplified version of the model proposed by Hutton (1988) to better understand ascent and emplacement processes of the Lavadores Granite: A diapiric ascent into the middle crust and intercepting a frozen intracrustal ductile shear zone (PTDSZ) lead to elongate plutons with slight late ballooning and tectonic deformation (Fig. 30).

The ascent and emplacement dynamics of granitic magma have been studied in laboratory for a long time (e.g. Nickel et al., 1967; Ramberg, 1981).

The magma emplacement dynamics inside the magmatic chamber is influenced by convective mechanism such as fluid, thermal and chemical convection. The solid

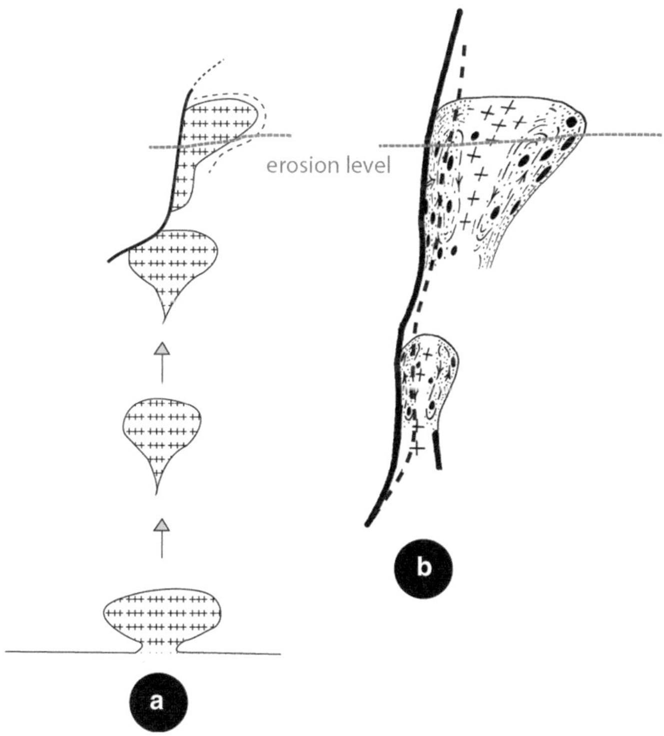

Fig. 30 Model of ascent and emplacement of Lavadores Granite (adapted from Ayrton, 1988; Hutton, 1988; model 4). **a** General sketch from genesis to ascent and emplacement. **b** Detail of emplacement phase, highlighting the role of shear zone plan on enclaves and magmatic flow distribution (adapted from Hutton, 1988)

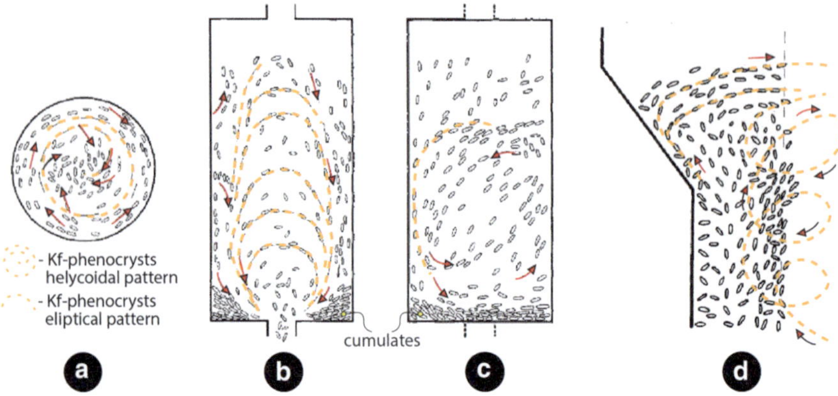

Fig. 31 Magmatic foliation patterns and associated magmatic structures generated by experimental work. **a** helicoidal ascentional patterns; **b** the planar flux structures—magmatic foliation (S_{mag})—are marked by elongate particles in circular to elliptical patterns; **b** and **c**—gravitic concentration of particles—buoyancy of Kfs-phenocrysts; **c** and **d**—quasi-parallelism of flux structures to the experiment cell walls central conduit; **d**—vortex patterns of experiment cell walls central conduit (adapted from Nickel et al., 1967)

component of magmatic fluids (e.g. Kfs-phenocrysts, Bt crystals and enclaves) is the convection-pattern markers.

In Nickel's et al. (1967) pioneering experiments (Fig. 31) he generated different patterns such as helicoidal ascentional patterns (Fig. 31a). Other patterns are planar flux structures—magmatic foliation (S_{mag}), marked by elongate particles in circular/elliptical patterns (Fig. 31b) and gravity caused concentration of particles (Fig. 31b, c). Buoyancy effects result in Kfs-phenocrysts, quasi-parallelism of flux structures to the experiment cell walls (Fig. 31c, d) and a central conduit with vortex patterns (Fig. 31d).

Any models of magma segregation, ascent, and emplacement must also take all parameters that control the magmatic systems into account such as bulk chemical composition, temperature, H_2O, CO_2 and O_2 fugacity (f_{O2}).

The entire evolution of a magmatic system occurs between the *solidus* and the *liquidus* of the system. The *solidus* of system (0% melt fraction) is the point at which the system produce the very first melt on an anatectic pathway, or on the other end, crystallization is completed in a magmatic differentiation process. The *liquidus* of a system (100% melt fraction) is the point at which all the system is liquid. Thus, the melt system of a magmatic system is highly scale depend and heterogeneous. Looking e.g. on the entire magmatic body it can never reach the *liquidus* because at some sub volume a certain crystal volume will be already or still present. On the smaller scale, however, 100% melt batches may exist in the overall granitic mush.

Beside the mineral composition, the evolution of a granitic magmatic system between its *liquidus* and *solidus* depends on several factors (McByrney, 1984). Increasing H_2O content lowers the *liquidus* temperature, facilitates the melting and lowers viscosity. This is one main factor controlling the segregation, ascent and emplacement of a granitic rock. Another, the CO_2 increases the bulk viscosity andf_{O2}favors higher crystal/melt ratio than reduced conditions.

Ascent and emplacement of a granitic magma—and associated acid and basic magmas hybridization mechanisms—depends on the bulk system rheological behavior (Fig. 32). The hybridization of more felsic and mafic magmas occurs only within a narrow temperature range (Castro et al., 1995). The upper limit is the higher temperature that a more felsic magma (e.g., granitic) can reach in nature (granite *liquidus* temperature). The lower limit is the lower temperature at which the more mafic magma (e.g. gabbroic) can mingling in nature (Arzi, 1978; Van der Molen & Paterson, 1979—definition of critical melt fraction, CMF; Fig. 32).

Following a magmatic differentiation path until the crystal fraction (Ø) reaches the value of Ø $\approx$ 22–32%, all systems behave like a Newtonian liquid (Field I). This is the field of initial magmatic mixing.

At crystal fraction values Ø $\approx$ 22–32% a major change in rheological properties on the magmatic system take place and is called First Rheological Threshold (RT1), when the Non-Newtonian behavior appears (Field II) (Fernandez & Gasquet, 1994). The amount of melt decreases and crystal volume fraction increases resulting in pseudoplastic behavior (Fernandez & Gasquet, 1994). Here, the apparent viscosity depends on the strain rate. This non-Newtonian behavior ends at Ø $\approx$ 65–72% when suspension (magmatic liquid + crystals fraction) acquires a yield stress (internal cohesion) achieving the Second Rheological Threshold (RT2). From this point towards the system behaves like a Bingham-type body (Field III) (Fernandez & Gasquet, 1994). Magmatic material's viscosity issue was partly reviewed in Mukherjee and Mulchrone (2012).

Following a partial melting path, once an interconnected network forms, the source is permeable and melt could migrate at a melt fraction of 2–10%, achieving at the First Percolation Threshold (PT1) (Vigneresse et al., 1991). Melting is going on and accumulates in the magma chamber, then the Second Percolation Threshold (PT2), equivalent to the CMF, is reached for melt fractions of approximately 28–35%. At CMF the strength of the rocks will decrease rapidly and the system began to behave like magma. The CMF values change between 10 and 50% depending on the rock system composition, textures, grain sizes and geometry, and others (Arzi, 1978; Wickham, 1987). An alternative pathway at PT1 is the migration of the first felsic melts if there are favorable conditions: a driving force and a favorable space to ascent.

The structures observed in magmatic systems greatly depend on their rheological behavior. The relative fraction of liquid versus solid controls the bulk rheology of magmatic system and dynamically changes during the temperature-decreasing pathway. Consequently, each magmatic structure correlates to a specific rheological field (Fig. 32 and Table 3).

Typical magmatic structures develop in Field I, where gravitational processes are dominant on a low viscosity highly convective system. Examples are cumulates, aggregations of crystals and enclaves by sinking or by buoyancy and schlieren, which are aggregates of mafic minerals aligned on main flow plane. Magmatic layering, the alternation of non-miscible layers with different mineralogical composition into de main flow plane and mesoscopicly difficult to observe SPO of minerals are other

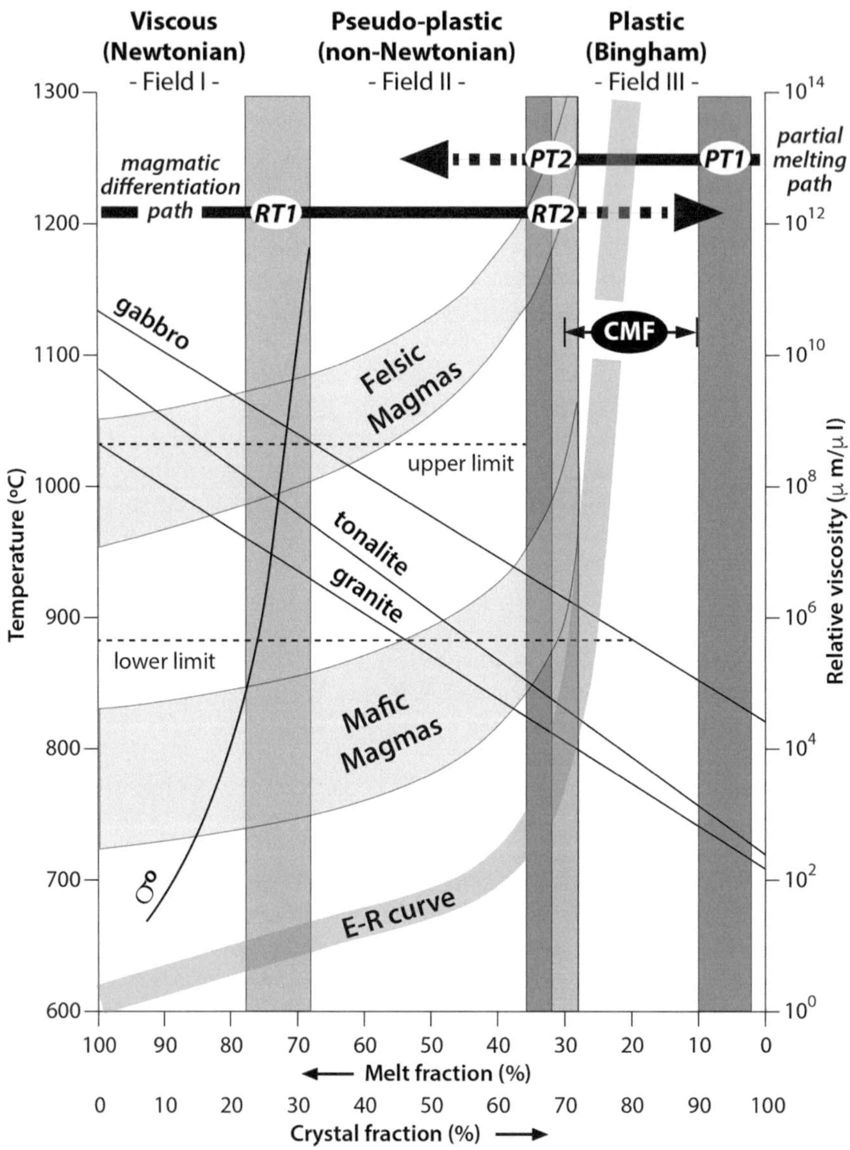

Fig. 32 Relationship between melt fraction (%), relative viscosity (μm/μl), temperature (°C) and magma composition (Granite–Tonalite–Gabbro). CMF—Critical Melt Fraction; PT1—First Percolation Threshold; PT2—Second Percolation Threshold; RT1—First Rheological Treshold; RT2—Second Rheological Treshold; E-R curve—Einstein-Roscoe viscosity curve; σ_0—Evolution of yield stress from Soo (1967)'s equation (adapted from Castro et al., 1995; Fernandez & Gasquet, 1994)

Table 3 Structures and microstructures generated in different rheological fields (Field I—Newtonian; Field II—non-Newtonian; Field III—Bingham)

Structures and microstructures	Rheological field		
	Field I	Field II	Field III
• Cumulates	——		
• Schlieren	——		
• K-feldspar zoning	——		
• Felsic magma and enclaves with solid-state HT plastic deformation			——
• Felsic magma and enclaves with microfractures			——
• Weakly SPO	——		
• Strong SPO on felsic magma and enclaves		——	
• Fabric parallelism across boundaries	——	——	
• Refraction of fabric across boundaries			——
• Mafic enclaves with chilled margins	——	——	
• Mafic enclaves with diffuse margins	——		
• Mafic enclaves with sharp margins	----	----	——
• Mafic “pillow-lavas” enclaves	——		
• Magmatic layering	——		
• Mafic magma dikes	----	——	——
• Planar and non-planar dyke-like enclave swarms	----	——	
• Magmatic shear zones		----	——
• Folded mafic dikes			——
• Sheared mafic dikes, boudins			——
• Tension gashes and shear fractures			——

Fields I, II and III: Solid lines—more frequent; Dashed lines—less frequent, No lines—no occurrence. Data sources: Castro et al (1995), Fernandez and Gasquet (1994), Nédélec and Bouchez (2015), Petford, 2003, Tobisch et al. (1997), present work

typical structures in Field I. Intrusion of more mafic magmas into more felsic ones result in enclaves, drop-like to rectangular shaped mm-hundred meter sized structures. Mafic enclaves with chilled margins are the result of a narrow fine-grained layer of quickly frozen crystal around the enclave’s boundary (high temperature contrast), whereas mafic enclaves with diffuse margins show a narrow hybrid compositional margin on the enclaves (low temperature contrast). K-feldspar zonation is another typical structure in Field I and develops by internal, congruent or incongruent alignments of mineral inclusions.

In Field II, the solid particle fraction increase and therefore, a relative viscosity develop. This leads to the rotation of rigid particles in a low viscosity matrix, generating locally a strong SPO of anisotropic shaped crystals (e.g. feldspar and biotite), as well as paralyzed enclaves along flow planes. Strong SPO magmatic fabric occur inside enclaves, (i) parallel to the host felsic magma (low viscosity contrast), and (ii) non-parallel to the host, more felsic magma with a refraction across the boundary of the enclave, or an enclave rotation causes pressure shadows where melt may accumulate (high viscosity contrast). Other typical structures in Field II are more mafic magma dikes intruded into the more felsic parts and planar dyke-like and non-planar heterogeneous dyke-like enclave swarms (Tobisch et al., 1997), with the basic geometry of mafic magma dike desegregation generates enclaves with chilled and sharp boundaries.

In Field III, an abrupt increase in relative viscosity of the magma causes a Bingham behavior of magma chamber and the generation of following magmatic structures: magmatic shear zones are planar narrow zones with high temperature deformation features (Castro et al., 1995). Fabric angle misorientation between different magma compositions lead to refraction patterns of the fabrics and across the compositional boundaries (Fernandéz & Gasquet, 1994). Mafic dikes occur were more mafic magmas intrude into open fractures within the more felsic magma chamber, and folded and sheared mafic dikes (e.g., boudins) are plastically deformed mafic magma dikes with high viscosity contrast to the more felsic host (Nédeléc & Bouchez, 2015; Petford, 2003). Other structures in Field III are mafic enclaves, which are widely scattered, or accumulated in swarms, but which are more angular with sharp boundaries and with a larger variation in size (Tobisch et al., 1997). Also typical, are local areas within the felsic magma and enclaves that underwent high temperature, solid-state plastic deformation and, microfratures in rigid minerals with limited propagation and often filled with melt of granitic composition (Tobisch et al., 1997). Tension gashes and shear fractures are often secondarily filled with felsic magma and occur typically in high angle to the magmatic fabric (Castro et al., 1995).

3.2.2 Lavadores Granite Magmatic Flow Structures

The Lavadores Granite has a porphyritic texture with centimeter sized pink Kfs-phenocrysts in a medium to coarse-grained matrix of quartz, plagioclase and biotite (Fig. 33).

On in equigranular textures of magmatic plutonic rocks, the large crystal grains (the megacrysts on a porphyritic texture) could have a genetic relationship with the smallest grain generation, (the groundmass or matrix) that is a point of controversy. Among these, the distinction between phenocrysts and porphyroblasts in granitic rocks is a long-term discussion among the petrologists (e.g. Dickson, 1995; Rong & Wang, 2016; Vernon, 1986; Vernon and Paterson, 2002). On Lavadores Granite the Kfs are typical euhedral, cm-sized and are notable marksof the magmatic flow. It is recognized their igneous origin, rather than the metamorphic one, so we call herein the large K feldspar megacrysts as Kfs-phenocrysts.

Microgranulares enclaves and Bt-schlieren are very common in the Lavadores Granite and are often concentrated in planar trails with biotite and enclave long-axes aligned along the flow direction (Fig. 33). These structures, together with domains of well-aligned Kfs-phenocrysts define a magmatic flow foliation and lineation.

Fig. 33 Outcrop of Lavadores Granite showing its most remarkable mesoscopic feature: pink Kfs-phenocrysts and mafic microgranular enclaves. Red dashed line mark flow plane

3.2.3 Kfs-Phenocrysts Classification and Measurements

Each textural, chemical or microstructural feature provides part of the geological history about the process of melting, ascent and emplacement of the magma. The Kfs-phenocrysts are potential testimonies of the synmagmatic stage of the Lavadores Granite.

Commonly, the study of Kfs-phenocrysts in granitic rocks focuses on its mineralogical and crystallographic characteristics such as the geometric and spatial variation of Kfs, the perthitic character and tryclinicity, and the chemical composition.

The Lavadores Granite Kfs-phenocrysts on a close observation reveals a complex and fascinating set of enigmatic characteristics.

What a petrologist expects from a plutonic feldspar megracrystal, according the nucleation and grow of magmatic laws (e.g. Kirkpatrick, 1975), is the developing of a nice and perfect polyhedral shape with prismatic habit controlled by the development of {010} pinacoid and {110} 3rd prism faces. The prism termination is made by {001} 3rd order pinacoid and {101} 2nd order prism.

The Lavadores Granite Kfs-phenocrysts are mainly orthoclase (Canilho, 1975) with tabular habit, sometimes perthitic, exhibiting Carlsbad twin with $Or_{73\text{-}91}$ composition (Silva, 1995), were blessed by the nature with an almost ubiquitous Carlsbad twin (Fig. 34).

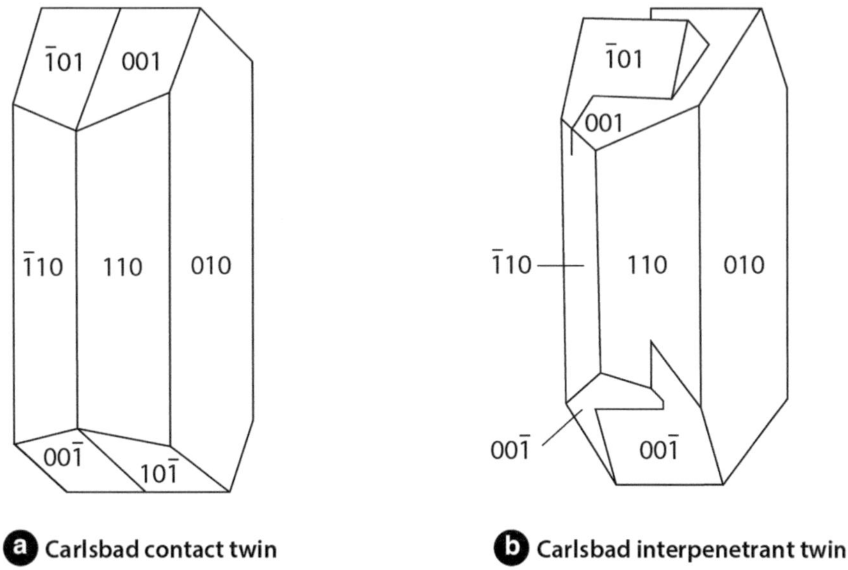

Fig. 34 Characteristics of perfect Carlsbad twins—A and B (after Klein and Hurlbut, 1977)

In the Lavadores Granite Kfs-phenocrysts show several interesting features that illuminates geological history of this granite. The geometric and spatial relationship between phenocryst and enclaves, their shape, zoning, outer rim and color are the most relevant features.

Schmidt (2014) proposed a morphological classification based on petrological characteristics codifying by numbers. He used an 8 numbers string to classify each Kfs-phenocryst individually. Each number represents a single attribute: (1) enclave relationship, (2) shape, (3) zoning or inclusion, (4) 1st zoning, (5) 2nd zoning, (6) 3rd zoning, (7) outer rim, and (8) color (see Table 3 for detailed information). As a result, each Kfs-phenocryst is represented by 8-digit number, which clearly defines all these attributes of the crystal. To the Schmidt (2014) code we add a supplementary attribute, the length, in millimeters that is represented by three numbers, preceding the 8 numbers code inside the straight brackets.

Out of this Kfs-phenocryst classification scheme, the attitude of (010) plane and of the <010> axis (c-axis) are measured and computed.

(a) Attribute (nnn)—Length

The size of a crystal is the measure of its volumetric space and can be defined in several different ways (Higgins, 2006). Here, we adopt a field expeditious measurement using the longest sectional axis, along <001> if identifiable, in order to get a linear parameter of the size. We added this length value in millimeters as a 3-digit number to the crystal attribute string, preceding the 8 numbers code inside the straight brackets (Table 4). We can then perform a crystal-length distribution analysis, which is a first approach to a classic crystal size distribution (CSD) analysis. Note that the standard measurements for CSD studies must take in account a volumetric dimension based on 2D data.

(b) Attribute 1—Kfs-phenocryst—enclave relationship

Four different classes of enclave-feldspar contacts were defined (Fig. 35).

The first class [0] shows no relationship with an enclave. There is no contact between the feldspar-crystal shape and a single mafic enclave. Most of these feldspars are idiomorph and most likely have never been in touch with the mafic rock material. The second class [1] is characterized by a simple and regular contact without any evidence of mechanical interference. The third class [2] implies physical interaction between Kfs-phenocrysts and the enclaves. It is defined by a minimum of one face of the crystal being inside an enclave (Ksf-phenocryst "riding" on enclave boundary). It's not known if the Kfs-phenocryst is "coming in or coming out" of the enclave, but the Kfs-phenocryst portion partly inside the enclave is usually marked by dissolution

Table 4 Kfs-phenocryst attributes classification (1–8) and classe numbers (0–4) (Modified after Schmidt, 2014)

	Attributes for Kfs-phenocryst code: nnn [12345678] [nnn—length in millimeters (mm)]							
Class Number	1 Enclave relationship	2 Crystal shape	3 Zoning inclusions	4 1st zoning	5 2nd zoning	6 3th zoning	7 Outer rim	8 Color
0	No		No	No	No	No	No	
1	In touch	Euhedral	Inclusion	Euhedral	Euhedral	Euhedral	punctual, discrete	Pink
2	Partly inside	subhedral (One round face)	One zoning	One round face	One round face	One round face	Less than half	Colored rim
3	Full inside	subhedral (One euhedral face)	Two zonings	One euhedral face	One euhedral face	One euhedral face	Half or more	Patchy
4	With tail	Anhedral	Three or more zonings	Anhedral	Anhedral	Anhedral	Full	White

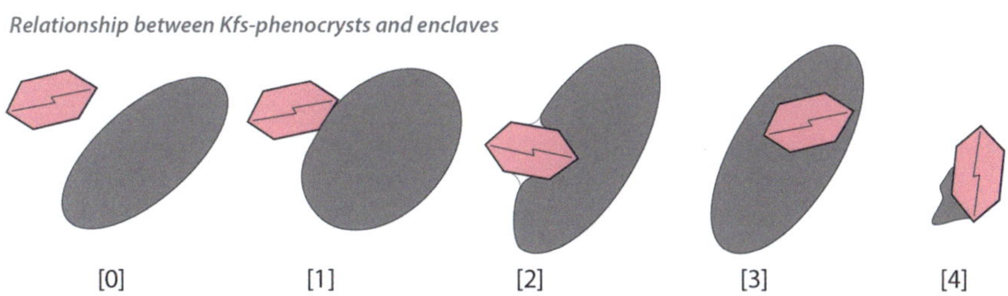

Fig. 35 Enclave—Kfs-phenocryst relationships features codification (attribute 1)

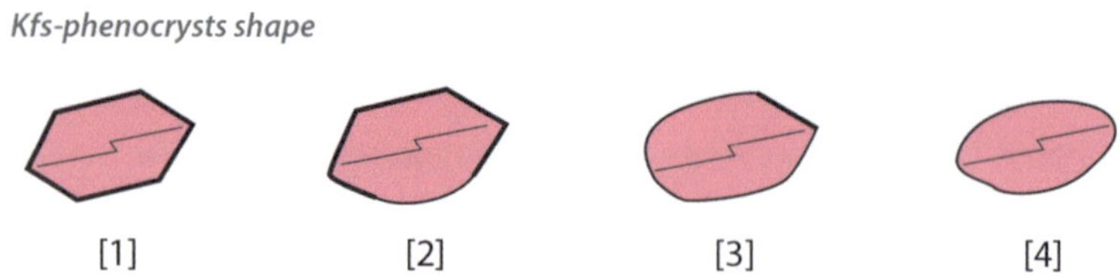

Fig. 36 Kfs-phenocrysts shape crystal codification (attribute 2)

and seems to be became more or less rounded. In the fourth class [3] the Kf-phenocryst is completely inside an enclave. It is often more or less rounded and shows a lot of dissolution features. This type of Kfs-phenocryst is in particular interesting because it shows how enclaves might change the Kfs-phenocryst appearance in the Lavadores Granite. The fifth class [4] is defined by a little part of enclave, which is connected to the K-feldspar phenocryst. The enclave looks like a tail. In most cases, the contact zone between mafic rock and feldspar shows some kind of dissolution. This type does not need a near enclave to occur. The enclave material is directly connected to the crystal and does not belong to a specific enclave.

(c) Attribute 2—Kfs-phenocryst shape

The shape of crystals is typically defined as euhedral for perfect grown crystal faces, anhedral for crystal that show nearly no face and subhedral for crystals with some grown faces. Here, we decided to define different sub-types of a subhedral shape to better distinguish the majority of subhedral ones (Fig. 36).

Type one [1] is characterized by a euhedral perfect crystal-shape. This crystal had enough space and chemical material to grow. The second type [2] is defined by only one rounded face, therefore, a subhedral crystal grown under almost ideal conditions. The third type [3] Kfs-phenocrysts have grown in less ideal magmatic conditions (it could have been limited on a growth process by either less space or missing material; or on a dissolution process by a chemical erosion of crystal faces), and typically show only one well-defined crystal face. This seems to be a very common type for subhedral crystals, which are inside an enclave. The fourth type [4] is the typical anhedral shaped Kfs-phenocryst that has no recognizable crystal faces. It is rounded and the shape often is marked by dissolution patterns. This is also a very common type for inside enclave feldspars.

(d) Attribute 3, 4, 5 and 6—Kfs-phenocryst zoning and inclusions

A special attribute of Lavadores Kfs-phenocrysts is a very common appearance of small-grain inclusions, usually biotite (Fig. 37; [1]). Sometimes these inclusions form typical congruent zonation inside the phenocryst (Fig. 12; [2], [3], [4]).

Regarding attribute 3, a crystal without any biotite inclusion is defined as zero [0]. A crystal with mineral inclusions is defined as one [1]. Nevertheless, [1] indicates that there are biotite minerals inside the Kfs-phenocryst, but they do not form a discrete zonation.

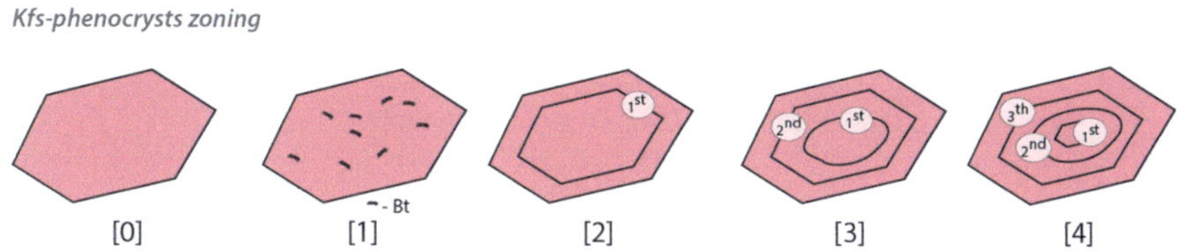

Fig. 37 Kfs-phenocrysts zoning and inclusion (attribute 3)

If there is a zonation formed by biotite grains inside the Kfs-phenocryst there could be until three or more additional attributes on codification: [2] 1 zonation; [3] 2 zonations; and [3] 3 or more zonations. This means that there are Kfs-phenocrysts crystals with at least up to three or more different zonings at Lavadores Granite.

Each of these zonations, attribute 4, 5, 6, is further coded according to the morphology of zoning (Fig. 38).

The classes of the attribute 2 classify every zonation similar to the classification of the phenocryst external shape (Figs. 36 and 38). Hence, a completely round zonation is classified as [4] or a perfect euhedral congruent zonation is defined as [1]. The [0] class is reserved for crystals without any zonation (Fig. 36).

(e) Attribute 7—Outer rim

Kfs-phenocrysts at Lavadores Granite often show biotite rims. They follow the shape of the crystal without any space between biotite and feldspar (Fig. 39). This attribute is used for the classification, as so far there has been no full explanation about this microstructure that is very special and common at these outcrops.

Four classes define the outer rim. The first one [0] defines a rim-free Kfs-phenocryst. The phenocryst still can contain biotite zonation or inclusions but without

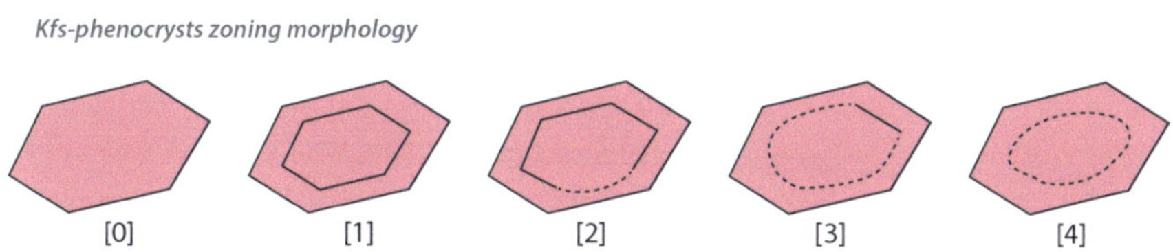

Fig. 38 Kfs-phenocrysts zoning morphology (attribute 4, 5 and 6)

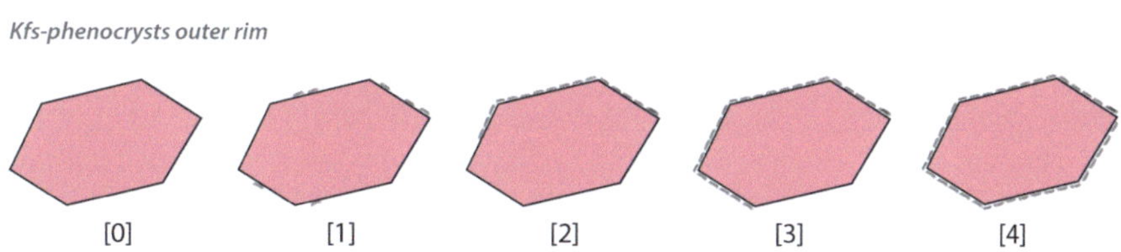

Fig. 39 Kfs-phenocrysts outer rim (attribute 7)

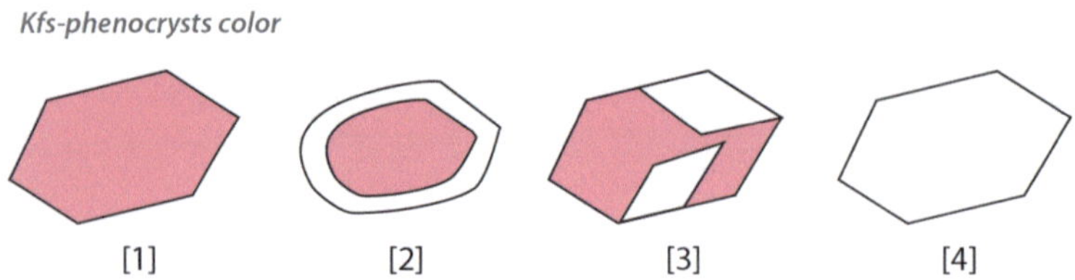

Fig. 40 Kfs-phenocrysts color (attribute 8)

any biotite around its outer shape. In case, biotite is only attached locally at the feldspar outer side it got the attribute [1]. The third classe [2] labels a crystal with less than a half rim of biotite around its external shape. Sometimes the surrounded part of the crystal shows little dissolution but the biotite always follows the shape. The fourth one [3] is similar to [2] but here the rim is more than half surrounding the phenocryst. The rim is dominant and biotite surrounds greater parts of the crystal. The last class [4] shows a complete outer rim of biotite, which follows the whole shape of the crystal. The feldspars looks like dunked in a melt of biotite and pulled out later.

(f) Attribute 8—Color

Feldspar group minerals usually occur in different colors depending on their chemistry, structural state, inclusions, etc. Kf-minerals of granitic rocks are less white and often pale yellow, lightly reddish or pinkish. In this color classification four different types of Kf-phenocrysts are observed (Fig. 40). Most of the Lavadores Granite Kfs-phenocrysts have a characteristic pink color, but sometimes they look blotty with white and pinkish parts.

Kfs-phenocrysts, which are completely pink, are classified as [1]. The second type [2] comprises of a one-color core and a different colored outer rim. Both is possible, a Kfs-phenocryst with a white core and pink rim or with a pink core and a white rim. The third type [3] shows a slightly blotty or patchy coloring. It is also formed by parts of white and pink feldspar but there is no real order. The different parts are randomly distributed in the crystal. The last one is a completely white crystal [4]. The white Kfs-phenocrysts are rare at Lavadores Granite.

(g) Examples of Lavadores Granite Kfs-phenocrysts classification (codes)

At Lavadores Granite outcrops are identified Kfs-phenocrysts exhibiting almost all the features considered on this classification (Fig. 41).

A systematic study of Kfs-phenocrysts at the Lavadores Granite outcrops revealed several interesting results (Fig. 41). The most frequent characteristic of Kfs-phenocrysts recorded by Schmidt (2014) has the code **39[01100001]**. Most of the feldspars are not in contact to enclaves (97.3%), and with 1.1% partially inside enclaves. The majority is euhedral (78.8%) with 20.2% being subhedral showing some rounded faces. Almost all have Bt inclusions (92.7%), with 4.4% comprising of one or two internal zonation. About half of the phenocrysts have no outer rim (53.6%) and 32.4% of all phenocrysts are surrounded by a half- or full-rim of Bt

Fig. 41 Examples of Kfs-phenocryst codes nnn [12345678] from Lavadores Granite

minerals. The majority of Kfs is pink (95.1%), the rest is either white (2.8%) or has a white rim—rapakivi texture (1.3%).

Taking into account the length variation of Kfs-phenocryst, we constructed a crystal length curve (Fig. 42a), which is interpreted as an approximation to a CSD curve. The lengths of Kfs-phenocrysts of the Lavadores Granite ranges from 10 to 100 mm, with 30 mm being the most frequent grain size class. The divergence in size distribution for crystals < 30 mm (dotted line) could be attributed to accumulation phenomena in the magma chamber (Fig. 42a).

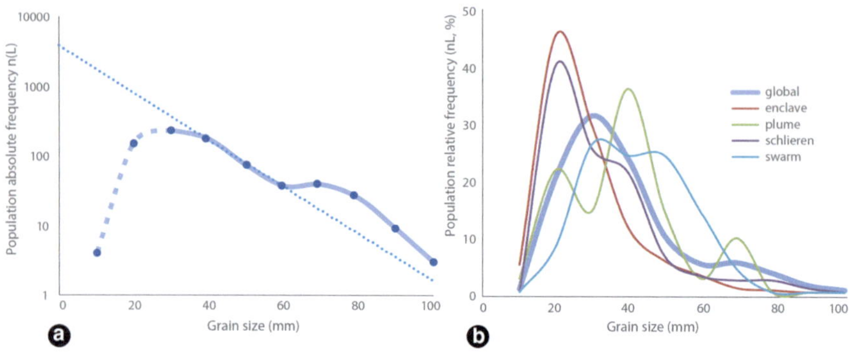

Fig. 42 Crystal length distribution of Lavadores Granite Kfs-phenocrysts. **a** The initial dashed portion of the curve is underestimated because the matrix Kfs grains weren't measured. The dotted line is the correspondent CSD straight line with a negative slope of −1/Gt. **b** Relative frequencies of Kfs-phenocrysts grains size related with enclaves, plumes, schlieren and swarms

In the field, we put another focus on the influence of enclaves, plumes, schlieren and swarms on the Kfs-phenocrysts nucleation and growth characteristics (Figs. 42b and 43).

Kfs-phenocrysts related to enclaves and/or schlieren have a modal grain size < 20 mm compared to non-related ones (Fig. 42b).

Swarm Kfs-phenocrysts and those inside plume structures are in mean larger compared to the global population. The related curves in Fig. 42b are shifted to the right. The bi- and tri-modal distribution indicated from this curves could be related to mixing effects between different feldspar populations or, heterogeneous nucleation and growth during emplacement (Fig. 42b).

The comparisons of attributes support several conclusions. Most of the Kfs-phenocrysts, which are in contact to enclaves are fully inside. Only in case of cumulated Kfs, a higher amount appears to be partly inside the enclaves, but still the majority is fully inside (Fig. 43a).

Kfs-phenocrysts inside of schlieren are mainly euhedral, following the global distribution tendency. However, the majority of crystals in plume structures are subhedral (with 1 round face), and anhedral grains are most frequent inside enclaves and swarm populations (Fig. 43b).

Almost all Kfs-phenocryst grains have biotite inclusions. Internal zonation is rare and only a small frequency of a one-internal zonation pattern for phenocrysts from plume structures and swarms can be observed (Fig. 43c). About half of the Kfs-phenocrysts inside the main granite and schlieren do not have an outer Bt rim. In contrast, most grains from enclaves, plumes and swarm are fully surrounded by biotite (Fig. 43d). Most of all phenocrysts are pinkish. However, there are some white grains related to schlieren and very distinctive pink grains with a white rim, rapakivi texture occur inside the plume structures (Fig. 43e).

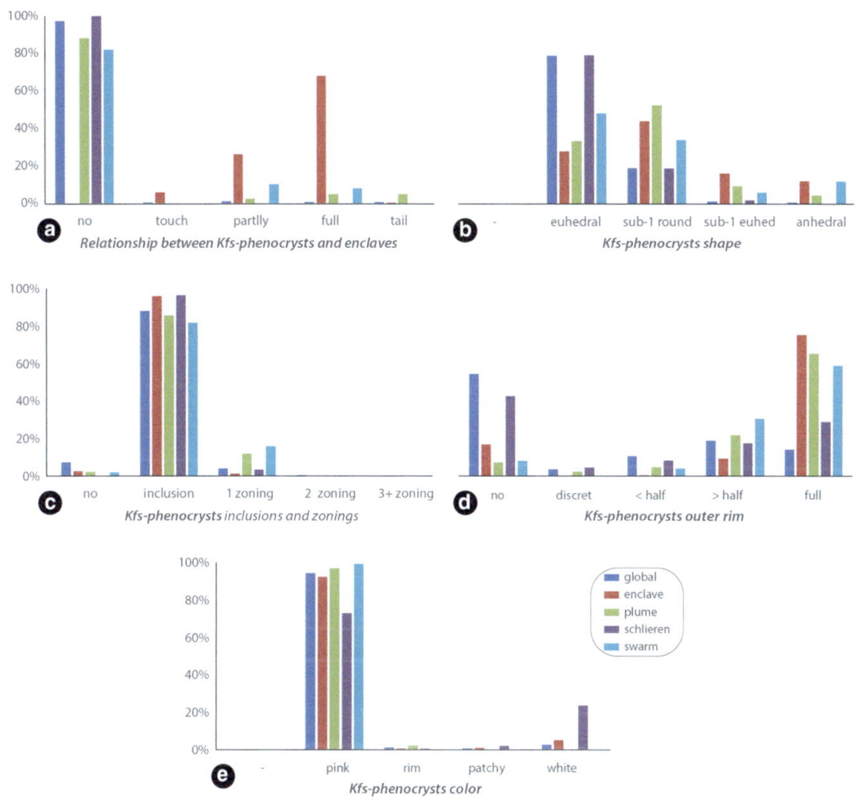

Fig. 43 Characteristics of Kfs-phenocryst related with enclaves, plumes, schlieren and swarms (data from Schmidt, 2014) with histograms showing the relationship with enclaves (**a**), shape (**b**), inclusions and zonation (**c**), outer rim (**d**) and color (**e**)

(h) Attribute Kfs-phenocrysts attitude—Planar and linear orientations

Kfs-phenocrysts of the Lavadores Granite are good magmatic flow markers (Fig. 44a). The (010) plane is parallel to the magmatic flow and the Kfs <001> direction (*c*-axis) defines the flow foliation-related lineation (Fernández-Catuxo, 1995). (010) is identified as the Carlsbad twin plan, parallel to the "big polygonal crystal face" and it is measured like a foliation plan (strike; dip; sense). The <010> (*c*-axis) is often not so easy to identify and it is measured like a lineation (trend; plunge; Fig. 44b).

In order to gain statistically significant data, we used a sampling grid (1 m^2) that is divided in 25 small squares (Fig. 44c).

For each square and if possible at least one Kfs-phenocryst foliation and related lineation must be measured. The measurement data (Fig. 44b) is transferred into stereographic projections as foliations (010) and lineations <001> .

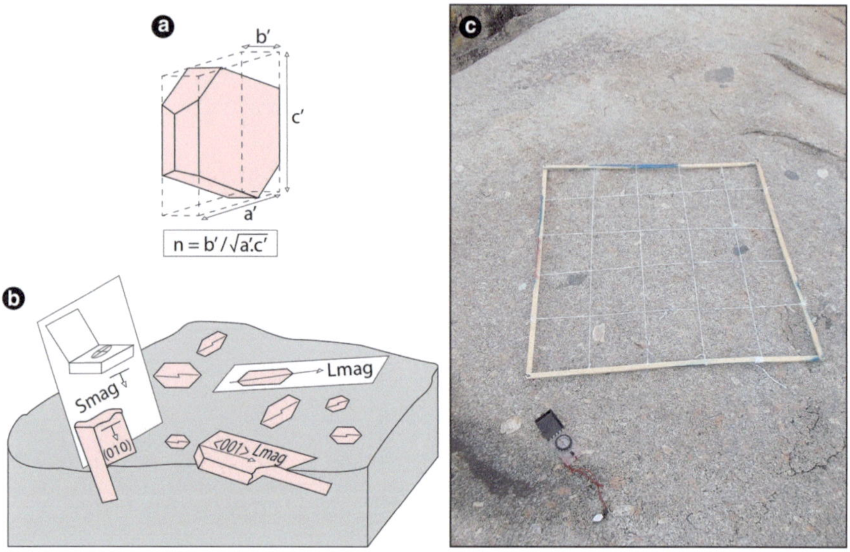

Fig. 44 **a**—Tabular habit of Kfs-phenocrysts; **b**—Magmatic flow structures—magmatic flow foliation and lineation, and measurement techniques (adapted from Laporte, 1987 in Fernández-Catuxo, 1995); **c**—One square-meter divided into 25 small squares used for statistical orientation measurements of Kfs-phenocrysts

3.2.4 Magmatic Flow Pathways

Based on the theoretical model for the Lavadores Granite, magmatic flow structures derived from flow processes inside the magmatic body were relatively less viscous volumes of mush have been forced along a volume of relative higher viscous material inside the magma chamber and/or the contact to the host rock.

A rotational deformation resulting from flow results on a shortening along the tectonic Z-axis (S_{mag}) linked to an elongation of the same amount along the tectonic X-axis (L_{mag}). Thus, L_{mag} represents the maximum extension of the magma and defines the direction of flow at the time of emplacement (Marre, 1986). Using the AMS data (Martins et al., 2011; Sant'Ovaia et al., 2014) and the Kfs-phenocrysts subfabric (Lorenz, 2014), it is possible to re-construct the magmatic flow and flow pathways from the Lavadores—Madalena granites (Fig. 45). Looking at the lineation at Lavadores beach, a contrasting pattern to the generally sub-horizontal lineation of the Lavadores—Madalena granites appears at the western boundary (generally, around Lavadores beach). Here, the Lmag is sub-vertical (Fig. 45).

The magmatic flow plane is generally sub-vertical in all the granite outcrops. Exceptional values of sub-horizontal flow planes were only recorded at the helicoidal pathway (Fig. 45), and they are probably related with convective cells. Locally at the Lavadores beach, two orientations of the S_{mag} are found (Fig. 20). Such patterns give an indication of the sort of processes that can intervenes during ascent and once magma has reached its final position.

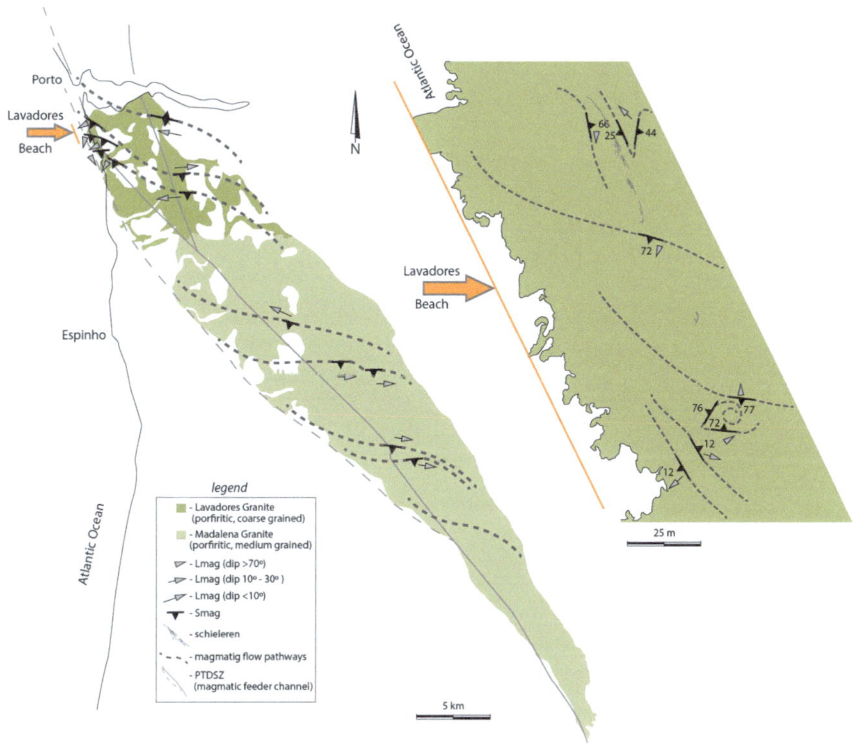

Fig. 45 Lavadores Granite magmatic flow structures and pathways (data from Lorenz, 2014; Martins et al., 2011, Sant'Ovaia et al., 2014)

The general magmatic flow pattern of the Lavadores-Madelena granites has a S-curved shape, subparallel to the contacts and strikes anticlockwise in the central parts. This seems to be the result of a granite emplacement during the very last episode of shear zone movement. The vertical flow plane and the vertical lineation at the western boundary could identify the feeding zone accommodated by the PTDSZ. The trend of the sub-horizontal lineation identifies the magmatic flow inside the eastern sector (Fig. 45).

3.2.5 Magmatic Cumulates

Textural features of certain igneous rocks provide clear evidence that the crystals of theses rocks were concentrated by accumulation. Rocks of this type are referred, therefore, as cumulates (Irvine, 1982; Wager et al., 1960). This process was initially thought to be the result of a gravity driven sinking of crystal on to the floor of a magma chamber. However, the accumulation of minerals floating at the top and at the outer parts of a magmatic convention cell is also admitted for granitic rocks

Fig. 46 Kfs-phenocrysts identified as real cumulates. **a**—Concentration against a magma chamber wall boundary (migmatites on the picture bottom); **b**—Concentration near an aplitic-pegmatite body; **c**—Embedding with enclaves on a swarm; d—Local concentration of euhedral crystals with a more evolved "intercumulus" material

(Schermerhorn, 1987; Vernon & Collins, 2011). Typically, cumulates have a texture formed by two contrasting types of mineral grains (Irvine, 1982): the *cumulus* grains formed mainly by euhedral grains defining a framework structure and the *intercumulus* grains, which grow in the empty space between the cumulus framework. The cumulated rocks are classified according the volume of postcumulus material: orthocumulates (50–25%), mesocumulates (25–7%) and adcumulates (<7%). Cumulate structures inside the Lavadores Granite are characterized by cm-sized, mainly euhedral Kfs-phenocrysts, which are highly concentrated along the magma chamber, and associated with enclaves on swarms (Fig. 46).

3.2.6 Field Activity/Geological Challenge

Our challenge on this outcrop is to classify the Kfs-phenocrysts to gain information about the emplacement and crystallization kinematics and dynamics of the Lavadores Granite. The field data sheet has twosides. **(i)** The front-side (Fig. 47) allows the data record and classification of each phenocryst, and it has three data zones. The first one is for general information about the geological stop (local, GPS coordinates, the granite type and specific outcrop characteristics). The second one is for a sketch of the one square meter window-grid (Fig. 44c), where the numbering and the position of all the Kfs-phenocrysts can be recorded. Here, you can also make a geological sketch (If it **is** not possible to use the window-grid, it **is** necessary to draw it on the outcrop). The third sheet is for recording the attributes of each phenocryst (Table 3—Use this table in the field activity—, Figs. 41 and 44b). **(ii)** The backside (Fig. 48) is used to analyze all the characteristics for each Kfs-phenocrysts using the classification (attributes) scheme proposed on this text. The attributes are: dimension (nnn), [1] enclave relationship, [2] crystal shape, [3] zoning inclusions, [4], [5] and [6] zoning morphology, [7] outer rim, [8] color, magmatic flow plan (S_{mag}) and magmatic flow lineation (L_{mag}) (Wulf net, low hemisphere). The analyses sheet, together with the projection of magmatic foliation and lineation on a Wulf net, is designed to be a first brief statistical evaluation.

3.3 STOP 2—Schlieren

3.3.1 Geological Layering in Granitic Rocks—Schlieren

Geological streaks are common in all geological environments, as it is the case of schlieren in granitic rocks. "Schlieren" (plural of "schliere") is a German word used to describe in homogeneities in various substances, such as streaks of varying colors in cement as a product of incomplete mixing (McCuish, 2001).

According to Bates and Jackson (1987) the term schlieren is used in geology to refer a structure that have the same general mineralogy as the plutonic rocks, that could be darker or lighter and with transitional boundaries with the host rock. Some schlieren could be modified enclaves or segregations of minerals.

The schlieren are also been described as being composed of thin wavy streaks of dark minerals, some of them truncated and cross bedded. Although not usually rhythmic, they may be finely banded (e.g. Barriére, 1981; McBirney & Noyes, 1979).

The ring schlieren were exotic magmatic structures, generally described as "alternating melanocratic and leucocratic bands in granite forming open to closed, nested, circular to elliptical, eccentric to concentric, prolate to oblate structures in which cross-cutting relations suggest a younging direction toward the centre" (Tweedale, 2012).

The genesis of schlieren is an open question inside the petrological community (e.g. Barbey, 2009). These preserved heterogeneities could reveal information about the chemical and physical processes of formation, and can also provide information about the final stages of magma emplacement (Pitcher, 1993). The schlieren generally suggest dynamic environments during the final stages of flow and crystallization.

For the origin of schlieren structures several processes have been proposed as explained in Barbey (2009), Clarke et al. (2013) and Weinberg et al. (2001).

Weinberg et al. (2001) proposed an interpretation to schlieren genesis, which is suitable to Lavadores Granite case study. In their model, the schlieren development

#1 - Field data sheet to Kfs-phenocrysts - individual data record and classification

General information

Local: Lavadores Beach - near migmatite contact

GPS Coordinates: 41°7'48,60'N; 8°40'10"W

Outcrop Characteristics: granite with presence of Bt-schlieren and swarms

Granite Type: Lavadores Granite, biotitic, porphyritic texture, centimetric pink Kfs-phenocrysts

Geological sketch

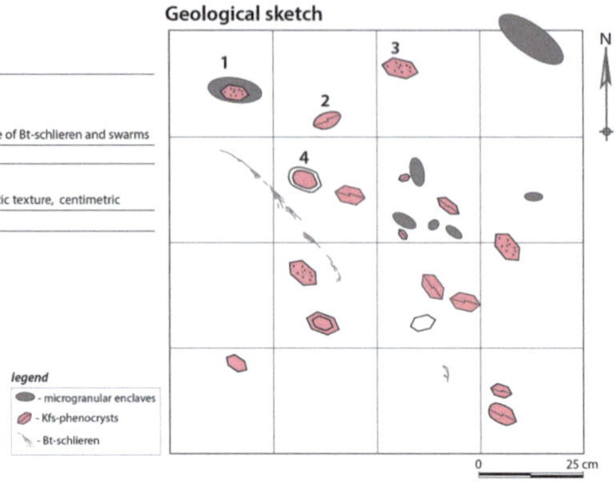

Kfs-phenocrysts individual data record and classification

crystal number	crystal lenght (mm)	Attributes [1]	[2]	[3]	[4]	[5]	[6]	[7]	[8]	Smag (strike/dip, sense)	Lmag (trend/plunge)	Observations
1	35	3	1	1	0	0	0	0	1	098°/76°S		
2	47	0	4	0	0	0	0	0	1	075°/80°S	165°/80°	
3	52	0	1	1	0	0	0	0	1	106°/74°S		
4	64	0	3	0	0	0	0	0	2	120°/70°S	205°/67°	rapakivi
5	(...)											
6												
7												
8												
9												
10												
11												
12												
13												
14												
15												
16												
17												
18												
19												
20												
21												
22												
23												
24												
25												

crystal average lenght (mm) =

Fig. 47 Field data sheet front-page (#1)—stop geological information, Kfs-phenocrysts individual data record and classification

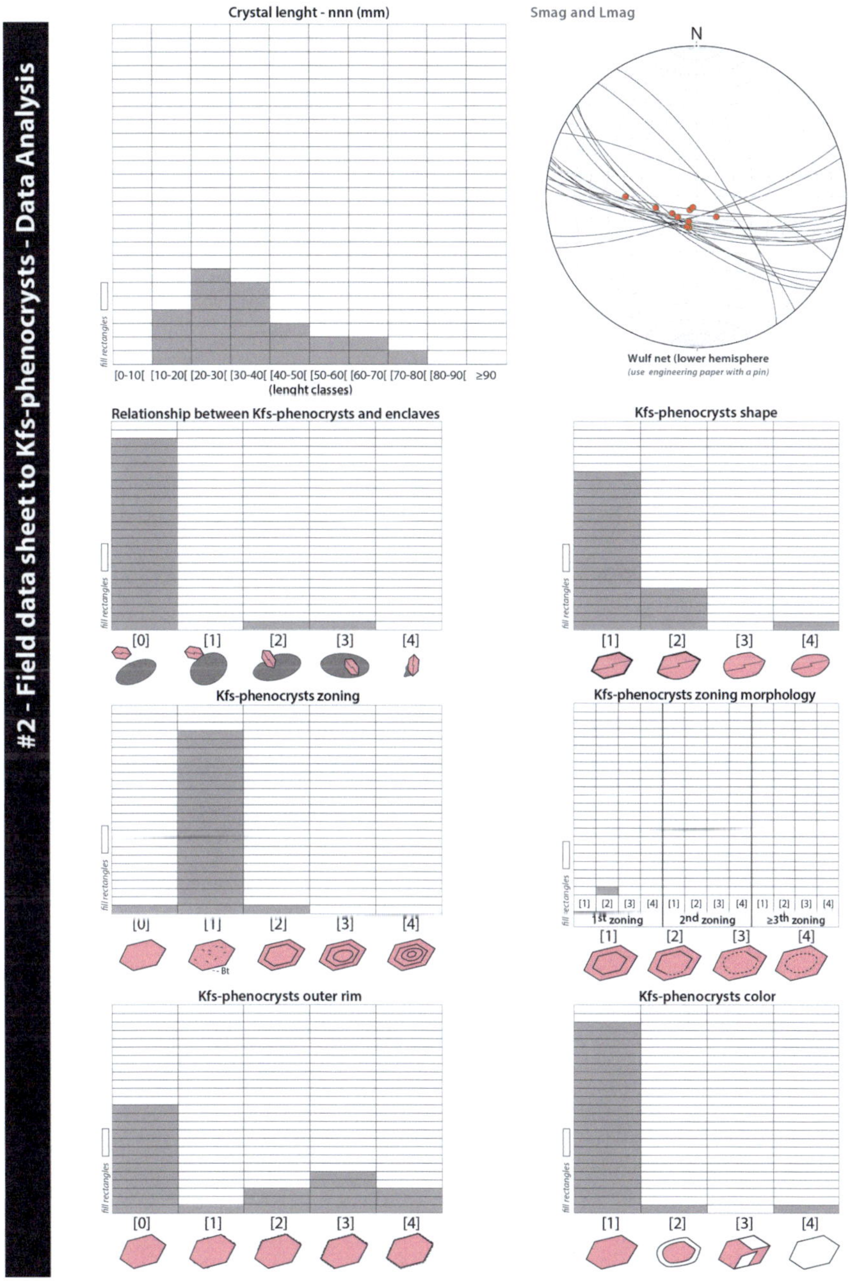

Fig. 48 Field data sheet back-page (#2)—Kfs-phenocrysts data analysis (example)

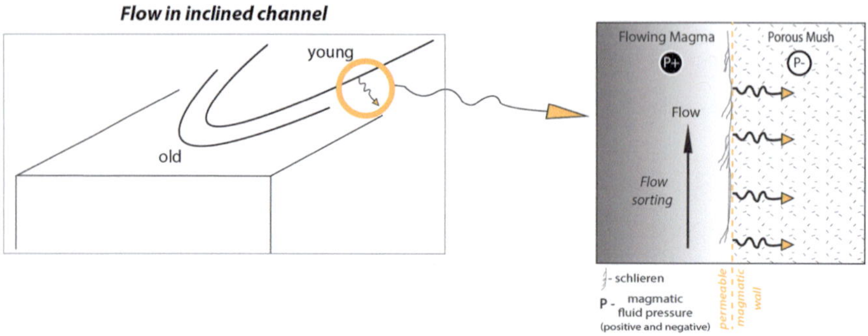

Fig. 49 Sketch of schlieren development. Interstitial melt migrates from the flowing magma into the porous mush wall. The concavity points to the younger layers. By an increasing distance to the wall the ability for melt migration decreases gradually. Adapted from Figs. 9 and 10—Paterson (2009); Weinberg et al. (2001)

is the result of shear sorting of a flowing magma (e.g. Barriére, 1981) in combination with a migration of interstitial melt out of the flowing magma to the porous and/or permeable magmatic wall—this is assumed to be a porous mush when the schlieren developed (Fig. 49).

Thereby, the pressure gradient between the flowing magma and the porous mush is of great importance. By a negative pressure gradient toward the walls a certain amount of melt is pressed out of the flowing magma. As a general rule, feldspar as quartz and other felsic minerals crystallize as the last mineral phases in a magmatic system and the mobile phases stay (Glazner & Johnson, 2013). By pressing out this phase, mostly the mafic minerals, which are already crystallized, form the schlieren along the contact to the porous wall. The pressure gradient is probably triggered by the hotter flowing magma and the mush that is relatively cooled down and at an advanced stage of crystallization.

Furthermore, the dynamic energy of the flowing magma could be an explanation for the better development of schlieren at more curved segments. By the impact of the flowing magma on the wall an overpressure or compaction of the biotite minerals could be generated, what may lead to additional filter pressing. At the direct vicinity of the wall more interstitial melt is able to migrate into the mush. The further away from the contact the more is the decreasing ability and the schlieren gradually disappear. Although, the magma and especially the schlieren is thought to be almost completed crystallized, they are supposed to be flowable anyway to align and laminate, in particular the mafic minerals along the porous mush walls. The flow direction of the mobile magma is thought to be to the top, almost parallel to the developed schlieren and the porous wall (Fig. 49).

The measurement data of the Lavadores Granite Kfs-phenocrysts reinforces this assumption to a certain degree. It shows an almost parallel layering and a distributed lineation along these schlieren layers. By taking a closer look on the schlieren, especially at the north-western part of the central schlieren outcrop, this makes it clear, that the biotite layers are branching always to the south. This could be an evidence for an ascending magma with a slight lateral, probably clockwise, movement.

On Lavadores Granite the schlieren are of the following types: (i) alternating gradated bands of felsic material and biotite; and (ii) concentrations of biotite-rich laminae associated with megacrysts.

3.3.2 Schlieren Outcrops

The Lavadores beach schlieren are exposed mainly in the northern and eastern part. These structures are relatively thin and long-drawn out. Figure 50 presents the schlieren outcrops.

Their longest continuous length—central outcrop—is ~50 m and the orientation is NNW–SSE, sub-parallel to the shoreline. In northern direction—northern outcrop—the schlieren are slightly curved to NW where in ~15 m distance additionally two smaller schlieren are exposed. To the south—southern outcrop—some further schlieren crop out. These are ~35 m away from the main schlieren outcrop and their position is approximately in continuity and parallel to it. The area between these two structures was unapproachable and it was not clear if they are connected by further schlieren structures. In the following, these three outcrops are considered as the northern, the main and the southern schlieren outcrops (Fig. 50).

The schlieren itself has a relatively sharp border to NE and it was possible to take measurements on their dip direction. The central outcrop shows that the schlieren dips at 64°–74° to the W.

The outer sharp border, which is facing NE, shows a convex form. Otherwise, its SW border has a more diffuse appearance and it is concave. A closer look reveals that the schlieren of the main outcrop are not perfectly aligned. In places, they are laterally shifted and do not always match the adjacent schlieren layers. This points to the evolution of separated segments of several meters, which are pieced to one whole structure. In detail, the schlieren itself are laminated aggregates of mafic minerals in the felsic Lavadores Granite. Partially, layers with a higher amount of felsic minerals are intercalated. The darker parts are made up of biotite arranged along the schlieren layers.

A thin section of schlieren evidence a relatively obvious arrangement of biotite crystals indicated by dashed lines (Fig. 51). Almost all laminae are subparallel at the outer, convex shaped border. The further away from the outer border of the schlieren the more the layers are branched out and normally curved to a certain degree where they disappear.

The schlieren sections at the whole outcrop are not equally evolved. From section to section their intensity of evolution may change. A remarkable feature is that schlieren generally are better and broader evolved if they are curved. Additionally, the outer border is sharper in this case and the color contrast to the host rock is very distinct.

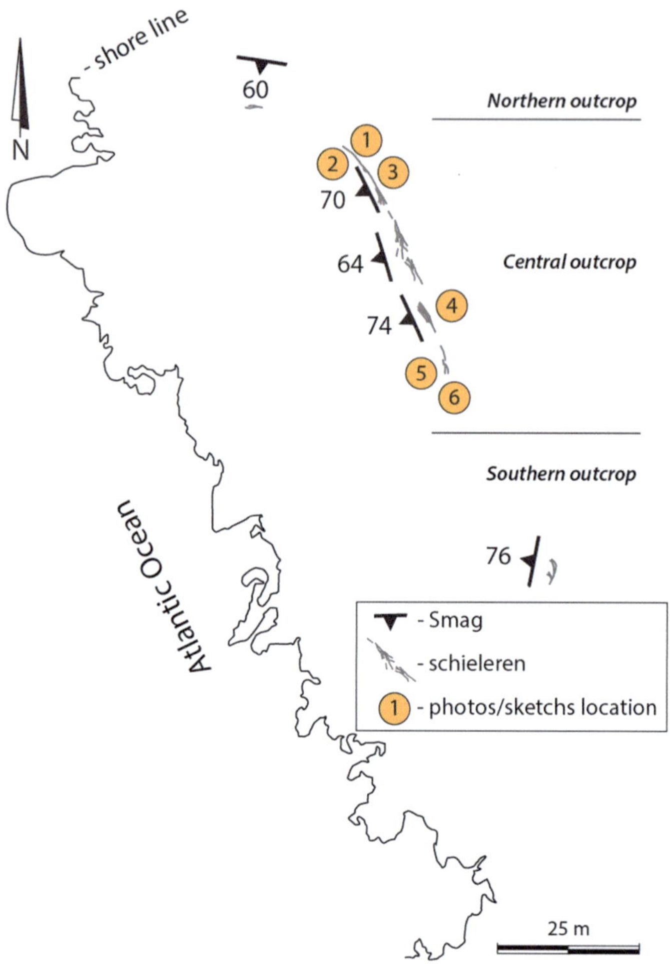

Fig. 50 Sketch of schlieren outcrops. Numbers 1–6 indicate the location of outcrops with photographs and field sketches that will be shown further in this text (adapted from Lorenz, 2014)

Schlieren—Outcrops Locations 1 and 2

The schlieren section at the northern part of the central outcrop is slightly curved and well-develop (Figs. 52 and 53). The outer border is very sharp and the schlieren show a clear color contrast to the Lavadores Granite. The Kfs-phenocrysts, which are aligned by the schlieren, seem to have an effect on the schlieren, as well. The detailed sketch of the border shows biotite layers, which are apparently curved by the interaction with Kfs-phenocrysts (Fig. 52). The feldspar crystals seem to be pressed out by the flowing magma into the granite and the biotite layer had been bent on the same way. Additional attention should be paid to the white color of

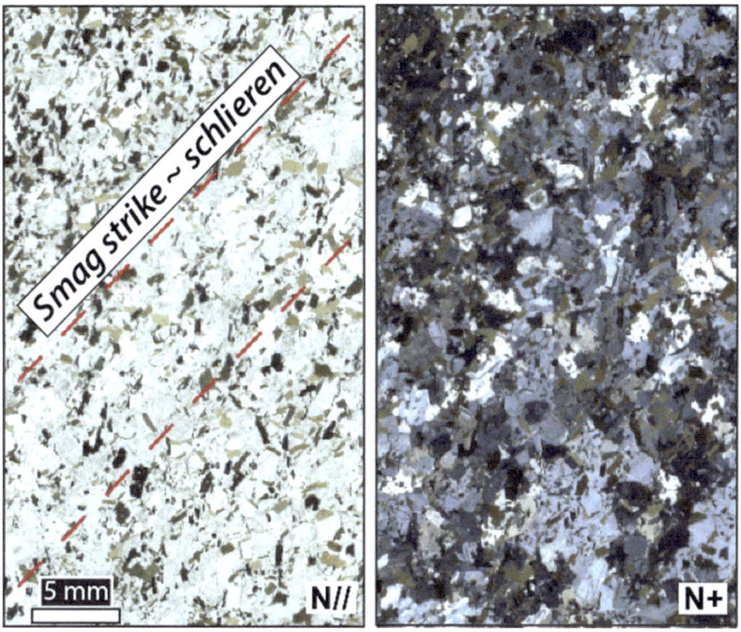

Fig. 51 Thin section from Lavadores Granite crosscutting Bt-schlieren. The red dashed lines indicate the orientation of biotite on schlieren. This orientation matches approximately the strike of magmatic foliation (Smag). N// Plane polarized light (left microphoto) and N+ cross-polarized light (right microphoto), same scale for both microphotos

the crystals inside the schlieren and the pink color just beyond the transition to the granite (detail on Fig. 52). To the concave side of the schlieren this color change is more diffuse and no clear border is visible, just like happen on the schlieren itself. In some places, K-feldspar had been cumulated on the convex border and in its vicinity. The forms of these accumulations are different, generally they are of slight thickness and long-drawn out along the transition layer. Two common forms had been observed. These are: (i) accumulations with recognizable euhedral crystals and (ii) accumulation masses where crystals can scarcely be recognized. This enrichment of Kfs is a possible explanation to the migration of the interstitial melt into the granitic porous mush (Weinberg et al., 2001) where it crystallized either to euhedral megacrysts or to K-feldspar masses. Dashes lines ("///—Kfs mass") mark such slight K-feldspar masses in the upper right corner of Figs. 52 and 53. Further to the South, the branching of the single biotite layers starts directly at the outer border and no parallel layers are visible. Probably, this results from a less force that flowing magma exerts on the wall. Especially in the southeastern half of this part of the schlieren are less curved and they are developed not as well as at the previous one. There is also no clear border and no obvious color change. Solely the more frequent presence of bigger Lavadores Granite Kfs-phenocrysts indicating a possible transition.

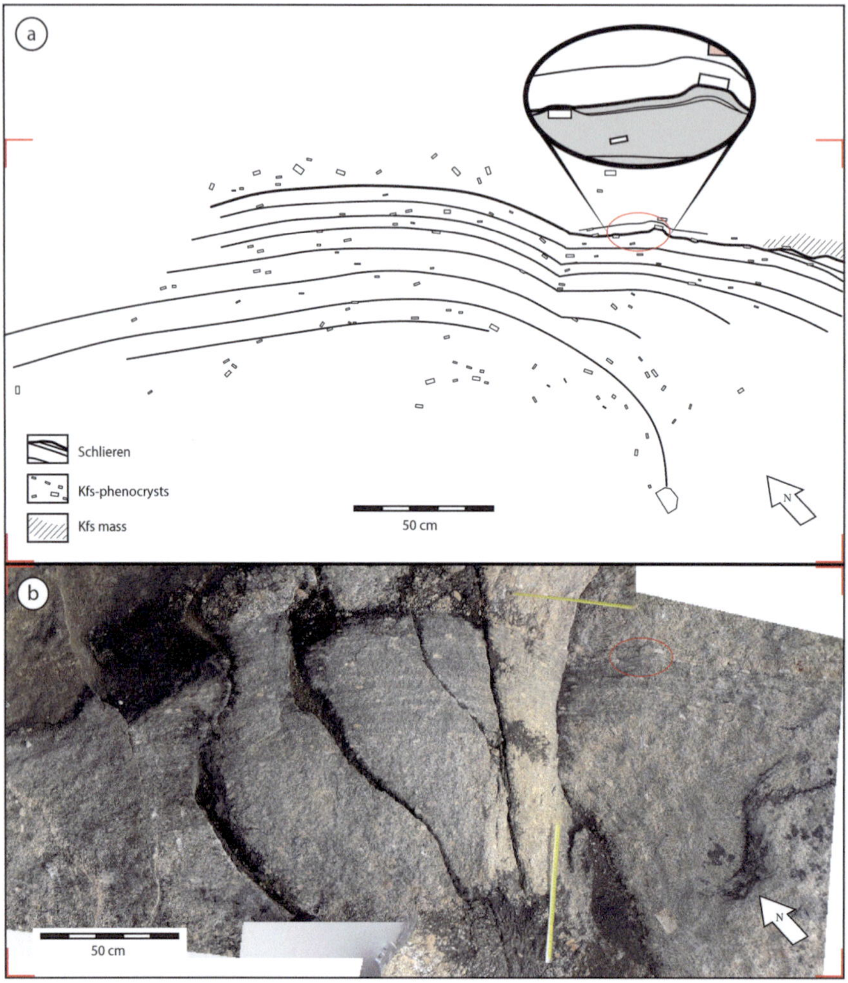

Fig. 52 Sketch (**a**) and photograph (**b**) of well-developed schlieren location 1 within the central outcrop (Fig. 3.2). The red marks locate the observed area in (**a**) and (**b**). The detailed sketch (see red ellipse) shows the outer convex border of the schlieren what is also the transition zone of white and pink K-feldspars. In addition biotite layers seem to be influenced by Kfs-phenocrysts

This lack of color of pink Kfs-phenocryst inside the schlieren carry on its genetic model some constrains. Inside all schlieren structures scattered Kfs-phenocryst are present. They are, unlike the Lavadores Granite Kfs-phenocrysts, almost all whitish coloreds and of minor length (3–4 cm), which also differs less. Generally, the color of minerals is caused by major components or by impurities (Nassau, 1978). The white color of the K-feldspar crystals inside the schlieren can be explained by the migration of interstitial melt into the porous granitic mush during schlieren formation and the loss of required components or impurities, which are responsible for the

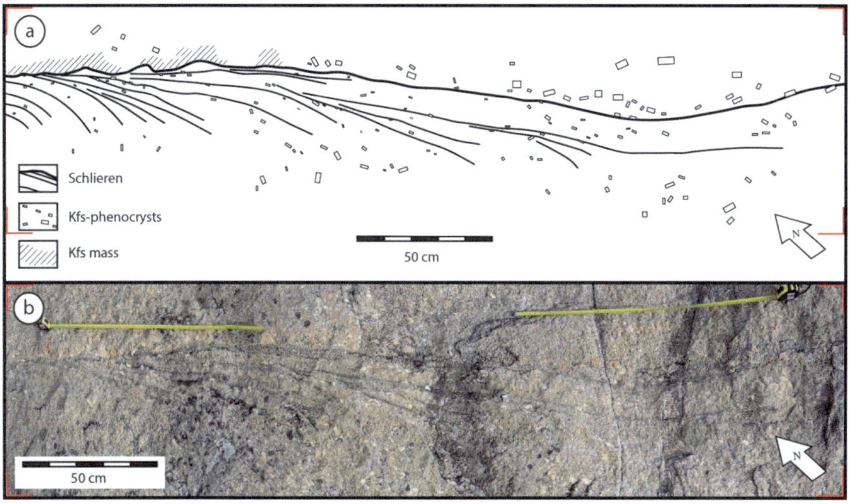

Fig. 53 Sketch (**a**) and photograph (**b**) of schlieren (Fig. 50, location 2 within the central outcrop). Red marks locate the observed area in (**a**) and (**b**). The schlieren are not as well developed as the schlieren of the location 1 shown in Fig. 52

pinkish appearance of K-feldspar. The loss of melt also explains the minor size of the Kfs-phenocryst, as the unavailability of melt was not able to provide further crystallization. Kfs-phenocryst, which are surrounded by the schlieren, are perfectly aligned along its layers. This is an evidence of early crystallization of the Kfs-phenocrysts that was stopped during schlieren development and the corresponding melt loss. Feldspars in nearer adjacencies are affected by these structures as well. They also are aligned along these nearby schlieren layers.

Schlieren—Outcrop Location 3

At the next location (Fig. 54) the schlieren are even less well developed and another outstanding feature had been analyzed. White and pink colored Kfs-phenocrysts and cumulates are clearly separated from each other, but as distinguished from the previous location this color change is not necessarily connected to the schlieren itself. A barrier ("white/pink feldspar boundary"—Fig. 54) had been developed irregularly with white colored feldspars at the central long-drawn part, where the crystals have a certain degree of alignment along this separation layer. The thickness of this band varies and, in some places, it seems that the barrier had been pushed out. At this bulges K-feldspar had been crystallized in masses what gives more emphasis to the barrier. At the southern half of the location of this schlieren, an accumulation of euhedral Kfs-phenocrysts is exposed inside the Lavadores Granite and several examples very close to the border seem to be aligned (Fig. 54, upper right corner). Compared to the other sections, the amount of Kfs-phenocrysts is much higher. Weinberg et al. (2001),

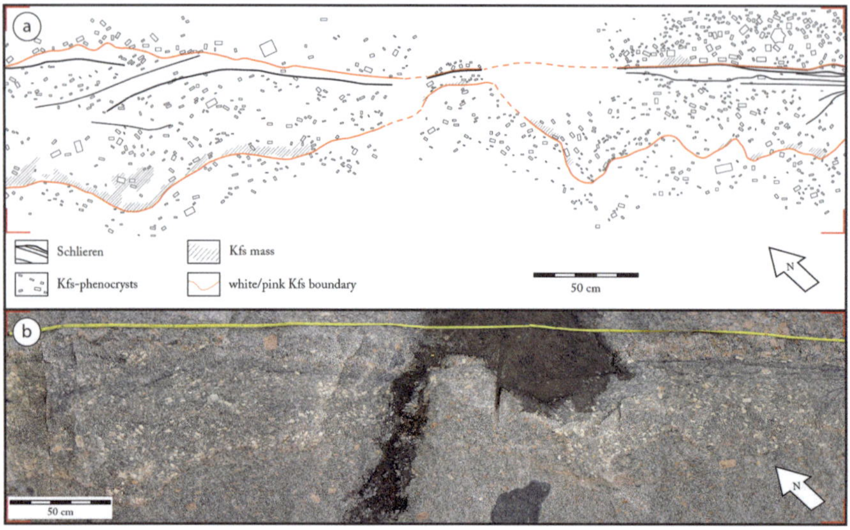

Fig. 54 Sketch (**a**) and photograph (**b**) of schlieren of location 3 within the central outcrop (Fig. 50). The red marks locate the observed area in (**a**) and (**b**). Schlieren are not well developed and only barely visible. The orange line in (**a**) marks the border among white and pink feldspars

described similar observations on granite-granodiorite Tavares Plutonand interpreted them as a kind of log jamming caused by megacrysts.

Schlieren—Outcrop Location 4

Further to the south (Fig. 50, location 4 of central outcrop), some structures enables to theorize about the consistency of magma during the development of schlieren. Figure 55 shows a relatively thick package of almost perfectly parallel-aligned biotite layers. These layers are interrupted by other structures, which seem to be oozed out of a certain layer and propagating through the schlieren package. The severed biotite layers had been dragged forming "drag-folds" that are used to determine the propagation direction. The direction is always from the outer convex border towards the inner concave border. The biggest example of these oozed out structures had propagated through the entire schlieren package. It seems that the transition zone to the granite at the inner border where a small enclave is visible had stopped it. The structure, severaldm-sized next to it, also contains a supposed lighter colored enclave. All these oozed out structures tell us that after the build-up of schlieren structures the magmatic mush is still with high mobility. Due to the absence of any further deformation on these structures by flowing magma, it is obvious that this flow process was already almost "frozen". Wiebe et al. (2007), describe similar structures. The authors interpret them as sinking enclaves producing steep schlieren by propagation through a crystal mush. The described enclaves at this outcrop are

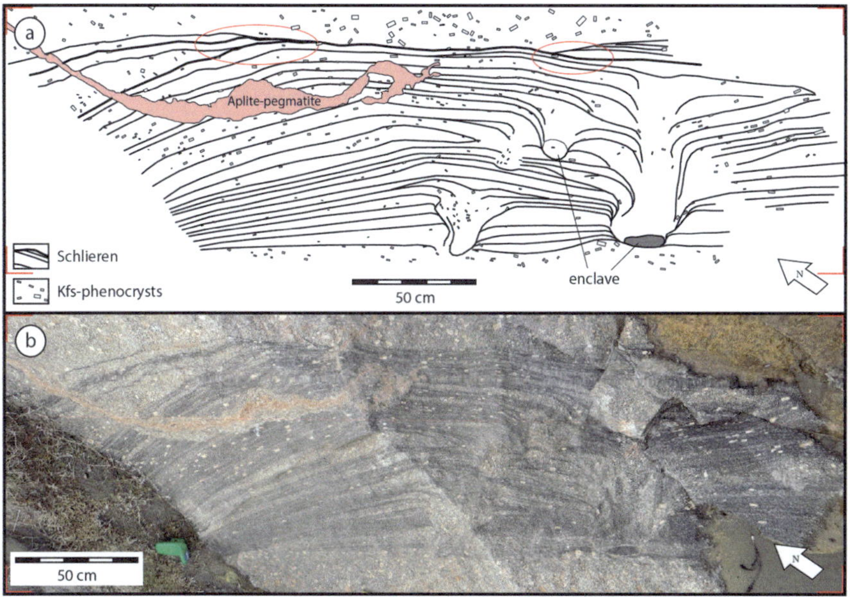

Fig. 55 Sketch (**a**) and photograph (**b**) of very well developed schlieren of location 4 within the central outcrop (Fig. 50). The red marks locate the observed area in (**a**) and (**b**). Some oozed out structures indicate a relatively fluid consistency of the magma shortly after schlieren development. At the top some kind of "crossbedding" structures are visible (red lines ellipses)

a very good evidence for such structures. An alternative to a purely gravitational mechanism is that the Bt-schlieren layer could be disrupted by internal pressure of felsic mush that drags anisotropies (Fig. 55).

On the eastern margin of the schlieren a kind of "crossbedding" had been observed (encircled by red lines in Fig. 55). The small schlieren coming from NW and from SE seem to be cut off by the slightly central curved schlieren package, which indicates the relative age of these two structures.

According to this observation, they can be classified as the older structures in the E and the younger structures in the W. Again this lets us assume that schlieren formation is a process of several stages.

The youngest structure at this location is the aplite-pegmatite that cut through all the other structures. At the center of this thin aplite-pegmatite structure, feldspar and quartz, partially with crystal faces, had been crystallized. The aplite-pegmatite is embedded by a microgranular rim that shows a more basic composition.

Schlieren—Outcrop Location 5 and 6

Schlieren with completely different shape were formed at the southernmost part of the schlieren central outcrop (Fig. 50, locations 5 and 6). The whole segment

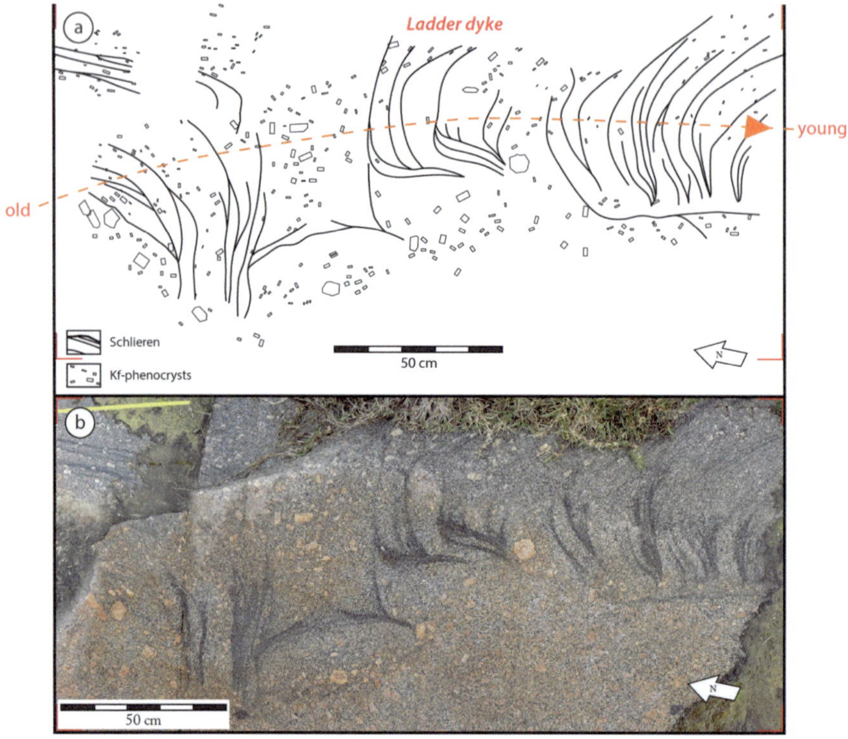

Fig. 56 Sketch (**a**) and photograph (**b**) of schlieren of location 5 within the central outcrop (Fig. 50). The red marks locate the observed area in (**a**) and (**b**). This kind of schlieren is interpreted as magma pathways of ladder dikes, which propagated in southern direction. Each biotite layer shows a position of one magma pulse

has still the same orientation like it was already described before (strike NNW–SSE; dip 64°–74° W). The orientation of the several biotite layers changes from the beforehand-described normal appearance of schlieren layers into curved structures with tapered schlieren towards the inner western side—bottom of figures (Figs. 56 and 57). The Kfs-phenocrysts, which are in direct contact with the schlieren are aligned along their layers. In the center of the figure, which is formed by differently oriented biotite layers, this is not the case where the composition is more felsic and the Kfs-phenocrysts are randomly distributed. The adjacent schlieren in southern direction (Fig. 57) shows an interesting structure. It is wedge-shaped and clearly limited at the margins to the granite. Several schlieren run transversal inside this structure.

Weinberg et al. (2001) described these schlieren as ladder dikes. These structures propagate and grow in the direction of the concave side of the biotite layers. The movement of the schlieren can be described either by a slight displacement of the source that supports the dike with hot ascending material or a linear source that releases material step-by-step in one direction. Each biotite layer shows the position

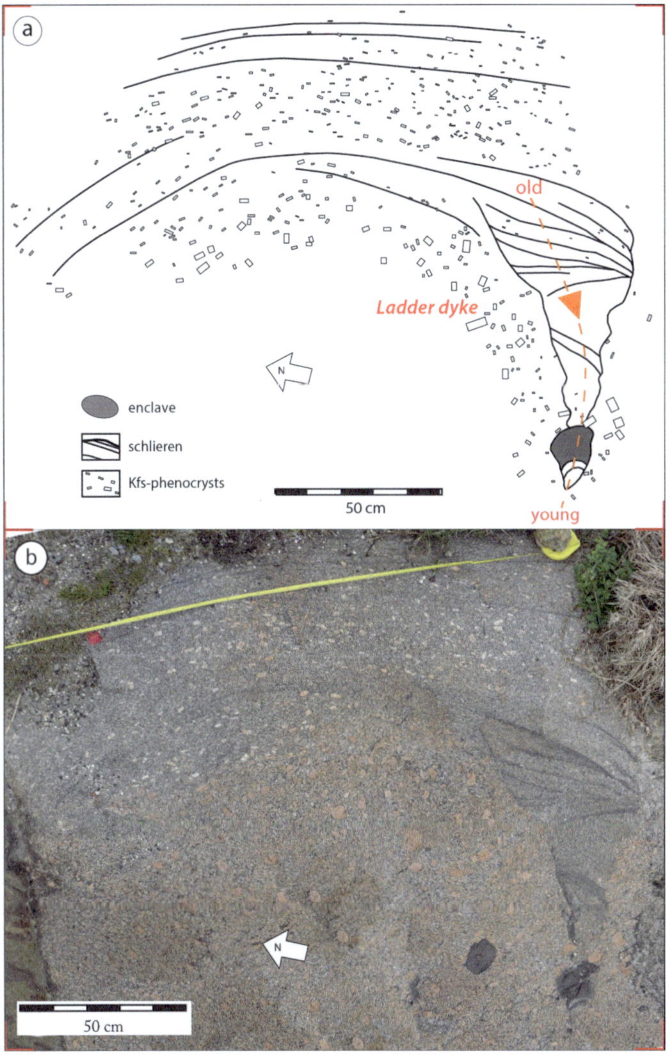

Fig. 57 Sketch (**a**) and photograph (**b**) of schlieren of location 6 within the central outcrop (Fig. 50). The red marks locate the observed area in (**a**) and (**b**). The cone-shaped structure is probably a ladder dike that propagated in western direction. The last magma pulse was only of a small diameter and of a clearly mafic composition

of the dike until it propagates. The outcrop sketched on Fig. 56 is considered to be the transition zone between the "normal" schlieren and the ladder dike. A possible explanation for the arrangement of these two adjacent structures is a second, probably smaller, heat and magma source that provided the development of the ladder dike. This dike propagated first in southern direction (Fig. 56) and in the next location it proceeds in western direction where the source probably dries up and the movement stops and a mafic enclave remains (Fig. 57).

3.4 *STOP 3—Microngranular Enclaves*

3.4.1 Enclaves Classification

The identification of an enclave traditionally must verify the existence of any kind of contrast between two portions of a rock, one in major percentage (host rock) that completely involves the other minor portion (enclave). The relationship between the host and the enclave is generally not unequivocal. This relationship, which was evident in the volcanic rocks studied by Lacroix (1903), is particularly enigmatic in plutonic rocks with special emphasis on granitic ones.

The most accepted classification of enclaves was proposed by Didier (1973) who established terms such as microgranular enclave (or mafic microgranular enclaves—MME) and surmicaceous enclave in addition to the terms xenoliths and schlieren, already widely used.

The origin of granitic enclaves can occur by (Fig. 58):

- The enclave has origin in the granitic host rock itself (host-rock);
- The enclave has origin in the magmatic sequence of the host granitic (magmatic suite);

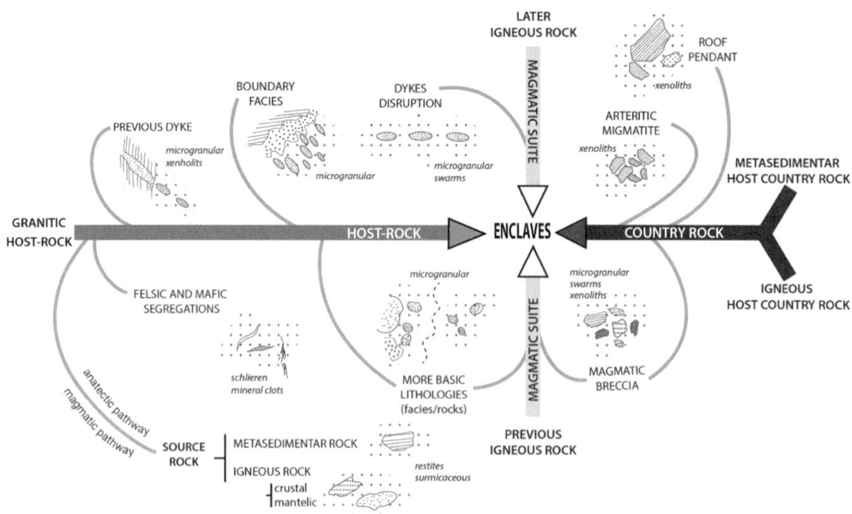

Fig. 58 Genetic links for the origin of the granitic rocks enclaves (after Rodrigues, 1989). At each of the three origins are associated the more common geological processes. In italic are identified the enclaves that are more frequent in each origin

- The enclave has origin in the metasedimentary or igneous country rocks (country rock).

Enclaves originated directly from host granite rock can be the result of different geological processes: they may be (*i*) fine-grained terms of boundary facies; (*ii*) result from the dismantling of early dikes not fully digested in the granitisation process; (*iii*) biotitic (schlieren) or other mineralogical segregations or concentrations; or (*iv*) materials originating from protholit (sedimentary, crustal igneous or mantellic igneous) not fully digested, in an anatectic or a magmatic genetic process (Fig. 58).

The most studied source of enclaves has been the magmatic suite of the host granite rock. The less differentiated lithologies, more basic magmatic terms, generate fascinating and ubiquitous mafic microgranular enclaves (MME). On the other hand, the more differentiated (more acidic) magmatic terms may, although rarely, generate enclaves by dismantling late emplaced dikes (Fig. 58).

Finally, enclaves of xenolith type are linked to the metasedimentary or igneous host country rock, results from roof-pendent, arteritic migmatites and magmatic breccia (Fig. 58).

3.4.2 Enclave Swarms

Enclave swarm is used to describe an abnormal concentration of the same or different enclave lithologies and xenoliths in a limited area within a granitic host-rock magmatic environment. These structures can show five basic types of swarm geometry (Tobisch et al., 1997): dyke; small raft; lens; pipe/vortex and large massif (Fig. 59)

3.4.3 Lavadores Beach Enclaves and Swarms

The enclaves of Lavadores Granite are mentioned in the bibliography for the first time by Alves (1966), and were object of a more complete investigation in the 90's (e.g. Silva, 1995, 2010; Silva & Neiva, 1998).

The most abundant and representative enclaves were classified as mafic microgranular enclaves, petrographically classified as diorite, quartz-diorite, quartz-sienite with alcaline feldspar, quartz-sienite, monzodiorite, quartzo-monzodiorite, tonalite and granodiorite (Silva, 2010).

Geochemical studies carried on the mafic microgranular enclaves led to the conclusion that the Lavadores Granite host enclaves belong to several geochemical affinities, basic, alkaline and even Mg–K (Silva, 1995; Silva & Neiva, 1998). They represent magmas generated by partial fusion of enriched mantle, which repeatedly intruded the partial crystallized magmatic granitic chamber (Silva, 2010).

The enclave types that outcrop in Lavadores beach are mainly mafic microgranular enclaves, tonalitic to granodioritic compositions, migmatitic and double enclaves (Fig. 60).

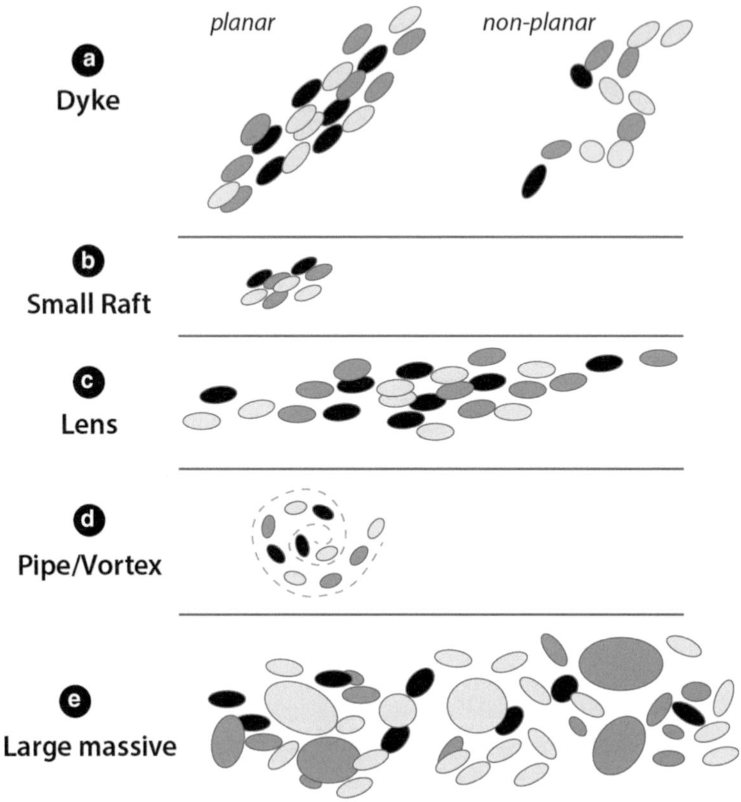

Fig. 59 Swarms enclave types: **a** dyke; **b** small raft; **c** lens; **d** pipe/vortex: **e** large massif (adapted from Tobisch et al., 1997)

Field relationships and macroscopic textures provide clues to genetic processes of mafic microgranular enclave swarms in the Lavadores Granite pluton. The swarm mafic microgranular enclaves are heterometric and typically have lobate shape, sometimes narrow chilled margins, and exhibit compositional variation (Fig. 61).

The most developed swarm of enclaves in Lavadores Beach shows an elliptical shape corresponding to a swarm enclave geometry of large massive type (Fig. 59e), being possible to follow it for about 10 m. A characteristic of this swarm enclave is the concentration of accumulated Kfs-phenocryst and various types of enclaves. The limit of the swarm enclaves with the host granite is only defined by the higher concentration of the enclaves, and there is no physical border at all. There are also very gradual transitions that can be described, merely, as zones of greater occurrence of these structures. The genesis of these swarms may be linked to the establishment of convective cell limits, as we will see below (see Sect. 3.5).

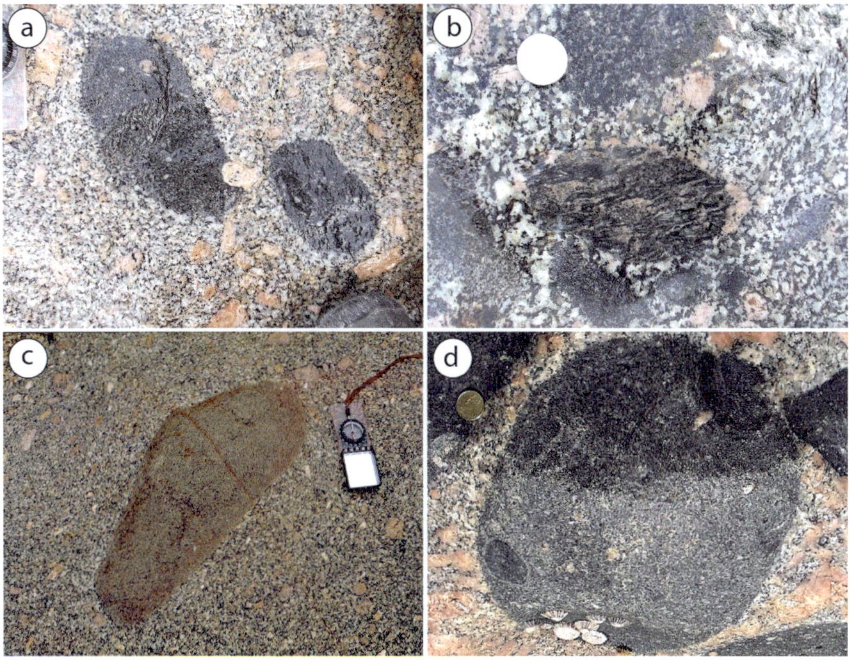

Fig. 60 Lavadores enclave types: **a** tonalitic mafic microgranular; **b** migmatitic; **c** granodioritic microgranular; **d** double

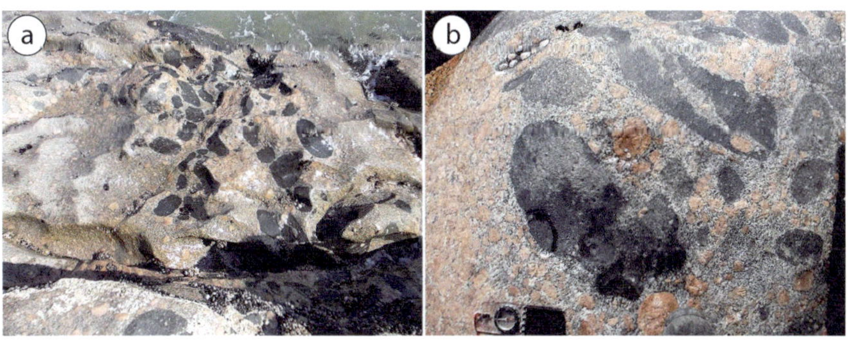

Fig. 61 Lavadores Granite swarms: **a** microgranular non-planar dike swarm; **b** detail of a microgranular enclave swarm with Kfs-phenocrysts filling the space between enclaves

3.5 STOP 4—Interaction Between Acid and Basic Magmas

3.5.1 Enclaves Relationship with Granite Host-Rock

Lavadores Granite is a showcase of granitic magmas microstructures, mafic microgranular enclaves and Kfs-phenocrysts growth features (see Sect. 3.2.3: Table 3; Figs. 34, 35, 36, 37, 38, 39. 40, 41, 42, 43 and 44).

The big petrological question is: what is common process that can link all these "microstructures"?

The logical answer is that they testify sequential stages of the dynamic's interaction process between acid magmas (granitic microstructures and K-feldspars) and basic magmas (mafic microgranular enclaves).

This text will introduce an empirical model to explain all these relationships, beginning by the interaction between Kfs-phenocrysts and mafic microgranular enclaves.

In the model proposed to explain this interaction, for simplicity reasons, it is constructed on the concept of a static enclave versus dynamic Kfs-phenocrysts—K-feldspar crossing thought out enclaves. However, the opposite situation can also occur in the nature: static Kfs-phenocrysts versus enclaves convectively moving all around in magmatic chamber—enclaves passing through the Kfs (Fig. 62). This dynamic interaction between Fk-phenocrysts and enclaves is also admitted to occur on enclaves with different dimension and composition.

The Lavadores Granite Kfs-phenocrysts exhibit several microstructures that are not usually on common igneous megacrysts and points to a complex genetic process. The Kfs ubiquous characteristic is its shape, habit and dimensions: they are euhedral tabular centimetric crystals (Figs. 62 and 1).

One interesting microstructure is the Kfs-phenocrysts transecting the boundaries of enclaves: some of them are "coming-in" and others "coming-out" the enclave.

Sometimes, these "transecting" crystals are observed "coming-in" the microgranular enclaves (Figs. 62 and 2). They are fully euhedral and "drags" the enclave boundary—especially when an outer rim exists—in the sense of the entry movement. Inside the enclaves, the Kfs could exhibits rounded morphologies, subhedral and anhedral—more or less, ten times more abundant than in the host granite—, as well as euhedral, what points to some kind of dissolution process (Figs. 62, 3a, b, c).

The "coming-out" movement of Kfs is still more interesting. Most frequently, these Kfs are fully rounded (Figs. 62 and 4) and "drags" the enclave boundary to the outside. However, more rarely it was observed a complex "riding" Kfs with half rounded shape inside the enclave and a half euhedral overgrowth masking an older rounded shape outside the enclave (Figs. 62 and 6a). Once the Kfs are outside the enclave the general rule is the crystal growing. They are in equilibrium with the granitic melt crystallization process and stable crystal faces growth is improved (Figs. 62 and 5). Sometimes an astonishing tail ornaments partially the Kfs faces what can be interpreted as testimony of the carrying behind of the enclave stuff

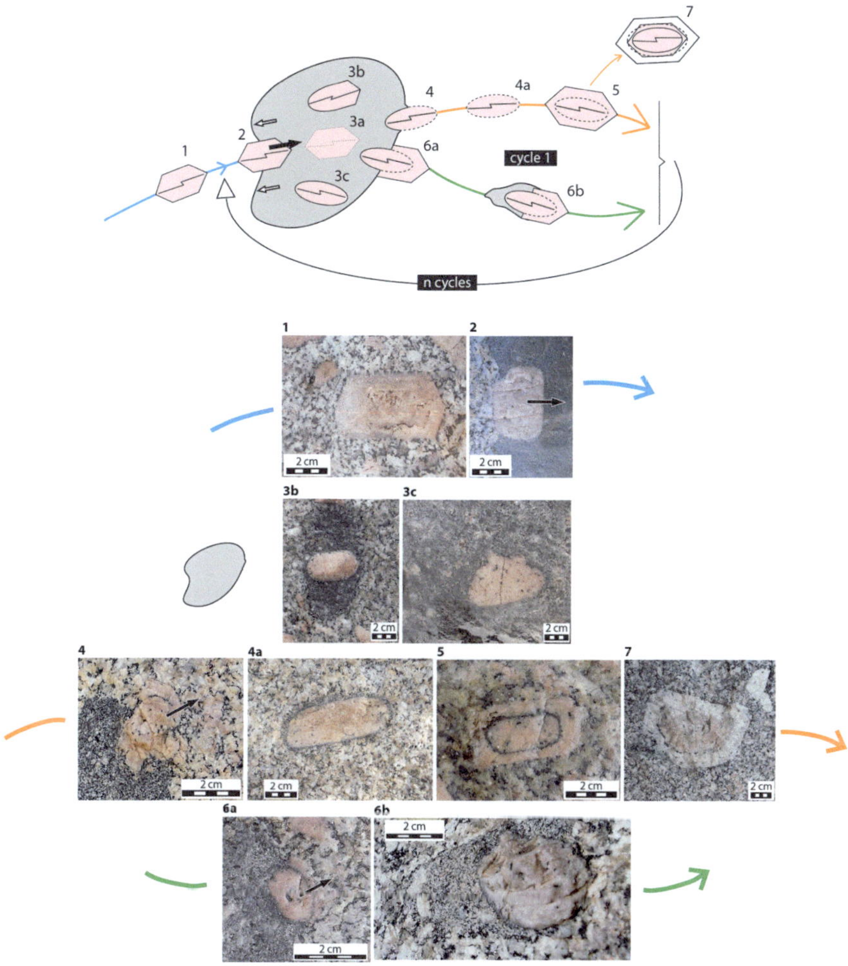

Fig. 62 General mechanism showing the sequential stages of the interaction between Kfs-phenocrysts and enclaves inside a granitic magmatic chamber. Black arrow (inside Kfs 2) shows the movement sense in the hypothesis of static enclave and dynamic Kfs; and white arrow (inside enclave) shows movement sense in the hypothesis of dynamic enclaves and static Kfs

material (Figs. 62 and 6a). Later on, rapakivi growth could happen (Figs. 62 and 7) over anhedral or euhedral Kfs (Figs. 62, 4a, 5 and 6b) in ~2% of the crystals.

All of these Kfs microstructures can be organized on a logical sequence as explained on Fig. 62. This is a cyclic mechanism and can be reset at any of the types of Kfs-phenocryst described (Figs. 62, 4a, 5 and 6b). Whenever these Kfs are out or on the vicinity of an enclave they can be caught, at any time, in a new dynamic Kfs-enclave cycle.

This mechanism is based on detailed field observations, but like it was previously referred, it is important to give relevance that the sense of convectively moving of

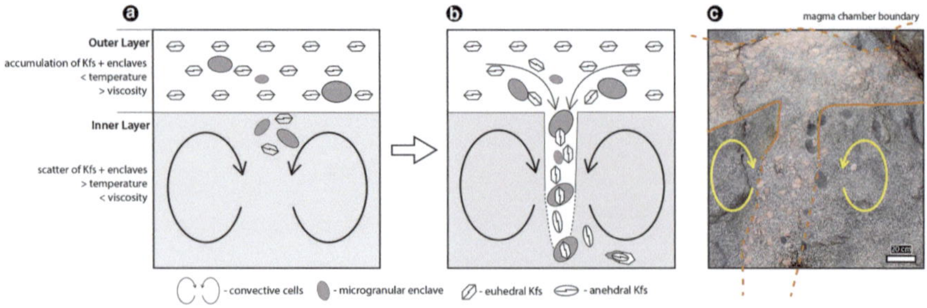

Fig. 63 The sketch shows the mechanism that generates the interaction between two layers inside a magmatic chamber with different viscosities and temperature (outer layer—OL, and inner layer—IL). **a** Initial process with larger accumulations of Kfs and enclaves on the OL and the development of convective cells in the IL; **b** More effective action of convective cells on IL will break the boundary OL–IL, promoting the absorbing and dispersing of solid load of OL (granitic mush, accumulations of Kfs-phenocryst and enclaves) throughout the IL—generating the observed microstructures with the interaction between Kfs-phenocrysts and enclaves; **c** Lavadores Granite outcrop that records this mechanism

the Kfs-phenocrysts and enclaves all around in a magmatic chamber is relative—and it could be the enclaves passing through the more statics Kfs. We believe that this inversion of the "flowsense", definitely do not change de mechanical and geochemical processes of the interaction between Kfs-phenocrysts and enclaves.

The detailed observation of Lavadores Granite allows the assembly of a general mechanism that explains all these microstructures in relationship with the thermodynamic stratification on the convection cells boundaries, as could happen near the magmatic chamber border (Fig. 63).

The thermodynamic stratification in the border of the magmatic chamber enhanced the development of a dynamic interface splitting two different magmatic mushes (outer layer—OL and inner layer—IL): one, OL, with a higher content of crystals and enclaves solid charge (< temperature and > viscosity) and other, IL, with higher content of magmatic liquid (> temperature and < viscosity). McBirney (1984) proposed de concept that on the inner layer (IL) of magmatic systems convective cells will be developed.

As a result of convective cells in the inner layer, at the limits of convective cells, a heterogeneity nucleation point will occurs: the interface between the two layers is broken, allowing the absorbing and dispersing of outer layer solid load—Kfs-phenocrysts, enclaves, granitic mush—throughout the inner layer (Fig. 63). This mechanism is proposed to be the keyinteraction between acid and basic magmas that could explain the features observed in Lavadores Granite.

3.5.2 Enclaves Hybridization

The existence of rounded mafic microgranular enclaves within an igneous host of granite composition, could indicate different origins for the enclave: it may result from mixing and mingling processes involving more basic lithologies; it may be the result of dismantling an early or syn-magmatic dike; it may be a border facies; among others possibilities (for more details see Sect. 3.4.1, Fig. 58).

Lavadores Granite mafic microgranular enclaves are interpreted as the result of the mingling between basic and acid magmas. Silva (2010) proposed that they are the result of repeatedly intrusion of a partial crystallized granitic magmatic chamber by partial fused enriched mantle magma.

After the intrusion episode, perfect rounded mafic microgranular enclaves will develop and if they are in thermodynamic equilibrium within the host granite they will freeze, and nothing more happen. This story is very short: at outcrops scale the rounded shape indicates that both magmas had a liquid like behavior at the same time (e.g. Fernandez & Gasquet, 1994; Fernandez et al., 1997; McBirney, 1984) and chilled margin of enclaves will testify the suddenly crystallization.

However, at Lavadores Granite the history is more complex. The primary rounded mafic microgranular enclaves (Fig. 64, P) became instable and a transient state will occurs (Fig. 64, t). The reason to this destabilization was already explained (see Fig. 63). On the proposed genetic sketch (Fig. 64) the concept of transient enclave (t) was introduced for better understanding the cyclic enclave hybridization mechanism. Once the original basic igneous term ("primary/stable" microgranular enclave—P) was introduced into the granitic magmatic chamber, a sequential and gradual change of basic composition starts towards a final more acid hybrid enclave state. To this intermediate compositional states—that sometimes never are revealed in outcrop—it is proposed the general term "transient" enclave.

As a consequence of the progression of this transient state, several cyclic situations can occur (Fig. 64): (h1)—The mafic microgranular enclaves solidify without any significant exchanges with host granitic rock, only biotic cooling rims identifies a freezing border; (h2)—The mafic microgranular enclaves suffer a hybridization process from outside in that is not complete: relicts of an older transient state will remain inside a hybridized granitic host generating double enclaves; when the hybridization process referred to h2 is complete, two end terms can occur—the hybrid mafic microgranular enclaves achieved some stability and a rounded shape with a cooling rim will develops (h3) or the hybridization process goes so far that there are no physical boundaries between enclave and host granitic rock and a diffuse rim will develops (h4).

Each one of these end terms (h1–h4) can re-enter this cycle of enclave hybridization in a non-limited number of times and generate all the enclave combinations that we see in the field.

3.5.3 Enclave Disruption Mechanism (EDM)

The nature records the intermediate steps between the mafic microgranular enclave transient state (t) and the end terms (h1–h4) above described (Sect. 3.5.2, Fig. 64), on a process called herein as enclave disruption mechanism (EDM)—Fig. 65—that will be described on the following.

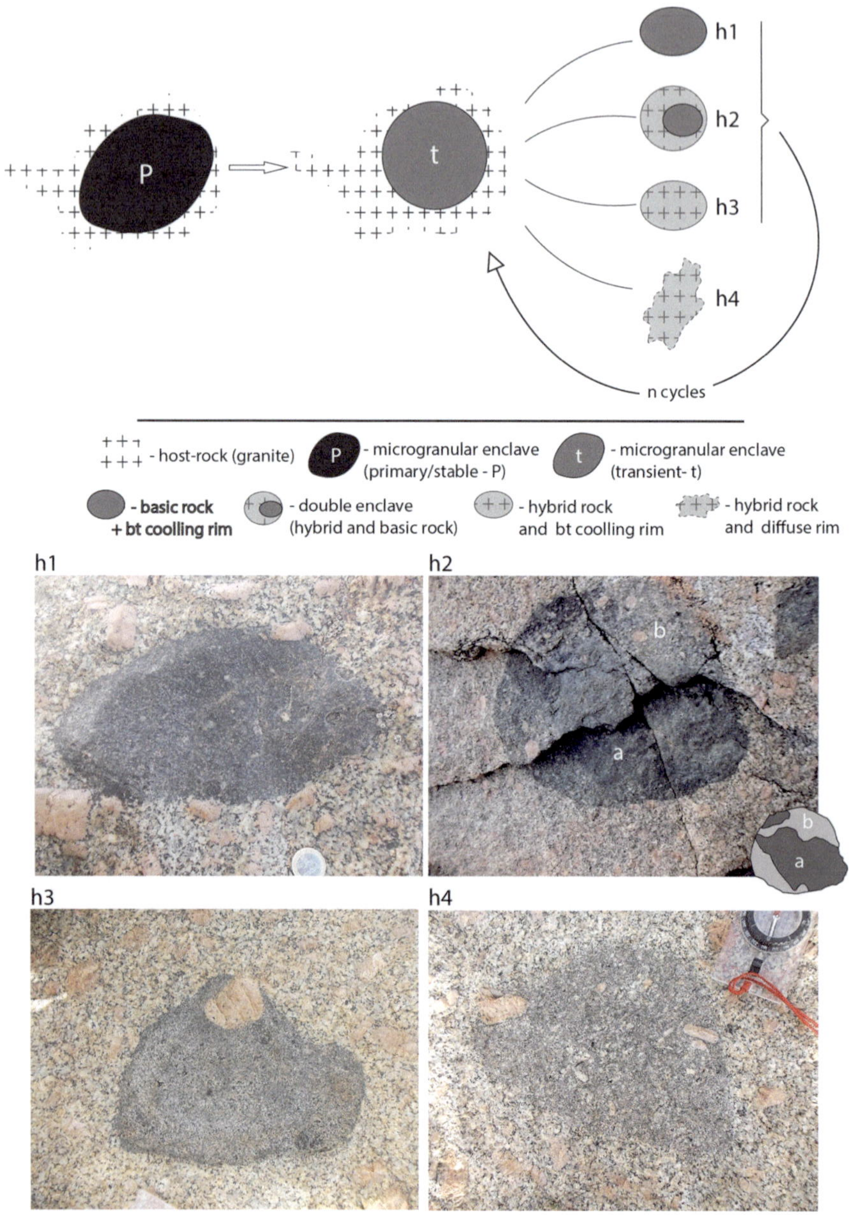

Fig. 64 Cyclic enclave hybridization mechanism. h1—basic rock with Bt cooling rim; h2—double enclave of hybrid and basic rock; h3—hybrid rock with Bt cooling rim; h4—hybrid rock with diffuse rim

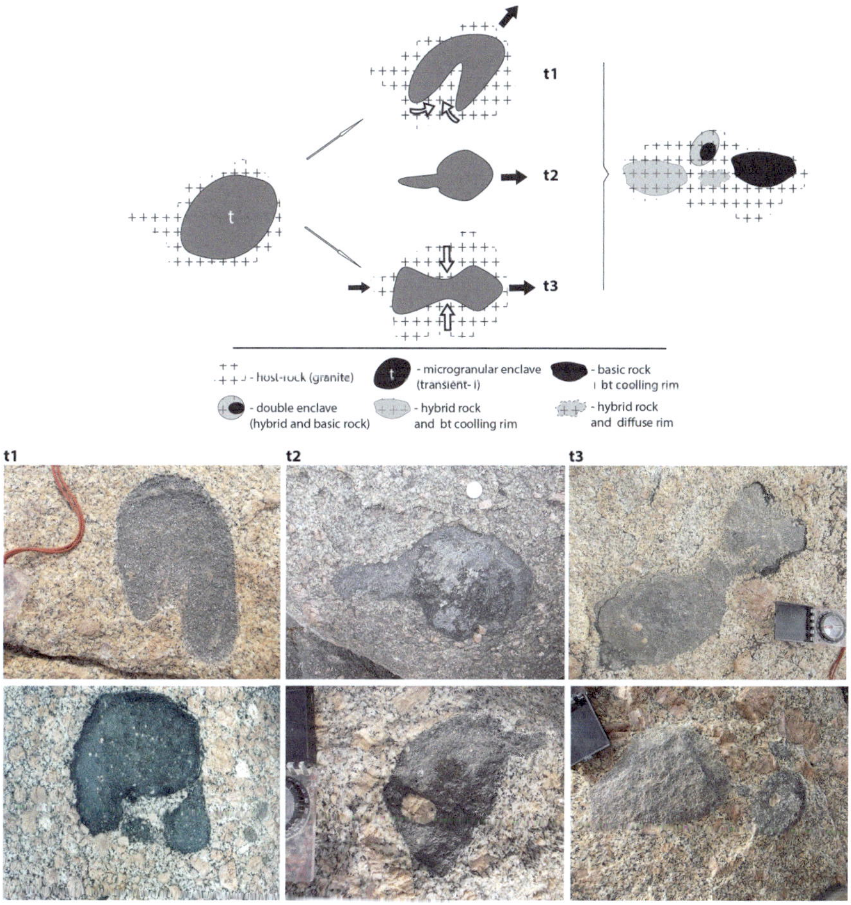

Fig. 65 Enclave Disruption Mechanism (EDM). **a** and **b**—t1. The granitic host-rock moving in the sense of arrows penetrate the enclave; **c** and **d**—t2. The enclave leaves a tail behind; **e** and **f**—t3. Granitic host-rock that is in its final cooling stage generates squeezing forces (porous pressing or some compressive tectonic effect) promoting the enclave disruption. White arrows: entrance of granitic magma sense; black arrows: relative motion of enclave

The convective movement into a magmatic chamber acts against the granitic mush and the enclaves imprinting on them relative velocities. This general magmatic dynamics generates several ways how the enclaves are disrupted inside the magmatic chamber. The most frequent that can be empirical deduced from Lavadores Granite observations is proposed on the illustration of Fig. 65a, b (t1). The motion of enclave (black arrow) generates negative pressures (white arrows) on the "back" of enclaves—usually identified by a more diffuse and transitional rim or even by a concavity border—that triggers the disruption of enclave boundary with mingling between enclave material and granitic mush.

Sometimes (Fig. 65c, d—2), the movement of enclave (black arrow) triggers a boundary disruption that leaves behind a tail promoting the mingling.

The enclave motion into a magma cell can also generate an enclave elongation (Fig. 65e, f—t3) resulting on a "pulling" head (big black arrow) with a "lazy" ending (small black arrow). This kind of movement induces negative pressure (white arrows) concavities that promotes the enclave disruption.

All the Kfs-phenocryst microstructures (Fig. 62) and mafic microgranular enclaves features (Figs. 64 and 65) can be observed along the beach outcrops of Lavadores Granite (Fig. 29).

3.5.4 Field Activity/Geological Challenge

On the outcrops of this Stop 4 the geological challenge is to the identify structures and microstructures, which allows, empirically, the identification of the enclave hybridization level (Fig. 64, and respective caption): four basic types of enclaves reveal different levels of hybridization (h1. Basic rock; h2. Double enclave of hybrid and basic rock; h3. Hybrid rock with cooling rim; h4. Hybrid rock with diffuse rim).

The activity/challenge proposed is the utilization of a field data sheet (Fig. 66) to record data, to identify and quantify the hybridization enclave levels.

3.6 STOP 5—The Feldspathic Plume Structure

3.6.1 Structural Features and Composition of the Lavadores Feldspathicplume

On the northeastern mapped sector (Lorenz, 2014) outcrops an association of an irregular mafic tonalitic layer with an exotic K-feldspar crystallization, plus a heterogeneous nucleation on the tonalite boundary and a dendritic and comb grown. This structure was designated as feldspathic plume.

The feldspathic plume structures represent an almost completely continuous structure. In the northern part of this outcrop its direction is to SW–NE and suddenly it turns to approximately N–S and after several small bends it gets thinner and disappears (Fig. 29). The plume structures in the center of the outcrop shows an attitude of approximately 148°/55°SW. It should be noted, that this attitude is very similar to migmatite wall rock boundary (146°/55°SW). In the northwestern part, two more isolated and relatively short segments of the plume structure are exposed. These are very close to the migmatite and they are not as well developed as the long continuous plume structures several meters away from the migmatite. The northern part is best developed and the feldspathic plume structure was better exposed (Fig. 67). The plume outcrops over a length of about 5.5 m.

On the western part of this small sector the plume has a thickness of > 10 cm, while on the northeastern part, examples of thickness > 30 cm are common.

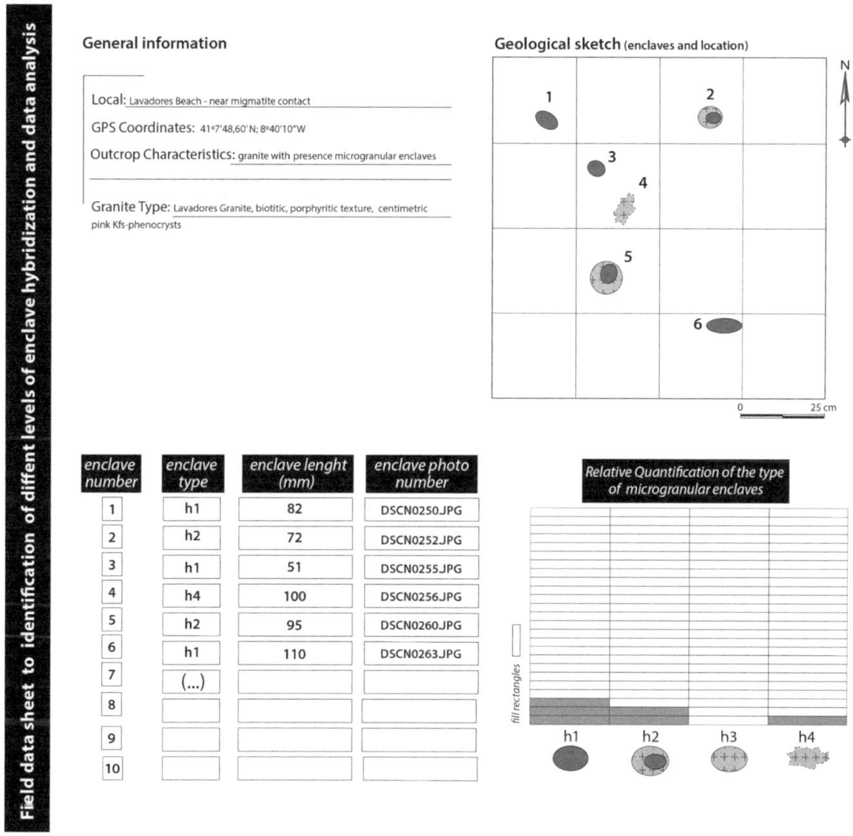

Field data sheet to identification of diffent levels of enclave hybridization and data analysis

General information

Local: Lavadores Beach - near migmatite contact

GPS Coordinates: 41°7'48,60'N; 8°40'10"W

Outcrop Characteristics: granite with presence microgranular enclaves

Granite Type: Lavadores Granite, biotitic, porphyritic texture, centimetric pink Kfs-phenocrysts

Geological sketch (enclaves and location)

enclave number	enclave type	enclave lenght (mm)	enclave photo number
1	h1	82	DSCN0250.JPG
2	h2	72	DSCN0252.JPG
3	h1	51	DSCN0255.JPG
4	h4	100	DSCN0256.JPG
5	h2	95	DSCN0260.JPG
6	h1	110	DSCN0263.JPG
7	(...)		
8			
9			
10			

Fig. 66 Field data sheet—stop geological information, identification of different level of enclave hybridization and relative quantification. On structural and microstructural properties was used classic petrographic terminology (e.g. Hibbard, 1995; Castro et al., 1990; Vernon, 2004). On this data sheet the mandatory properties are assigned with bold type. After fulfilling the boxes it is necessary to evaluate the type of hybridized enclave level (h1, h2, h3 and h4) that better match with the observed enclave

The feldspar crystals grew inside a band of tonalitic composition with similar size. The feldspar crystals can also nucleates in the granite and grew crosscutting the entire mafic band and sometimes throwing itself into the granite.

On the northeastern part the feldspar crystallization apparently had been stopped by the transition to the granite, or had been cut-off by a relative movement between the plume and the granite. Further evidence of this movement is the lateral shift of the mafic band (this structure is marked by the red arrow—Fig. 67).

Inside this thinned out passage (without signals of deformation) no feldspathic structures are present what, probably, is the result of later tonalitic matrix crystallization than the supposed movement.

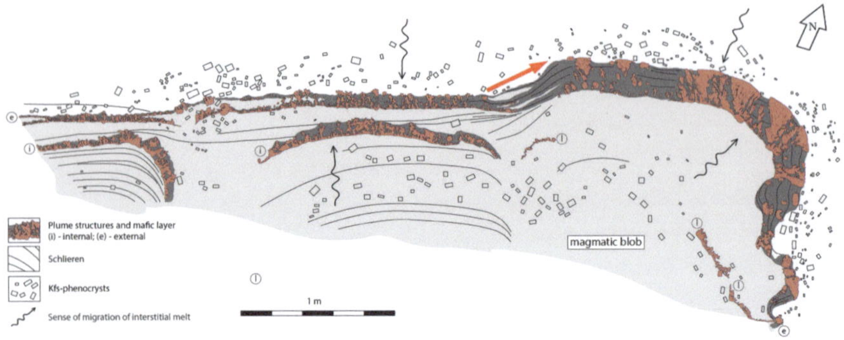

Fig. 67 Sketch of the feldspar plume structures (e—external, i—internal and lo—"lost" plume, see text for explanation). The schlieren shows the same orientation like the second layer (internal) of plume structures. The schlieren are bent and partially developed crossbedding. The red arrow marks a supposed movement, only on the NE part, where the plume structures show a more massive crystallization and probably a cut-off top. See text for more explanation

Fig. 68 Comparison between a dendritic development of plume structures (**a**) and a more massive development (**b**). The top of the plume structures in (**b**) is cut-off. Notice the scale in both pictures

The plume in this NE sector is completely different of all other plume structures in the mapped area. The thickness of the tonalitic layer is higher and the plume structures are developed massively. A dendritic branching is only visible by taking a closer look. A comparison between a dendritically developed plume structure (Fig. 68a) and a massively developed structure (Fig. 68b), which is stopped sharply by the transition to the granite, it is shown for better understanding.

3.6.2 Origin and Development of Plume Structures

Weinberg et al. (2001) described structures that associate feldspathic felsic pockets and mafic layers. The authors interpreted these structures as leucogranite pockets developed from an interstitial melt that migrated from a mush. In their description the structures have irregular shapes and "at grain scale they form irregular branches that grade into the felsic interstitial material". The structures are connected to mafic minerals especially biotite and sometimes even layers of this, but these layers are only at the lower half or below these structures, considering provable bottom-up direction for leucogranitic migration. Furthermore the authors state that narrow pockets of leucogranite also occur inside irregular sheets of mafic-rich granodiorite of varying width (1–10 cm).

In contrast to the structures described by Weinberg et al (2001), the feldspathic plume structures seem to crystallize completely inside a clear "open" layer and it seems they were able to crystallize freely and not only in the interstitial space of other, already crystallized mineral assemblage.

Similarly to the Weinberg et al (2001)'s leucogranite pockets, the plume structures also form irregular branches, which sometimes crystallize out of the mafic layer into the granite (Fig. 67—"lost" plume, lo). Dendritic crystallization also had been observed on K-feldspar megacrysts, which are surrounded by the mafic layer. The direction of crystallization is always perpendicular to the mafic band and towards the Lavadores Granite as well (Fig. 69).

Fig. 69 Plume structures growing surrounded by a mafic band. On the surface of Kfs-phenocrysts a dendritic crystallization in only one direction is visible. The mafic band shows forms that are similar to schlieren structures on the northern mapped area. The layers seem to be influenced/bent by Kfs-phenocrysts

3.6.3 Global Modelto the Lavadores Granite and Featuring Related Structures

Vermicular quartz grains (Qtz) grows inside the dendritically grown Kfs. Quartz grains shows the same extinction position over numerous grains indicating that these supposed several grains are just one bigger quartz crystal that branches out in different directions and only several branches are cut in the observation section (Fig. 70).

The plume structures are an intergrowth between vermicular Qtz and Kfs and not with Pl. This structure should be declared as symplectite, the general topic for intergrowths between different minerals. The exact process that is responsible for the appearance of the plume structures is yet unknown.

All these observations lead to the hypothesis that the dendritic plume structure crystallization is a result of specific conditions inside schlieren and in their vicinity.

However, the plume structures are an almost completely continuous band, especially in the southern part of the outcrop. They had formed only a single layer and their bends are also very sharp and irregular. Although, if it there are many analogies to the schlieren (see Sect. 3.3) thatdoes not match their appearance perfectly.

A possible explanation for this issue could be the so-called magma blobs (Weinberg et al., 2001). These are ascending magmatic structures, which are of a slight

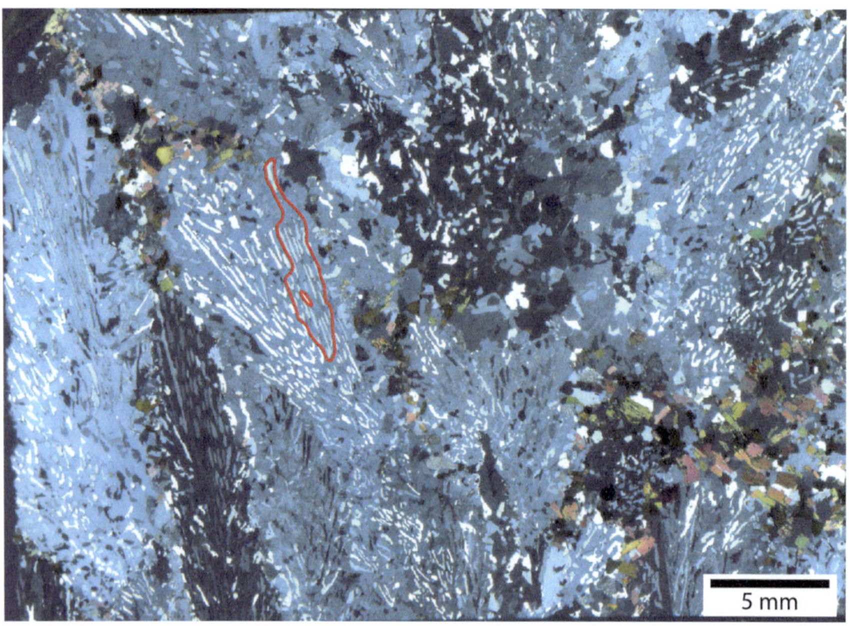

Fig. 70 Microphotography of a small part of feldspathic plume structure. The phase with vermicular shape are Qtz grains embedded in Kfs. The similar extinction of many (supposed) small Qtz grains indicates bigger Qtz crystals that branch out in different directions. The red line marks an example of Qtz with an absolutely identical extinction (Microphoto, N+)

different composition and/or fabric, than the surrounding granite. Commonly, they have a diameter up to several meters and they are bordered by irregular schlieren. The interior of such a magma blob is commonly rich in mafic enclaves, further schlieren and aggregates of Kfs-phenocrysts that are mainly arranged along the margins of the blob.

The fact that at Lavadores beach the area between the plume structures and the migmatite country rock contains accumulations of Kfs-phenocrysts, mafic enclaves and schlieren is a good evidence for such a magmatic blob case. An additional argument is the already mentioned similarity of the tonalitic layer and the mafic schlieren. According to this, the tonalitic layer and the plume structures are a transition zone between the magma blob and the typical Lavadores Granite (Fig. 71). However, compared to the magma blob described by Weinberg et al. (2001), the dimension of the proposed magma blob at the Lavadores Granite is quite wide and its delimitation by the schlieren and feldspathic plume structures is not completely continuous.

The open space necessary to the development of the feldspathic plume structure has a possible explanation given by Koopmann (2004). According to this author, quenched enclaves result in a reduction of their volumes. At this process "hollow

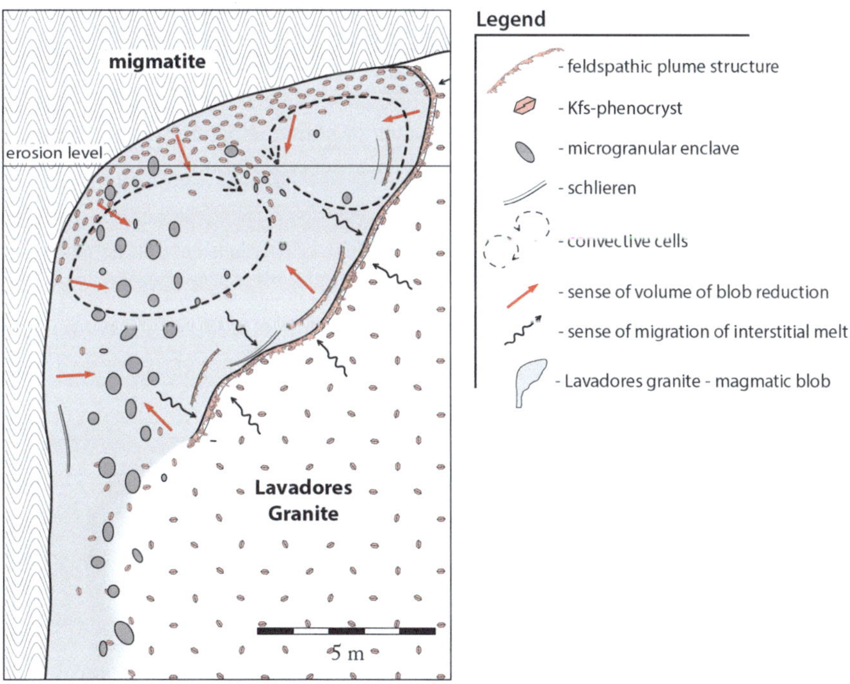

Fig. 71 Final state of the feldspathic plume structure (cooling phase and crystallization). The horizontal line shows the approximate erosion level on the actual outcrop (Modified after Lorenz, 2014)

space" is formed at their boundaries and due to a pressure gradient, melt can intrude into this void.

By projecting this process to the Lavadores magmatic blob it is assumed that after the injections stopped the superheated magma was quenched due to a high temperature gradient between the hot magmatic blob and the relatively cool Lavadores Granite and wall rock. Consequently, the volume of the blob had been reduced (red arrows in Fig. 71). The negative pressure gradient at the boundary zone led to a migration of an interstitial melt out of adjacent parts of the Lavadores Granite, and of residual K-feldspar rich fluids from blob magma, leading to a crystallization at this rheological boundary (sidled black arrows in Fig. 71).

The excessively supply of felsic melt could be a reason for the development of the plume structures, which consist completely of Kfs-phenocrysts and vermicular Qtz (Fig. 70). The plume structures seem to be crystallized freely inside the open tonalitic layer/schlieren and the model of "hollow space" at the transition zone would explain this free crystallization of the plume structures.

The blob melt migrated even behind the first schlieren layer and formed a second layer of plume structures but it is still linked to schlieren (Fig. 71). This link probably is explainable by the difference in rheological behavior of schlieren structures and the remaining phases. It might be more delicate to the decompression process and to the intake of melt. The parts of the outcrop where schlieren were possibly disturbed by the turbulent magma flow no plume structures had been recognized.

The nucleation of dendritic plume structure on magmatic blob wall indicates the colder wall of "hollow space" while the growth of plume structures in one direction is an indication for an intake of melt from the Lavadores Granite.

Acknowledgements The work of the author Jorge Pamplona is supported by national funding awarded by FCT—Foundation for Science and Technology, I.P., projects UIDB/04683/2020 and UIDP/04683/2020. Soumyajit Mukherjee (IIT Bombay) invited edited and reviewed this article. Mukherjee (2023) summarized this chapter.

References

Alburquerque, C. A. (1971). Petrochemistry of a series of granitic rocks from northern Portugal. *Geological Society of America Bullettin, 82*, 2783–2798.

Almeida, A., Leterrier, J., Noronha, F., & Bertrand, J. M. (1998). U-Pb zircon and monazite geochronology of the Hercynian two-mica granite composite pluton of Cabeceiras de Basto (Northern Portugal). *Comptes Rendul Acadamic Science, Paris, 326*, 779–785.

Almeida, A., Santos, J. F., & Noronha, F. (2014). Contribuição dos sistemas isotópicos Sm-Nd e Rb-Sr para o estudo petrogenético do maciço granítico peraluminoso de duas micas da cidade do Porto (NW Portugal). *Comunicações Geológicas, 101*(I), 27–30.

Altmeyer, T., Köpping, J., & Mengert, M. (2014). *Tectonic Evolution of the Praia dos Beijinhos (NW-Portugal). (Unpublished Bachelor Thesis)* (p. 94). University of Mainz.

Alves, C. M. (1966). Os encraves granulares do Granito de Lavadores (Vila Nova de Gaia). *Rev. Fac. Ciências Lisboa, 2ª série* (C)14, 51–60.

Ameglio, L., Vigneresse, J. L., & Bouchez, J. L. (1997). Granite pluton geometry and emplacement mode inferred from combined fabric and gravity data. In J. L. Bouchez, D. H. W. Hutton, & W. E. Stephens (Eds.), *Granite: From segregation of melt to emplacement fabrics* (pp. 200–214). Kluwer Academic Publishers.

Andrade, M. M., Borges, F. S., Marques, M. M., Noronha, F., & Pinto, M. S. (1983). *Contribuição para o conhecimento da faixa metamórfica da Foz do Douro (Nota prévia).* Sumários do I Congresso Nacional de Geologia.

Antunes, I. M., Neiva, A. M., Silva, M. M., & Corfu, F. (2009). The genesis of I- and S-type granitoid rocks of the Early Ordovician Oledo pluton, Central Iberian Zone (central Portugal). *Lithos, 111*, 168–185.

Arenas, R., Díez-Fernández, R., Rubio-Pascual, F. J., Sánchez-Martínez, S., Martín-Parra, L. M., Matas, J., Tánago, J. G., Jiménez-Díaz, A., Fuenlabrada, J. M., Andonaegui, P., & Garcia-Casco, A. (2016). The Galicia—Ossa-Morena Zone: Proposal for a new zone of the Iberian Massif. Variscan Implications. *Tectonophysics, 681*, 135–143.

Arzı, A. A. (1978). Critical phenomena in the rheology of partially melted rocks. *Tectonophysics, 44*, 173–84.

Ayrton, S. (1988). The zonation of granitic plutons: The "failed ring-dyke" hypothesis. *Schweizerische Mineralogische Und Petrographische Mitteilungen, 68*, 1–19.

Ballèvre, M., Martínez-Catalán, J. M., López-Carmona, A., Pitra, P., Abati, J., Díez-Fernández, R., Ducassou, C., Arenas, R., Bosse, V., Castiñeiras, P., Fernández-Suárez, J., Gómez-Barreiro, J., Paquette, J. L., Peucat, J. J., Poujol, M., Ruffet, G., & Sánchez-Martínez, S. (2014). Correlation of the nappe stack in the Ibero-Armorican arc across the Bay of Biscay: a joint French–Spanish project. In K. Schulmann, J. R. Martínez Catalán, J. M. Lardeaux, V. Janoušek, & G. Oggiano (Eds.), *The Variscan Orogeny: Extent, Timescale and the Formation of the European Crust* (vol. 405, pp. 77–113). Geological Society London, Special Publication.

Barbarin, B. (1999). A review of the relationships between granitoid types, their origins and their geodynamic environments. *Lithos, 46*(3), 605–626.

Barbarin, B. (1996) . Genesis of the two main types of peraluminous granitoids. *Geology, 24*, 295–298.

Barbey, P. (2009). Layering and schlieren in granitoids: A record of interactions between magma emplacement, crystallization and deformation in growing plutons. *Geologica Belgica, 12*, 109–133.

Bard, J. P., Capdevila, R., & Matte, P. (1971). Structure de la chaîne Hercynienne de la Meseta Ibérique: Comparaison avec les segments voisins. In *Symposium Histoire Structurale du Golfe de Gascogne* (vol. 22(4), pp. 1–68). Publications de l'Institute Français du Pétrole, Collection Colloque et Séminaire.

Barriére, M. (1981). On curved laminae, graded layers, convection currents and dynamic crystal sorting in the Ploumanac'h (Brittany) subalkaline granite. *Contributions to Mineralogy and Petrology, 77*, 214–224.

Bates, R. L., & Jackson, J. A. (1987). *Glossary of Geology* (p. 788). 3rd Ed. American Geological Institute, Alexandria, Virginia.

Bea, F., Montero, P., Talavera, C., & Zinger, T. (2006). A revised Ordovician age for the Miranda do Douro ortogneiss, Portugal. *Zircon U-Pb ion microprobe and LA-ICPMS dating.Geologica Acta, 4*, 395–401.

Berthé, D., Chokroune, P., & Gapais, D. (1979a). Orientations prèférentielles du quartz et orthogneissification progressive en régime cisaillant: l'example du cisaillement sud- armoricain. *Bulletin De Mineralogie, 102*, 265–272.

Berthé, D., Chokroune, P., & Jegouzo, P. (1979b). Orthogenesiss, mylonite and non-coaxial deformation of granites: The example of South Armoricain shear zone. *Journal of Structural Geology, 1*, 31–42.

Bose, N., Dutta, D., & Mukherjee, S. (2018). Role of grain-size in phyllonitisation: Insights from mineralogy, microstructures, strain analyses and numerical modeling. *Journal of Structural Geology, 112*, 39–52.

Bose, N., & Mukherjee, S. (2020). Estimation of deformation temperatures, flow stresses and strain rates from an intra-continental shear zone: The Main Boundary Thrust, NW Himalaya (Uttarakhand, India). *Marine and Petroleum Geology, 112*, 104094.

Burg, J. -P., Iglesias, M., Laurent, Ph., Matte, Ph., & Ribeiro, A. (1981). Variscan intracontinental deformation: the Coimbra-Cordoba shear zone (SW Iberian Peninsula). *Tectonophysics, 78*, 161–177.

Canilho, M. H. (1975). Contribuição para o conhecimento do granito de Lavadores. *Boletim Sociedade Geológica Portugal, 19*, 173–195.

Capdevila, R., Corretgé, G., & Floor, P. (1973). Les granitoides varisques de la Meseta Ibérique. *Bulletin de la Societé Géologique de France, 7*(153, 3/4), 209–228.

Carrington da Costa, J., & Teixeira, C. (1957). *Carta Geológica de Portugal na escala de 1/50000 e Notícia Explicativa da folha 9-C (Porto)* (p. 38). Serviços Geológicos de Portugal.

Carter, N. L., & Tsenn, M. C. (1987). Flow properties of continental lithosphere. *Tectono- Physics, 136*, 27–63.

Castro, A., De la Rosa, J. D., Fernández, C., & Moreno-Ventas, I. (1995). Unstable flow, magma mixing and magma-rock deformation in a deep-seated conduit: The Gil-Márquez Complex, southwest Spain. *Geologische Rundschau, 84*, 359–374.

Castro, A., De La Rosa, J., & Stephens, W. E. (1990). Magma mixing in the subvolcanic environment: petrology of the Gerena interaction zone near Seville, Spain. *Contributions to Mineralogy and Petrology, 105*, 9–26.

Chappel, B. W., & White, A. J. R. (1992). I- and S-type granites in the Lachlan Fold Belt. *Transactions of Royal Society of Edinburgh: Earth Sciences, 83*, 1–26.

Chaminé, H. I., Leterrier, J., Fonseca, P. E., Ribeiro, A., & Lemos de Sousa, M. J. (1998). Geocronologia U/Pb em zircões e monazites de rochas orto-derivadas do sector Espinho–Albergaria-a-Velha (Zona de Ossa Morena, NW de Portugal). In: Azerêdo, A. (coord.), Proceedings of V Congresso Nacional de Geologia (Vol. 84, issue (1), pp. B115–B118). Comun. Inst. Geol. Min., Lisboa.

Chaminé, H. I., Gama Pereira L. C., Fonseca P. E., Moço, L. P., Fernandes, J. P., Rocha, F. T., Flores, D., Pinto de Jesus, A., Gomes, C., Soares de Andrade, A. A., & Araújo, A. (2003). Tectonostratigraphy of middle and upper Palaeozoic black shales from the Porto–Tomar–Ferreira do Alentejo shear zone (W Portugal): new perspectives on the Iberian Massif. *Geobios, 36*, 649–663.

Clarke, D. B., Grujic, D., McCuish, K. L., Sykes, J. C. P., & Tweedale, F. M. (2013). Ring schlieren: Description and interpretation of field relations in the Halifax Pluton, South Mountain Batholith, Nova Scotia. *Journal of Structural Geology, 51*, 193–205.

Costa, M. M., Neiva, A. M., & Azevedo, M. R. (2006). Geoquímica dos granitóides variscos tardi-a pós-D3 da região de Aguiar da Beira: Os maciços de Pera Velha e Ferreira de Aves. *Res. VII Congresso Nac. Geologia, 1*, 61–64.

Dias, G., Leterrier, J., Mendes, A. C., Simões, P. P., & Bertrand, J. M. (1998). U-Pb zircon and monazite geochronology of post-collisional Hercynian granitoids from the Central Iberian Zone Northern Portugal. *Lithos, 45*, 349–369.

Dias, R., Ribeiro, A., Coke, C., Pereira, E., Rodrigues, J., Castro, P., Moreira, N., & Rebelo, J. (2013). Evolução estrutural dos sectores setentrionais do Autóctone da Zona Centro-Ibérica. In R. Dias, A. Araújo, P. Terrinha, & J. C. Kullberg (Eds.), *Geologia de Portugal* (vol I, pp. 73–148). Geologia Pré-Mesozóica de Portugal. Escolar Editora.

Dickson, F. W. (1995). Orthoclase porphyroblasts in gneisses and granites: Cordilleran Section. *Geological Society of America Abstracts with Programs, 27*(5).

Didier, J. (1973). *Granites and their enclaves* (p. 393). Elsevier.

Faure, M., Lardeaux, J.-M., & Ledru, P. (2009). A review of the pre-Permian geology of the Variscan French Massif Central. *Comptes Rendus Geoscience, 341*, 202–213.

Faure, M., Leloix, C., & Roig, J. Y. (1997). L'évolution polycyclique de la chaîne hercynienne. *Bulletin de la Société Géologique de France, 168*, 695–705.

Fernández-Catuxo, J. (1995). *Geología Granítica del Macizo del Confurco (Galicia, España). (Tese de Doutoramento)* (p. 298). Universidade de Oviedo.

Fernandez, A. N., & Gasquet, D. R. (1994). Relative rheological evolution of chemically contrasted coeval magmas: Example of the Tichka plutonic complex (Morocco). *Contributions to Mineralogy and Petrology, 116*, 316–326.

Fernandez, C., Castro, A., De la Rosa, J. D., & Moreno Ventas, I. (1997). Rheological aspects of magma transport inferred from rock structures. In J. L. Bouchez, D. H. Hutton, & W. E. Stephens (Eds.), *Granite: From segregation of melt to emplacement fabrics* (pp. 75–91). Kluwer Academic Plubishers.

Fernandez, F. J., Chaminé, H. I., Fonseca, P. E., Munhá, J. M., Ribeiro, A., Aller, J., Fuertes-Fuentes, M., & Borges, F. S. (2003). H-T fabrics in a garnet-bearing quartzite from Western Portugal: Geodynamic implications for the Iberian Variscan Belt. *Terra Nova, 15*, 96–103.

Ferreira, J., Martins, H. C. B., & Ribeiro, M. A. (2014). Geocronologia (U-Pb) e Geoquímica do granito do Pedregal. *Comunicações Geológicas Tomo, 101*(I), 89–92.

Ferreira, N., Iglésias, M., Noronha, F., Pereira, E., Ribeiro, A., & Ribeiro, M. L. (1987). Granitóides da Zona Centro-Ibérica e seu enquadramento geodinâmico. In F. Bea, A. Carnicero, J. C. Gonzalo, M. Lopes Plaza, & M. D. Rodriguez Alonso (Eds.), *Geologia de los Granitoides y Rocas Asociadas del Macizo Hesperico, Libro de Homenaje a L. C. Garcia de Figuerola* (pp. 37–52). Editorial Rueda.

Franke, W. (1989). Variscan plate tectonics in Central Europe—current ideas and open questions. *Tectonophysics, 169*, 221–228.

Franke, W. (2000). The mid-European segment of the Variscides: Tectonostratigraphic units, terrane boundaries and plate tectonic evolution. In W. Franke et al. (Eds.), *Orogenic processes: Quantification and modelling in the variscan belt. Geological Society of London* (vol. 179, pp. 35–61). Special Publication.

Glazner, A. F., & Johnson, B. R. (2013). Late crystallization of K-feldspar and the paradox of megacrystic granites. *Contributions to Mineralogy and Petrology, 166*, 777–799.

Gonçalves, A., Sant'Ovaia, H., & Noronha, F. (2019). Emplacement mechanism of Caria-Vila da Ponte Pluton (Northern Portugal): Building and internal magmatic record. *Journal of Structural Geology, 124*, 91–111.

Hatcher, R. D. (1989). Tectonic synthesis of the U.S. Appalachians. In R. D. Hatcher, W. A. Thomas, & G. W. Viele (Eds.), *The appalachian-ouachita orogen in the United States* (pp. 511–535). Geological Society of America, Boulder.

Hibbard, M. J. (1995). *Petrography to petrogenesis* (p. 587). Prentice Hall.

Higgins, M. D. (2006). *Quantitative textural measurements in igneous and metamorphic petrology* (p. 265). Cambridge University Press.

Hutton, D. H. (1988). Granite emplacement mechanisms and tectonic controls: inferences from deformation studies. *Transactions of the Royal Society of Edinburgh: Earth Sciences, 79*, 245–255.

Hutton, D., Dempster, T., Brown, P., & Becker, S. (1990). A new mechanism of granite emplacement: intrusion in active extensional shear zones. *Nature, 343*, 452–455.

Irvine, T. N. (1982). Terminology for layered intrusions. *Journal of Petrology, 23*, 127–162.

Kirkpatrick, R. J. (1975). Crystal growth from the melt: A review. *American Mineralogist, 60*, 798–814.

Klein, C., & Hurlbut, C. S., Jr. (1977). *Manual of Mineralogy* (after J.D. Dana). 20ª Ed. (p. 596). John Wiley & Sons.

Koopmann, A. (2004). *Magma mingling: Die hydrodynamische genese magmatischer dispersionen.* (Unpublished PhD thesis). Universität Würzburg.

Kruhl, J. H. (2003). Analysis of microfabrics in geomaterials. In J. H. Kruhl (Ed.), (Organizer), *Second European Workshop: Analysis of Microfabrics in geomaterials.* (not published).Tectonic and Materials Fabric Section, TUM, Munich.

Kurz, W., Fritz, H., Tenczer, V., & Unzog, W. (2002). Tectonometamorphic evolution of the Koralm Complex (Eastern Alps): constraints from microstructures and textures of the "Plattengneis" shear zone. *Journal of Structural Geology, 24*, 1957–1970.

Lacroix, A. (1903). *Notice sur les travaux scientifiques de M.A. Lacroix* (p. 126). Libraire Polytechnique Ch. Beranger, Paris.

Lancelot, J. R., Allegret, A., & Iglesias, M. (1985). Outline of the upper cambrian and the lower paleozoic evolution of the Iberian Peninsula according to U-Pb dating of zircons. *Earth and Planet Science Letters, 74*, 325–337.
Laporte, D. (1987). *Un exemple d'intrusion syntectonique: l'intrusion d'Ile-Rousse, Cor e du Nord-Ouest; étude pétrographique, minéralogique et géochimique; analyse structurale* (p. 422). Thèse nouveau régime, Univ. Saint-Étienne.
Lefort, J. P. (1981). Manaslu leucogranite: a collision signature of the Himalaya. A model for its genesis and emplacement. *Journal Geophysical Research, 86*, 10545–10568.
Leterrier, J., & Noronha, F. (1998). Evidências de um plutonismo calcoalcalino Cadomiano e de um magmatismo MORB no Complexo Metamórfico da Foz do Douro. *Com. Inst. Geol. Min., Lisboa, 84*, B146–B149.
Lexa, O. (2003). *PolyLX toolbox for MATLAB—reference manual*. Charles University, 2.0 edn.
Lisle, R. J. (1985). *Geological strain analysis: A manual for the Rf/ϕ Method* (p. 99). Pergamon Press.
Llana-Fúnez, S., & Marcos, A. (1998). Malpica-Lamego Deformation Zone: a major crustal-scale shear zone in the Iberian Variscan Belt (Galicia, N Portugal). In *Canadian tectonics group 18th annual meeting and geological association of Canada NUNA conference* in honour of P.F. Williams, Abstracts volume "Evolution of Structures in Deforming Rocks". Canmore, Canada.
Lorenz, A. (2014). *How does granite emplace in middle to upper crust—Structural study of the Lavadores granite, Portugal* (p. 102). (Unpublished Diplomat Thesis). Johannes Gutenberg-Universität.
Lotze, F. (1945). Zur gliederung der varisciden der Iberischen Meseta. *Geotekt Forsch, 4*(6), 78–92.
Marre, J. (1986). *The structural analysis of granitic rocks* (p. 123). North Oxford Academic Publishers Ltd.
Martins, H. C., Almeida, A., Noronha, F., & Leterrier, J. (2001). Novos dados geocronológicos de granitos da região do Porto: granito do Porto e granito de Lavadores. In T. Boski et al. (Eds.), *Actas do VI Congresso de Geoquímica dos Países de Língua Portuguesa e XII Semana de Geoquímica* (pp. 146-148). Universidade do Algarve, Faro.
Martínez-Catalán, J. R. (1990). A non-cylindrical model for the northwest Iberian allochthonous terranes and their equivalents in the Hercynian belt of Western Europe. *Tectonophysics, 179*, 253–272.
Martínez-Catalán, J. R. (2011). Are the oroclines of the Variscan belt related to late Variscan strike-slip tectonics? *Terra Nova, 23*, 241–247.
Martínez-Catalán, J. R., Arenas, R., Abati, J., Sánchez Martínez, S., Díaz García, F., Fernández Suárez, J., Cuadra, P. G., Castiñeiras, P., Barreiro, J. G., Díez Montes, A., Clavijo, E. G., Rubio Pascual, F. J., Andonaegui, P., Jeffries, T. E., Alcock, J. E., Díez Fernández, R., & López Carmona, A. (2009). A rootless suture and the loss of the roots of a mountain chain: The Variscan belt of NW Iberia. *Comptes Rendus Geoscience, 341*, 114–126.
Martínez-Catalán, J. R., Arenas, R., Díaz-García, F., & Abati, J. (1997). Variscan accretionary complex of NW Iberia: Terrane correlation and succession of tectonothermal events. *Geology, 25*, 1103–1106.
Martins, H. C., Sant'Ovaia, H., Abreu, J., Oliveira, M., & Noronha, F. (2011). Emplacement of the Lavadores granite (NW Portugal): U/Pb and AMS results. *C. R. Geoscience, 343*, 387–396.
Martins, H. C., Sant'Ovaia, H., & Noronha, F. (2009). Genesis and emplacement of felsic Variscan plutons within a deep crustal lineation, the Penacova-Régua-Verín fault: An integrated geophysics and geochemical study (NW Iberian Peninsula). *Lithos, 111*, 142–155.
Matte, P. (1986). Tectonics and plate tectonics model for de variscan belt of Europe. *Tectonophysics, 126*, 329–374.
Matte, P. (2000). The Variscan collage and orogeny (480–290 M.a.) and the definition of the Armorica microplate: tectonic approach. In *Variscan-Appalachian dynamics: the building of the Upper Paleozoic basement*. Basement Tectonics 15, A Coruña, Program and Abstracts, 12.
McBirney, A. R. (1984). *Igneous petrology* (p. 504). Freeman, Cooper & Company.

McCuish, K. L. (2001). *Schlieren in the South Mountain Batholith and Port Mouton Pluton, Meguma Zone, Nova Scotia.* (Unpublished Honours BSc. Thesis) (p. 101). Dalhousie University, Halifax, Nova Scotia.

Mehnert, K. R. (1968). *Migmatites and the origin of granitic rocks* (p. 393). Elsevier.

Mendes, F. (1968). Contribuition à l'étude géochronologique, par le méthode au strontium, des formations cristallines du Portugal. *Bol. Mus. Lab. Min. Geol., Universidade Lisboa, 11*(1), 159.

Mukherjee, S. (2013). Deformation microstructures in rocks. *Springer Geochemistry/Mineralogy* 1–111. ISBN 978-3-642-25608-0.

Mukherjee, S. (2014). Review of flanking structures in meso- and micro-scales. *Geological Magazine, 151*, 957–974.

Mukherjee, S. (2017). Review on symmetric structures in ductile shear zones. *International Journal of Earth Sciences, 106*, 1453–1468.

Mukherjee, S. (2023). Introduction to structural geology and tectonics field guidebook—Volume 2. In S. Mukherjee (Ed.), *Structural geology and tectonics field guidebook—Volume 2* (pp. xi–xiv). Springer Nature Switzerland AG. ISBN 978-3-031-19575-4.

Mukherjee, S., Bose, N., Ghosh, R., Dutta, D., Misra, A. A., Kumar, M., Dasgupta, S., Biswas, T., Joshi, A., & Limaye, M. (2020). *Structural geological atlas*. Springer. ISBN: 978-981-13-9825-4.

Mukherjee, S., Koyi, H. A. (2010). Higher Himalayan Shear Zone, Sutlej section: structural geology and extrusion mechanism by various combinations of simple shear, pure shear and channel flow in shifting modes. *International Journal Earth Science, 99*, 1267–1303.

Mukherjee, S., & Mulchrone, K. (2012). Estimating the viscosity and Prandtl number of the Tso Morari Gneiss Dome, western Indian Himalaya. *International Journal of Earth Sciences, 101*, 1929–1947.

Mukherjee, S., & Mulchrone, K. F. (Eds.). (2015). *Ductile shear zones: From micro- to macro-scales* (1st edn, p. 306). John Wiley & Sons.

Nassau, K. (1978). The origins of color in minerals. *American Mineralogist, 63*, 219–229.

Nédélec, A., & Bouchez, J.-L. (2015). *Granites petrology, structure, geological setting, and metallogeny* (p. 385). Oxford University Press.

Neiva, A. M., Williams, I. S., Ramos, J. M., Gomes, M. E., Silva, M. V., & Antunes, I. M. (2009). Geochemical and isotopic constraints on the petrogenesis of Early Ordovician granodiorite and Variscan two-mica granites from the Gouveia area, central Portugal. *Lithos, 111*, 186–202.

Nickel, E., Kock, H., & Nungässer, W. (1967). Modellversuche zur Fliessregelung in Graniten. *Schweizerische Mineralogische Und Petrographische Mitteilungen, 47*, 399–498.

Noronha, F., & Leterrier, J. (1995). Complexo metamórfico da Foz do Douro. Geoquímica e geocronologia. Resultados preliminares. *Mem. Mus. Lab. Miner. Geol. Fac. Ciênc. Porto, 4*, 769–774.

Noronha, F. P., Farinha Ramos, J. M., Rebelo, J. A., Ribeiro, A., & Ribeiro, M. L. (1979). Essai de corrélation des phases de déformation hercynienne dans le Nord-Ouest Péninsulaire. *Bol. Soc. Geol. Portug, 21*, 227–237.

Okudaira, T., Takeshita, T., Hara, I., & Ando, J. (1995). A new estimate of the conditions for transition from basal <a>to prism [c] slip in naturally deformed quartz. *Tectonophysics, 250*, 31–46.

Oliveira, J. T., Oliveira, V., & Piçarra, J. M. (1991). Traços gerais da evolução tectono-estratigráfica da Zona de Ossa Morena, em Portugal: Síntese crítica do estado actual dos conhecimentos. *Comunicações Dos Serviços Geológicos De Portugal, 77*, 3–26.

Pamplona, J., Gutiérrez-Alonso, G., & Ribeiro, A. (2006). Superposition of shear zones during orogenic development: An example from the NW Variscan belt (Viana do Castelo, NW Portugal). *Journal of Structural Geology, 28*, 1327–1337.

Pamplona, J., Rodrigues, B. C., Llana-Fúnez, S., Simões, P., Ferreira, N., Coke, C., Pereira, P., Castro, P., & Rodrigues, J. (2016). Structure and variscan evolution of malpica-lamego ductile shear zone (NW of Iberian Peninsula). In S. Mukherjee, & K. F. Mulchrone (Eds.), *Ductile shear zones: From micro- to macro-scales* (1st Edn, pp. 206–223). Wiley-Blackwell.

Passchier, C. W., & Trouw, R. A. J. (2005). *Microtectonics* (2nd ed., p. 366). Springer-Verlag.

Paterson, S. R., & Tobisch, O. T. (1992). Rates of processes in magmatic arcs: implications for the timing and nature of pluton emplacement and wall rock deformation. *Journal of Structural Geology, 14*(3), 291–300.

Paterson, S. R. (2002). Igneous origin of K-feldspar megacrysts in deformed granite of the Papoose Flat pluton. *Electronic Geosciences Electronic Geosciences*. https://doi.org/10.1007/s10069-002-005-3

Paterson, S. R. (2009). Magmatic tubes, pipes, troughs, diapirs, and plumes: Late-stage convective instabilities resulting in compositional diversity and permeable networks in crystal-rich magmas of the Tuolumne batholith, Sierra Nevada, California. *Geosphere, 5*, 496–527.

Pereira, E. (coord.). (1989). *Carta Geológica de Portugal, Folha 1, escala 1:200.000*. Serviços Geológicos Portugal.

Pereira, E. (1988). Soco Hercínico da Zona Centro-Ibérica—Evolução Geodinâmica. *Geonovas, 10*, 10–35.

Pereira, L. C., & Macedo, C. A. (1983). Sobre a idade dos granitos de Figueiró dos Vinhos, Pedrogrão Grande e dum pegmatito do Casal do Zote (Dornes) no sector da sutura ZOM—ZCI, a N de Tomar (Portugal Central); implicações geotectónicas. *Comum. Serv. Geol. Portugal, 69*(2), 265–266.

Pereira, M. F., Díez Fernández, R., Gama, C., Hofmann, M., Gärtner, A., & Linnemann, U. (2017). S-type granite generation and emplacement during a regional switch from extensional to contractional deformation (Central Iberian Zone, Iberian autochthonous domain, Variscan Orogeny). *International Journal of Earth Sciences, 107*(1), 251–267.

Pfiffner, O. A., & Ramsay, J. G. (1982). Constraints on geological strain rates - arguments from finite strain states of naturally deformed rocks. *Journal of Geophysical Research, 87*(B1), 311–321.

Petford, N. (2003). Rheology of granitic magmas during ascent and emplacement. *Annual Review of Earth Planetary Sciences, 31*, 399–427.

Petford, N., Cruden, A. R., McCaffrey, K. J. W., & Vigneresse, J. L. (2000). Granite magma formation, transport and emplacement in the Earth's crust. *Nature, 408*, 669–673.

Pinto, M. S. (1984). O granito gnáissico de Fânzeres (Porto, Portugal) – Idade e caracterização geoquímica geral. *Memórias e Notícias Coimbra, 98*, 231–242.

Pinto, M. S., Casquet, C., Ibarrola, E., Corretgé, L. G., & Ferreira, M. P. (1987). Síntese Geocronológica dos Granitóides do Maciço Hespérico. In F. Bea, A. Carnicero, J.C. Gonzalo, M. Lopes Plaza, M.D. Rodriguez Alonso (Eds.), *Geologia de los Granitoides y Rocas Asociadas del Macizo Hesperico, Libro de Homenaje a L. C. Garcia de Figuerola* (pp. 69–86). Editorial Rueda, Madrid.

Pitcher, W. S. (1979). Comments on the geological environments of granites. In M. P. Atherton, & J. Tarney (Eds.), *Origin of granites batholites, geochimical evidence* (pp. 1–8).

Pitcher, W. S. (1993). *The nature and origin of granite* (p. 387). Springer.

Ramberg, H. (1981). *Gravity deformation and the Earth's Crust: Experiments and geological applications* (p. 452). Academic Press.

Ramsay, J., & Huber, M. (1983). *Techniques of modern structural geology* (Vol. 1, p. 307). Academic Press.

Ribeiro, A., & Pereira, E. (1986). Flake tectonics in the NW Iberia Variscides. *Maleo, 2*(13), 38.

Ribeiro, A. (1974). *Contribuition à l'étude Tectonique de Trás-os-Montes Oriental. Memórias 24 (Nova Série)* (p. 168). Serviços Geológicos Portugal.

Ribeiro, A. (1984). Evolução Geodinâmica Da Zona Centro-Ibérica. *Geonovas, 1*(7), 145–146.

Ribeiro, A. (2013). II. Evolução geodinâmica de Portugal; os ciclos ante-mesozóicos. In R. Dias, A. A. Araújo, P. Terrinha, J. C. Kullberg. (Eds.), *Geologia de Portugal. Vol. I: Geologia Pré-Mesozóica de Portugal* (pp. 15–58). Escolar Editora.

Ribeiro, A., Munhá, J., Dias, R., Mateus, A., Pereira, E., Ribeiro, L., Fonseca, P., Araújo, A., Oliveira, T., Romão, J., Chaminé, H., Coke, C., & Pedro, J. (2007). Geodynamic evolution of the SW Europe Variscides. *Tectonics, TC6009*, 24.

Ribeiro, A., Pereira, E., Chaminé, H. I., & Rodrigues, J. (1995). Tectónica do megadomínio de cisalhamento entre a Zona de Ossa-Morena e a Zona Centro-Ibérica na região de Porto-Lousã. *Mem. Mus Lab. Min. Geol. Fac. Ciências Univ. Porto, 4*, 299–303.

Ribeiro, A., Pereira, E., & Dias, R. (1990). Part IV Central-Iberian Zone, allochthonous sequences, structure in the nortwest of the Iberian Peninsula. In R. D. Dallmeyer, & E. Martinez-Garcia (Eds.), *Pre-Mesozoic geology of Iberia* (pp. 220–236). Springer-Verlag.

Ribeiro, A., Pereira, E., & Severo, L. (1980). Análise da deformação da zona de cisalhamento Porto-Tomar na transversal de Oliveira de Azeméis. *Comunicações Dos Serviços Geológicos De Portugal, 66*, 3–9.

Ribeiro, M. L., Castro, A., Almeida, A., González-Menéndez, L., Jesus, A., Lains, J. A., Lopes, J. C., Martins, H. C., Mata, J., Mateus, A., Moita, P., Neiva, A. M, Ribeiro, M. A., Santos, J. F., & Solá, A. R. (2019). Variscan magmatism. In C. Quesada, J. Oliveira (Eds.), *The geology of Iberia: A geodynamic approach. Regional geology reviews* (pp. 497–526). Springer.

Rodrigues, B. C. (1989). *Os Encraves das Rochas Graníticas* (p. 137). (Unpublished MSc Thesis). Departamento de Geologia, Faculdade de Ciências da Universidade do Porto.

Rong, J., & Wang, F. (2016). *Metassomatic textures in granite: Evidence from petrographic observation* (p. 144). Science Press and Springer Science.

Sant'Ovaia, H., Ribeiro, M. A., Martins, H. C., Ferrão, F., Gomes, C., & Noronha, F. (2014). Estruturas e fabric magnético no maciço granítico de Lavadores-Madalena. *Comunicações Geológicas, 101*(Especial I), 313–317.

Schermerhorn, L. G. (1987). Granite fractionation by convective cumulation. *Revista Brasileira Geociências, 17*, 617–618.

Schmidt, A. (2014). *Development of feldspar classification from the Lavadores granite in Porto, Portugal* (p. 52). (Bachelor Thesis), University of Mainz.

Scholz, C. H. (1988). The brittle-plastic transition and the depth of seismic faulting. *Geologische Rundschau, 77*, 319–328.

Schulmann, K., Konopásek, J., Janousěk, V., Lexa, O., Lardeaux, J.-M., Edel, J.-B., Štípská, P., & Ulrich, S. (2009). An Andean type Palaeozoic convergence in the Bohemian Massif. *Comptes Rendus Geoscience, 341*, 266–286.

Schulmann, K., Martínez-Catalán, J. R., Lardeaux, J. M., Janoušek, V., & Oggiano, G. (2014). The Variscan orogeny: Extent, timescale and the formationof the European crust. In K. Schulmann, J. R. Martínez Catalán, J. M. Lardeaux, V. Janoušek, & G. Oggiano (Eds.), *The variscan orogeny: Extent, timescale and the formation of the European crust* (vol. 405, pp. 1–6). Geological Society of London, Special Publication.

Schulz, G. (1858). *Descripción geológica de la província de Oviedo.* Map to scale 1:400 000 (1857). Madrid.

Shaw, H. R. (1980). *The fracture mechanisms of magma transport from the mantle to the surface. Physics of magmatic processes* (pp. 201–264). Princeton University Press.

Silva, M. M. (1995). *Mineralogia, Petrologia e Geoquímica de Encraves de Rochas Graníticas de algumas Regiões Portuguesas* (p. 288). (PhD Thesis) Faculdade de Ciências e Tecnologia, Universidade de Coimbra.

Silva, M. M. (2010). O Granito de Lavadores e seus Encraves. In J. M. Cotelo Neiva, A. Ribeiro, M. Victor, F. Noronha, & M. Ramalho (Eds.), *Ciências Geológicas—Ensino e Investigação e sua História (I)—Geologia Clássica* (pp. 269–279). Associação Portuguesa de Geólogos e Sociedade Geológica de Portugal.

Silva, M. M., & Neiva, A. M. (1998). Geoquímica de encraves microgranulares e granitos hospedeiros da região de Vila Nova de Gaia, norte de Portugal. In *Actas do V Congresso Nacional de Geologia (Resumos Alargados), Lisboa.* Com. Inst. Geol. Min., Lisboa, 84, B35–B38.

Silva, M. M., & Neiva, A. M. (2001). Condições termobarométricas de cristalização de encraves de afinidade shoshonítica e granito hospedeiro de Lavadores. In T. Boski et al. (Eds.), *Actas do VI Congresso de Geoquímica dos Países de Língua Portuguesa e XII Semana de Geoquímica* (pp. 215–218). Universidade do Algarve, Faro.

Soo, S. L. (1967). *Fluid dynamics of multiphase systems. A blaisdell book in the pure and applied sciences* (p. 524). Blaisdell Publishing Company, Waltham, Mass.

Sousa, M., Sant'Ovaia, H., Tassinari, C., & Noronha, F. (2014). Geocronologia U-Pb (SHRIMP) e Sm-Nd do ortognaisse biotítico do Complexo Metamórfico da Foz do Douro (NWde Portugal). *Comunicações Geológicas, 101*(Especial I), 225–228.

Speer, J. A., McSween, Jr., H.Y., & Gates, A. L. (1994). Generation, segregation, ascent and emplacement of Alleghanian granitoid plutons in the southern Appalachians. *The Journal of Geology, 102*, 249–267.

Stipp, M., Stünitz, H., Heilbronner, R., & Schmid, S. M. (2002). The eastern tonale fault zone: A "natural laboratory" for crystal plastic deformation of quartz over a temperature range from 250 to 700 °C. *Journal of Structural Geology, 24*, 1861–1884.

Teixeira, C., Perdigão, J., & Assunção, C. T. (1962). *Carta Geológica de Portugal na escala de 1/50000 e Notícia Explicativa da folha 13-A (Espinho)* (p. 35). Serviços Geológicos de Portugal.

Tobisch, O. T., McNulty, B. A., & Vernon, R. H. (1997). Microgranitoid enclave swarms in granitic plutons, Central Sierra Nevada, California. *Lithos, 40*, 321–339.

Tweedale, F. (2012). *Occurrence and origin of ring schlieren in the South Mountain Batholith, Meguma Zone, Nova Scotia (Unpublished Honours BSc* (p. 82). Dalhousie University, Halifax, Nova Scotia.

Twiss, R. J. (1977). Theory and applicability of a recrystallized grain-size paleopiezometer. *Pure and Applied Geophysics, 115*, 227–244.

Valle Aguado, B., Azevedo, M. R., Schalteggerb, U., Martínez Catalán, J. R., & Noland, J. (2005). U-Pb zircon and monazite geochronology of Variscan magmatism related to syn-convergence extension in Central Northern Portugal. *Lithos, 82*, 169–184.

Van der Molen, I., & Paterson, S. (1979). Experimental deformation of partially-melted granite. *Contributions to Mineralogy and Petrology, 10*, 299–318.

Vernon, R. H. (1986). K-feldspar megacrysts in granites—phenocrysts, not porphyroblasts. *Earth Science Reviews, 23*, 1–63.

Vernon, R. H. (2004). *A pratical guide to rock microstructure* (p. 594). Cambridge University Press.

Vernon, R. H., & Collins, W. J. (2011). Structural criteria for identifying granitic cumulates. *Journal of Geology, 119*, 127–142.

Vigneresse, J. L. (1995). Control of granite emplacement by regional deformation. *Tectonophysics 249*, 173–186.

Vigneresse, J. L., Cuney, M., & Barbey, P. (1991). Deformation assisted crustal melt segregation and transfer. *Geological Association of Canada—Mineralogical Association of Canada Abstract, 16*, A128.

Voll, G. (1976). Recrystallization of quartz, biotite and feldspars from Erstfeld to the Leventina Nappe, Swiss Alps, and its geological significance. *Schweizerische mineralogische und petrographische Mitteilungen, 56*(3), 641–647.

Voll, G. (1980). Ein Querprofil durch die Schweizer Alpen vom Vierwaldst¨atter See zur Wurzelzone—Strukturen und ihre Entwicklung durch Deformationsmechanismen wichtiger Minerale. Neues Jahrbuch Geologie und Pal¨aontologie. *Abhandlungen, 160*, 321–335.

Wager, L. R., Brown, G. M., & Wadsworth, W. J. (1960). Types of igneous cumulates. *Journal of Petrology, 1*, 73–85.

Weinberg, R. (2003). Granite transport and emplacement: A review. In *Magmas to Mineralisation: The Ishihara Symposium* (pp. 125–127). GEMOC, Macquarie University, Geoscience Australia Record 2003/14.

Weinberg, R. F., Sial, A. N., & Pessoa, R. R. (2001). Magma flow within the Tavares pluton, northeastern Brazil: Compositional and thermal convection. *Geological Society of America Bulletin, 113*, 508–520.

Wickham, S. M. (1987). The segregation and emplacement of granitic magma. *Journal of the Geological Society of London, 144*, 281–297.

Wiebe, R. A., Jellinek, M., Markley, M. J., Hawkins, D. P., & Snyder, D. (2007). Steep schlieren and associated enclaves in the Vinalhaven granite, Maine: Possible indicators for granite rheology. *Contributions to Mineralogy and Petrology, 153*, 121–138.

Zeck, H. P., Whitehouse, M. J., & Ugidos, J. M. (2007). 496 ± 3 Ma zircon ion microprobe age for pre-Hercynian granite, Central Iberian Zone, NE Portugal (earlier claimed 618 ± 9 Ma). *Geological Magazine, 144*(1), 21–31.

Tectonically Significant Fault Zones in Central Europe (Germany, Czech Republic and Poland) and Their Surface and Subsurface Outcrops: Franconian Line, Hronov-Porici Fault, Sudetic Marginal Fault and Lusatian Fault

Lucie Novakova

Abstract This chapter describes and documents four main fault zones in central Europe and their outcrops in (sub)surface as a geological field guide. These up to 1500 km long often morphologically distinctive fault zones, including The Sudetic Marginal Fault, the Lusatian Fault, the Hronov-Porici Fault and the Franconian Line, are mainly oriented in the NW–SE direction. These fault zones played an essential role in controlling tectonics in Central Europe. They were reactivated during geological history, and some of them are also considered active. The fault zones cross three states, namely Germany, Czech Republic and Poland.

Keywords Fault zone · Central Europe · Bohemian massif · Slickensides · Tectonics · Reactivation

1 Introduction

The tectonic and geological history of Central Europe developed and influenced by Variscan and Alpine orogenesis. The key process controlling how Alpine tectonics became convolved with the older Variscan framework of Europe was the closure of the Palaeotethys and opening of Neotethys ocean systems and the development of an array of south Eurasian back-arc basins, followed or accompanied by the break-up of Pangaea and the early development of the Central Atlantic (Gee & Stephenson, 2006). Variscan Central Europe belongs to the most stable parts of the continent where the horizontal movements generally do not exceed 1 mmy^{-1} measured over

L. Novakova (✉)
Institute of Rock Structure and Mechanics, Czech Academy of Sciences, V Holesovickach 41, 182 09 Prague, Czech Republic
e-mail: lucie.novakova@irsm.cas.cz

Institute of Geology, TU Bergakademie Freiberg, Bernhard-Von-Cotta-Straße 2, 09599 Freiberg, Germany

S. Mukherjee (ed.), *Structural Geology and Tectonics Field Guidebook–Volume 2*, Springer Geology, https://doi.org/10.1007/978-3-031-19576-1_3

long distances (Nocquet et al., 2001). However, the seismicity of the area has been recorded nowadays, and the calculated magnitude is Mw = 6.5 (Štěpančíková et al., 2019).

The Bohemian Massif (BM) represents the largest exposure of Variscan basement in Central Europe. The BM is made up of several significant units: the Saxothuringicum, Moldanubicum, Tepla-Barandian Unit and the Lugian Unit (Fig. 1) and the nappe complexes on the west the Zone Erbendorf-Vohenstrauss (ZEV) and Münchberger Massif (MM). The most prominent fault zone is Elbe Fault System (EFS) is a WNW-striking zone extending from the southeastern North Sea to southwestern Poland along the present southern margin of the North German Basin and the northern margin the Sudetes Mountains (Scheck et al., 2002). This Fault Zone includes Sudetic Marginal Fault, Hronov-Porici Fault and Lusatian Fault described in this chapter. The zone is defined by interconnected faults associated with historical earthquakes (Špaček et al., 2006). The seismoactive zone on the NE margin of the Bohemian Massif is ~40–60 km wide and 150 km long. It comprises several NW–SE and NNW–SSE striking faults (Valenta et al., 2008), including Sudetic Marginal Fault and Hronov-Porici Fault.

Due to Quaternary cover in the Bohemian Massif, outcrops containing the direct record of the fault zone or the contact of rocks are quite limited. In this chapter, the most publicly accessible of them are documented and described. Geological significance, clarity and the access were the main criteria for the locality selection. Some of the localities provide regionally or even European important information.

2 Franconian Fault Zone

The Franconian Fault Zone (FFZ) (also called line or lineament) is one of the most significant fault lines in central Europe. The Franconian Line is located on the western margin of the Bohemian Massif near the German–Czech border. Franconian line is about 100 km long zone separating Cretaceous sediments on the west from the metamorphtites (amphibolites and paragneisses) in the east (Fig. 2). Coarse alluvial fan deposits in front of the FFZ attest to exhumation of the BM in the Early Triassic and again in the Late Cretaceous to Palaeocene (Zulauf & Duyster, 1997). The Triassic exhumation is dated by medium-temperature thermo-chronometers (zircon fission track: Jacobs et al., 1993; titanite fission track: Coyle & Wagner, 1998; sericite K–Ar: Wemmer & Ahrendt, 1997). The Cretaceous to Palaeocene exhumation is dated by low-temperature thermo-chronometers, e.g. apatite fission track by Wauschkuhn et al. (2015). The compressional intra-plate stresses driving the exhumation of the BM relaxed at the end of the Palaeocene owing to an uncoupling of the Central European foreland from the Alpine-Carpathian orogenic front (Ziegler & Dèzes, 2007), following which the BM was exposed to deep weathering and erosion, creating a regional peneplain. Morphotectonic studies of the KTB area show that the Palaeogene was a period of tectonic rest (Peterek & Schröder, 2010).

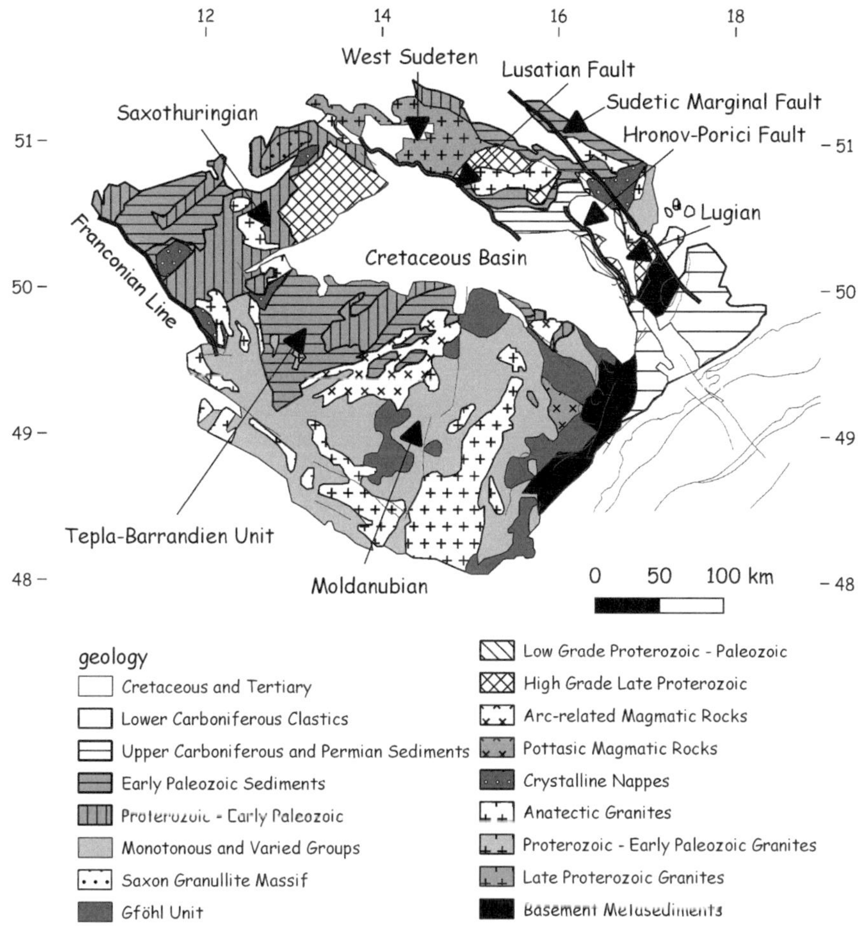

Fig. 1 Geological and tectonic situation of the Bohemian Massif (modified after Schulmann et al., 2009)

2.1 Locality KTB (Germany) 49°48′55″N; 12°07′14″E

The Kontinentale Tief Bohrung (KTB) (Fig. 3) was a continental deep drilling project in Bavaria from 1987 to 1995. The boreholes are located on the western margin of the Bohemian Massif, three km NW of town Windischeschenbach (Germany). The pilot hole is 4000 m deep, out of which 3276 m were cored. The main hole reaches 9101 m—the deepest straight borehole in the world. Amphibolites dominate the depth interval from 3300 to 7300 m. Felsic and mafic gneisses alternate in the section from 7300 to 9101 m (Hirschmann et al., 1997). The rocks are crosscut by cm to m-thick mafic to felsic late-Variscan dikes (Siebel et al., 1995). Its location near the crustal-scale fault system (Franconian Line) at the margin of the BM explains the strong

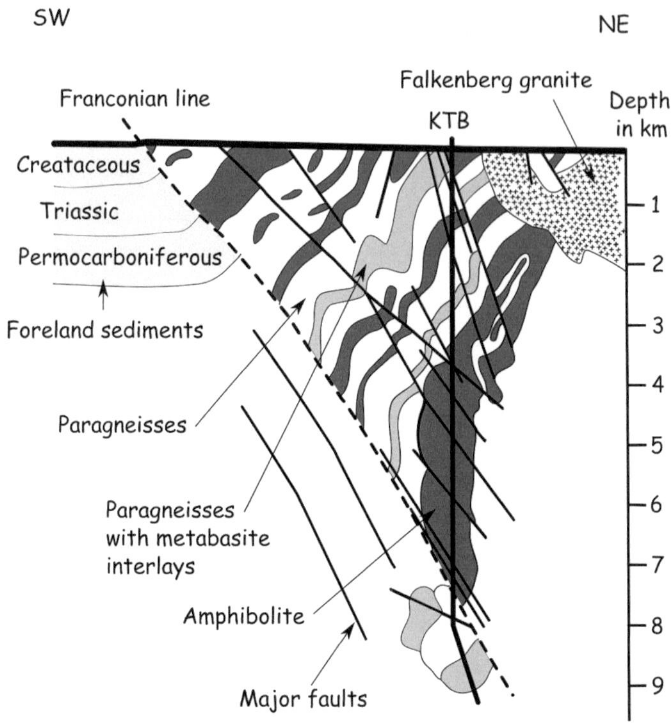

Fig. 2 Tectonic and geological situation of the Franconian Line, showing the intersection of KTB, the deepest fully cored borehole in the world (modified after Emmermann & Lauterjung, 1997)

brittle deformation observed in the KTB cores (Hirschmann et al., 1997; Zulauf & Duyster, 1997). The FL reverse fault zone intersects the KTB at 6850–7300 m depth below the surface; the faults dip 50–75° NE and accumulated ca. 6000 m of throw (Wauschkuhn et al., 2015).

2.2 *Locality Waldeck (Germany) 49°52′8.982″N; 11°57′9.803″E*

Ten kilometres NW from KTB, the Franconian Fault rises to the surface near the town, Waldeck. The outcrop is one of few places in which is the fault exposed. Tilted red and green sandy clay layers and fined grained sand and sandstones are exposed in the right part of the outcrop. The sediments are bordered by unweathered but very heavily broken metamorphic rocks of the basement (epigneisses in the left) (Fig. 4). The leap in time amounts to more than 200 million years. The epigneisses were thrust several kilometres onto the Keuper rocks. The thrusting occurred in a period between 60 and 100 million years ago.

Fig. 3 Drilling tower of the main borehole

3 Hronov-Porici Fault Zone

The Hronov-Porici Fault Zone (HPFZ) is located in the easternmost part of the Czech-Polish border. This zone lies in both countries—the Czech Republic and Poland. It is bounded on the north by the Vrchlabi line and on the south by the NovaPaka line (Stejskal et al., 2007). It is ~100 km long and up to 500 m wide zone of parallel fractures, dividing two important structural units—the Intra Sudeten Basin and the Gian Mts. foothills (Stejskal et al., 2009). The striking structure of the NW–SE has been created due to the post-Cretaceous bending composition and is filled with Upper

Fig. 4 Outcrop of the Franconian Fault

Cretaceous sediments (Valenta et al., 2008) and the Permo-Carboniferous volcano-sedimentary complex (Skácelová et al., 2009). HPFZ had a complicated tectonic development that began in the Late Paleozoic. Since then, several tectonic phases have occurred (Novakova, 2014; Valenta et al., 2008). The fault zone was developed from an asymmetric anticline due to regional reverse perturbation compression (Fig. 5) (Tásler et al., 1979). The major reversal disorder (traction) is accompanied by parallel or oblique normal or reversible disorders (Valenta et al., 2011). The NE block was relatively elevated (Kolínský et al., 2012). HPFZ reactivation is noted after Upper Cretaceous sedimentation during late Saxon tectogenesis. Previous phases were reactivated and local short orthogonal normal errors (Stejskal et al., 2006). According to Tásler et al. (1979), normal disorders in the HPFZ region represent younger tectonic elements, i.e. younger standard errors are morphologically more pronounced than older reversible disorders. Major dislocations in the Sudetenland such as SMF and HPFZ either follow a dominant shear direction or are oblique, which explains the different kinematics demonstrated by different segments of these disorders, from dip-slip to strike-slip transpressional faults (Wojewoda, 2007). Recent tectonic sinistral (oblique or horizontal) movements along steep faults of the NNE–SW are highly probable (Havíř & Špaček, 2004). The strongest historical earthquake was recorded on January 10, 1901, and reached M = 4.6 (Woldřich, 1901). Within 1705 to 2005, 30 earthquakes with macroseismic effects have been recorded within the HPFZ (Stejskal et al., 2006). A possible explanation for the current mobility of HPFZ was provided by Schenk et al. (1989). According to their geodynamic model, HPFZ, as a reverse

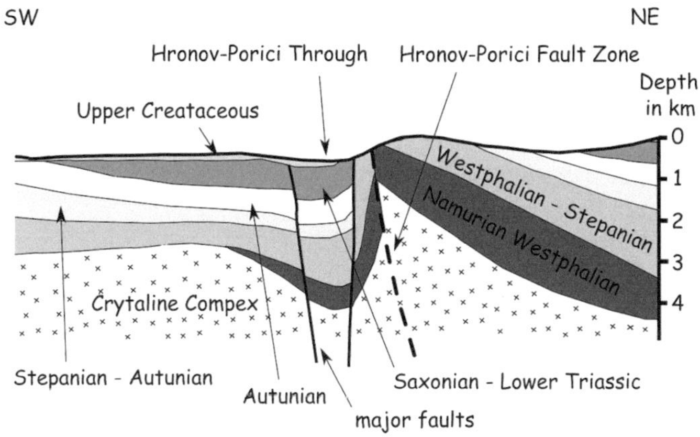

Fig. 5 Thrusting of the Hronov-Porici Fault Zone (simplified from Tasler et al., 1979)

fault, balances the compression caused by movements along the bounded faults and delimits HPFZ in the north and south.

3.1 Locality Male Svatonovice (Czech Republic) 50°32′14.809″N; 16°2′31.465″E

The outcrop exposes the main fault of the HPFZ. The rock wall is ~40 m long and 17 m high. Red-brown conglomerates and breccias of the Trutnov Formation are on the right-hand side of the wall. These are characterised by a variable percentage of basalt andesite clasts. Conglomerates are unconformably overlain by Cenomanian sandstones and siltstones of the Peruc–Korycany Formation, cropping out on the left-hand side of the wall (Fig. 6). Cretaceous sediments contain a mixture of glauconite and fine organic matter and a carbonised plant system at the base (dark grey colour). The dip of both Permian and Cretaceous sediments is steep (up to 89° SW) due to the close vicinity of the main fault line of the HPFZ. The dip of the marlstones is lower, grading to subhorizontal bedding at the axis of the Hronov-Porici Trough. Figure 7 presents a simplified sketch of the locality. The stereograms with the great circles are displayed in the equal area, the lower hemisphere. Arrows on the great circles determine the senses of movements on the fault planes.

We assume that the movements along the main fault have abandoned the lineations on the fault planes. The main fault is visible between two different geological formations. Units 1–3 belong to the Cretaceous sediments in the left-hand part of the wall, while units 4–6 belong to the Permian sediments on the right-hand side of the exposed outcrop. In Unit 4, there are mainly reverse faults with sub-vertical to almost vertical dipping.

Fig. 6 Outcrop of the Hronov-Porici Fault Zone

3.2 Locality Big Szczeliniec, National Park of the Table Mts. (Poland) 50°29′02″N; 16°20′38″E

The Table Mountains are the only example in Poland of mountains built of horizontally laid rocky layers. These rocks, called common sandstones, interspersed with marl, were formed in marine sediments that entered the area in the Late Cretaceous. It lies on even older ones, dating from the Permian, sandstones and conglomerates. The development of sculpture in the Table Mountains began after the Cretaceous

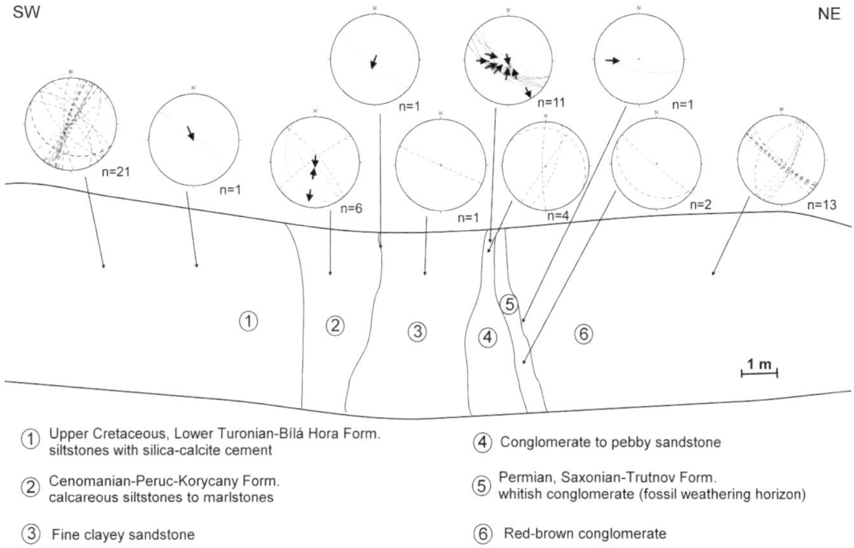

Fig. 7 Simplified sketch of the tectonic situation at the locality of Male Svatonovice. Stereonets (Schmidt projection, lower hemisphere) display fractures (dotted lines) and faults (solid lines) as great circles. The main fault intersects the Cretaceous sediments on the left (1–3) and the Permian sediments on the right (4–6). The arrows correspond to the slip direction and indicate the sense of the movement of the hanging-wall block. Reproduced from Fig. 6 of Novakova (2014)

Sea regressed ~70 million years ago. The period of tectonic calm, separated by three phases of tectonic movements, was formed by three steps with a flat top, rising to a height: 850–919 m—bastions with a flat top built of common upper sandstones—Big Szczeliniec (Figs. 8 and 9), Small Szczeliniec. 500–800 m—plateau Karlow and Lezyce, covered with marl. 400–500 m—the lowest south-eastern part. Water played an important role in shaping the Table Mountains, not only by surface erosion but also by its influence underground in the good zone. In the first case, chemical and mechanical erosion created fantastic rocky forms in exposed parts of common sandstones at the top of Big and Small Szczeliniec. In the second, removing eroded rock by groundwater at the contact of permeable sandstones with impermeable marls will create empty spaces inside the mountain. Then the sandstone shattered and sank. Changing climatic conditions over 7 million years had an impact. Rock cracks shaped geometries resembling buildings, mushrooms, animals and people. Table Mountain consists of two peaks: Big Szczeliniec (919 m asl) and a shallow saddle separated by Small Szczeliniec (870 m asl). Big Szczeliniec is located at <2 km from the Czech-Polish border. It is a massive plateau with a beautiful rock town, which rises above the surrounding terrain and is bordered on almost all sides by high sandstone walls. Big Szczeliniec is the highest peak of the Table Mountain National Park (Góry Stołowe). However, it is closed to tourists due to the numerous nesting sites of many protected bird species. Six hundred sixty-five steps lead to the top of the mountain and one can have nice views of the surrounding Czech and Polish mountains.

Fig. 8 Lookout to the Table Mountain—Big Szczeliniec

Fig. 9 Detailed photo of the sandstone rocks

3.3 Locality Red Rock, National Park of the Table Mts. (Poland)

The red colour of the granitic rock is caused by phenocrysts of feldspar, which were crystallized as the first minerals from magma during cooling (Fig. 10). It belongs to the part of Kudowa Massif, including granodiorites, tonalites and kataklazites. After the Variscan orogenesis during the Paleozoic, the Kudowa Massif was exhumated and heavily exposed to weathering, creating saprolites. During the Permian, the saprolites and the terrain were eroded. Later on, during the Cretaceous, the whole area was covered by the sea. The transgression of the see created sandstone rock formations, known as Table Mts. (Stolowe Gory). The reactivation of the fault zones was formed during Alpine orogenesis leaving the striations on the fault planes as documentation (Fig. 11).

4 Sudetic Marginal Fault Zone

The Sudetic Marginal Fault Zone (SMFZ) is a 250 km long fault zone, oriented NW–SE to NW (Fig. 12) significantly visible in the terrain's morphology. It is located in Poland and passes by the towns of Vápenná and Jeseník before graduation near the town of Opava. The northwestern part (150 km in length) is visible in terrain´s

Fig. 10 Outcrop of the granites on the east side of the Hronov-Porici Fault Zone

Fig. 11 Striations on the fault plane. Cover of the Canon camera with diameter of 56 mm for the scale

morphology (Skácel, 2004). The zone is accompanied by late Oligocene–Miocene basaltoid volcanism and numerous riverbeds and grabs (Badura & Zuchiewicz, 2008). Many authors have studied evolution and the history of SMF guilt (e.g. Štěpančíková, 2005; Novakova, 2015). SMF has traditionally been considered as an important structure border during the Paleozoic evolution of the Sudetenland (Cloos, 1922). An <200–1000 m offset has been estimated and documented from wells and determined. Skácel (2004) showed that the areas around SMF had to be more seismically active during Variscan and Saxon orogenic events. Maximum throw in the middle and SE part of the SMF in 1000–1200 m, range in the remaining segments between 400 and 800 m. The stair pattern of triangular facets shows a minimum of five episodes of footrest elevation (Badura & Zuchiewicz, 2008). Štěpančíková (2005) mentioned the reactivation of SMF during the Neogene. A clear difference in morphology of elevated blocks of the Sudeten Mountains and the decline of the Fore-Sudetic block reflects evidence of a long polygenetic history throughout the study area (Štěpančíková & Rowberry, 2008). The faults in the NE part of the Bohemian Massif have been reactivated at least four times since the Carboniferous. Similarly, during this period, NE–SW compression occurred (Novakova, 2015). Danišík et al. (2012) made a thermochronological reconstruction of the history of the Sudetes and faulting along the SMF based on zircon (U-Th)/He, apatite fission track and apatite (U-Th)/He dating. They have stated that the studied part of the SMF acted as a normal

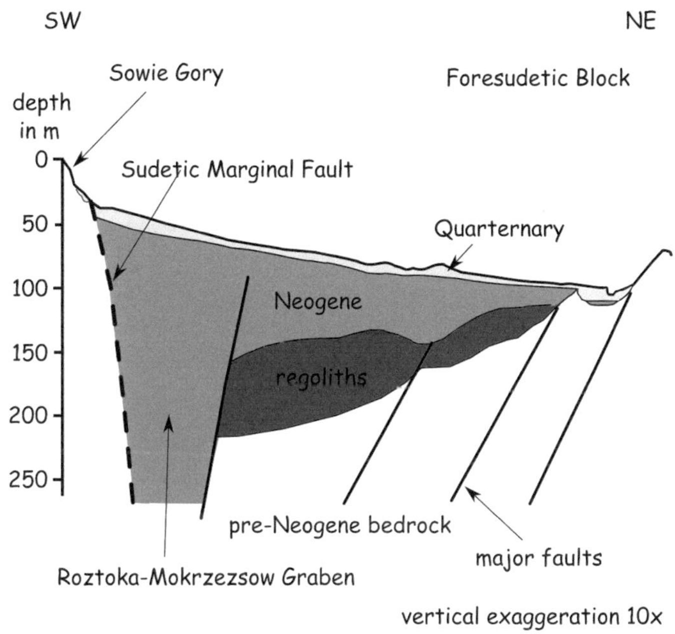

Fig. 12 Geological situation of the central part of the Sudetic Marginal Fault Zone (simplified from Badura et al., 2007)

fault during early Late Cretaceous burial, as a reverse fault during subsequent extensive European basins inversion and exhumation, and finally as a normal fault in the late Cenozoic.

4.1 *Locality Vápenná Quarries (Czech Republic) 50°16′36.172″N; 17°5′41.593″E*

A limestone deposit is located near a railway station ~500 m west of Vápenná village, ~10 km from the town of Jeseník in the Czech Republic. It was mined in several separate quarries up to that were 1500 m long and up to 300 m wide (Fig. 13). There are two or three levels in each quarry. The maximum high of the quarry face is ~80 m. Limestones are slightly karstified with the possibility to find up to 3 cm crystals of calcite. The terrain belongs to the Rychlebské Mountains. The altitude varies here from 430 to 520 m above water level. Limestone belongs to the northern end rock band Branná of the Devonian age. In the west, the group begins in the Bušín fault zone and continues east to its end on the Sudetic Marginal Fault in the village Vápenná. The rocks of the Branná Group are possibly divided into four distinct sets of layers. The second layer consists of siliceous graphitic slate, or sometimes 0.5 m thick graphitic layers. Crystalline black or grey limestone follow the slate. The limestone layer up

Fig. 13 Panoramatic view of the quarry

to 5 m wide contains a lot of graphitic pigment and often pyrite. Finally, the last layers are slightly folded white to grey crystalline limestone. Many calcite steps are possible to be found on the fault planes showing the directions of the movements (Fig. 14). This limestone forms a 5–10 m thick layer (Novakova, 2010).

Fig. 14 Calcite mineral steps indicating movements on the fault plane

4.2 Locality Na Špičáku Cave (Czech Republic) 50°17′00.600″N; 17°14′59.000″E

Na Špičáku Cave is one of the oldest documented caves in Central Europe. The cave is accessible for public. Guided tours cover 250 m of the caves. Over 400 m long underground labyrinth of corridors and cracks is created in Devonian 350–380 Ma marbles derived from marine sediments. Marble is part of a rock-forming sequence, newly arranged to cover the Desno unit (Vrbenská group). According to the NE-SW and NW–SE system of cracks, karst processes have developed, resulting in a system of low, cracked corridors with a total length of 400 m (Fig. 15). The main part of the cave is a 50 m long entrance corridor (2–3 m wide and 2 m high), followed by vertically oriented transverse corridors, which connect in the largest area of the cave. The karst process took place in the older Quaternary and was probably influenced by the melting waters of the continental glacier. In the younger Quaternary, the karst process ceased due to the overall deepening of the surrounding valleys.

Fig. 15 Marl caves in the vicinity of the Sudetic Marginal Fault

5 Lusatian Fault Zone

The Lusatian fault is of the NW–SE regional faults forming the Elbe Fault System (EFS). This particular fault belongs to the south-eastern part of the EFS, cutting through NW parts of the Bohemian Massif. Here the fault separates the Proterozoic and lower Paleozoic units of the Lusatian pluton and the Krkonoše–Jizera crystalline from the post-variscan sedimentary Permian and upper Cretaceous units of the Bohemian Cretaceous Basin and Permian basins (Fig. 16).

The Lusatian fault has a rich polyphase architecture, i.e. an internal structure (Coubal et al., 2014), filling a transversely variable, differently wide zone of this regional fault zone. The Lusatian Fault Zone is a higher-order conformal structure that comprises the main fault and associated zones, including the fault core and drug. It is the main failure of the contact of the pressed hangingwall and the footwall, on which the largest shift occurred. It includes fragile structures created in the process of its formation during the main formation. The core of the fracture, the brittle zone in the overlying and underlying blocks and the damage zone can be separated between these structures.

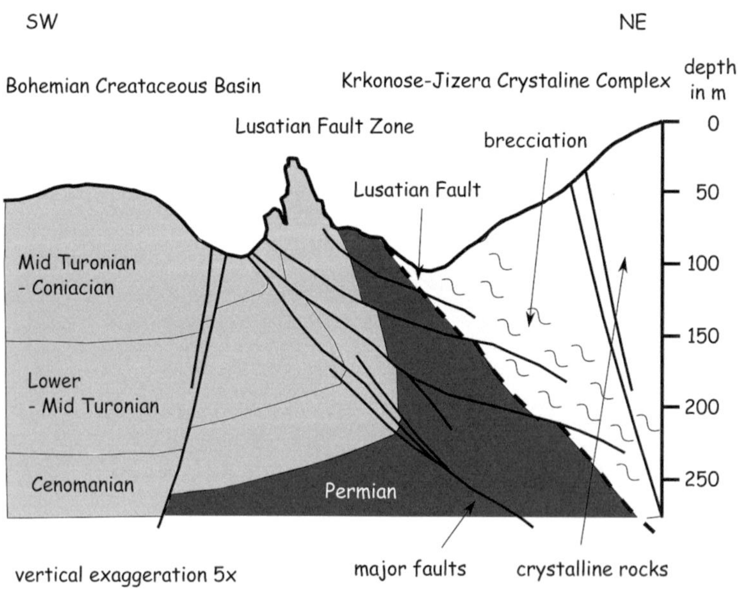

Fig. 16 Main units within the Lusatian Fault Zone. Simplified from Coubal et al. (2014)

5.1 Locality Dry Rocks (Czech Republic) 50°38′6.9″N; 15°12′49.09″E

The morphologically significant ridge of the Dry Rocks is formed by steep to slightly overturned layers of Cretaceous Cenomanian sandstones along the Lusatian fault. Upper Cretaceous sandy sediments of the Czech Cretaceous (sandstone and conglomerate of the Cenomanian sea) were elevated along the Lusatian fault during tectonic movements during the Tertiary and Quaternary. At the fault zone, quartz sandstones are in tectonic contact with Permian melafyrs and sedimentary rocks. Multiple tectonic mirrors (smooth layered surfaces) are evidence of tectonic movements and the "sliding" of individual layers one after the other. Sandstones are very variable in grain size and locally contain accumulations of oyster fauna. The originally horizontal layers thus found themselves in an almost vertical position and were exposed to a more intense drift. The oldest Upper Cretaceous sediments resisted the best of the draft—solid, quartzite Cenomanian sandstones, which were modelled in the form of a striking rocky ridge with sharp top rocks, towers and gorges (Fig. 17). Dry rocks are popular tourist site in the Czech Republic and also popular for climbing.

Fig. 17 Spectacular walls of the Dry rocks

5.2 *Locality Devil's Wall (Czech Republic) 50°40′27.41″N; 14°56′45.86″E*

The Devil's Wall is a true volcanic vein preserved from the sandstones of the Jizera Formation (Upper Cretaceous, Middle Turonian). It is located between the villages of Český Dub and Osečná in northern Bohemia. The top part of the ridge, which runs roughly parallel to the river Zábrdka in the NE–SW direction, has reached a height of up to 10 m. Since the beginning of the nineteenth century, rock has been mined and used for road gravel. From the original length of 12 km is preserved only incoherent remnants of a rock wall with a thickness of between 2 and a maximum of 4 m. Shallow ditches remained after the rock mining. Polyhedral, completely irregularly arranged blocks alternate with plate separation, sometimes forming fold-like formations, reflecting the process of viscous flow of the cooling melt (Fig. 18). The exposed walls of the Great Devil's Wall reveal at least two superimposed rock bodies, formed at different times and possibly even under different stress conditions. The vein is formed by a rock known as melilitic-olivine nephelinite, but reliable quantitative evidence is lacking. The Great Devil's Wall is part of a 5 km wide band formed by more than twenty roughly parallel veins, of which, for example, the Little Devil's Wall near Smržov is also known. Lusatian fault forming the northwestern boundary of this venous zone separates the sedimentary cover of the Upper Cretaceous basin and metamorphic formations of the Ještěd ridge, in which these veins are missing. The age of venous intrusions was determined using the K–Ar-method to ~55 to 80 Ma, i.e. for the Paleocene and Late Upper Cretaceous. The alkali content is low (30–35% SiO_2), and high CaO content up to 25% and MgO up to 20%.

There are also disputes about the origin of the melt. Some parts bear the features of the material composition of the upper mantle, while other xenoliths come from the upper part of the Earth's crust. Near the Red Hill, a pit has been preserved after historical mining by impregnating sandstone with iron hydroxides in the peripheral part of the vein (hence probably the name of the hill). About 1300 m long section of the vein with preserved remains of the rock wall called "Devil's head" and "Devil's chair" falls under the protection.

6 Summary

The Bohemian Massif is a large, consolidated relict of the Variscan orogeny located in Central Europe. Lately, when the Alpine orogenesis began south of the Bohemian Massif, old weakened zones were reactivated during the Cretaceous and Tertiary. Significant movements along these zones reshaped the landscape. Also, some of the reactivated zones evince remarkable seismic activity.

This article describes four of the most significant NW–SE fault zones within the Bohemian Massif—the Franconian Line, the Lusatian Fault, the Hronov-Porici Fault Zone and the Sudetic Marginal Zone. In addition to descriptions of brief geological

Fig. 18 Block of volcanic dyke surrounded by sandstones

situations and positions, relevant excursion sites are described to enable geology enthusiasts to explore the fault zones on their own.

Fault names	**Abbreviations**
Elbe Fault System	EFS
Franconian Line	FL
Hronov-Porici Fault Zone	HPFZ
Lusatian Fault Zone	LFZ
Sudetic Marginal Fault Zone	SMFZ

Acknowledgements Many thanks belong to my husband Petr for his support. I would also like to thank my friends and colleagues, namely Prof. L. Raschbacher, Dr R. Jonckheeree, Dr B. Wauschkuhn, Dr J. Malek and C. Aslanian for their helpful discussions. This work was carried out with the support of the long-term conceptual development research organisation RVO: 67985891. The research was funded by the EU/MEYS (CZ.02.2.69/0.0/0.0/19_074/0014756). Soumyajit Mukherjee (IIT Bombay) invited, handled and reviewed this article. Mukherjee (2023) summarized this chapter. Alexis Vizcaino and the proofreading team (Springer) are thanked.

References

Badura, J., Zuchiewicz, W., Štěpančíková, P., Przybylski, B., Kontny, B., & Cacoń, S. (2007). The Sudetic Marginal Fault: A young morphotectonic feature at the NE margin of the Bohemian Massif, Central Europe. *Acta Geodynamica Et Geomateria, 4*, 1–23.

Badura, J., Zuchiewicz, W., (2008). The Sudetic Marginal Fault as one of principal morphotectonic structures of the eastern portion of the European Cenozoic Rift System. 6–14th August 2008, International Geological Congress Oslo.

Cloos, H. (1922). Der Gebirgsbau Schlesiens und die Stellung seiner Bodenschätze, Berlin.

Coubal, M., Adamovič, J., Málek, J., & Prouza, V. (2014). Architecture of thrust faults with along strike variations in fault-plane dip: Anatomy of the Lusatian Fault, Bohemian Massif. *Journal of Geosciences, 59*, 183–208.

Coyle and Wagner (1998). Positioning the titanite fission-track partial annealing zone. *Chemical Geology, 149*, 1–2, 117–125.

Danišík, M., Štěpančíková, P., & Evans, N. J. (2012). Constraining long-term denudation and faulting history in intraplate regions by multisystem thermochronology: An example of the Sudetic Marginal Fault (Bohemian Massif, central Europe). *Tectonics, 31*, 1–19.

Emmermann, R., & Lauterjung, J. (1997). The German continental deep drilling program KTB: Overview and major results. *Journal of Geophysical Research, 102*(B8), 18179–18201. https://doi.org/10.1029/96JB03945

Gee, D. G., & Stephenson, R. A. (Eds.). (2006). European Lithosphere Dynamics. Geological Society, London, Memoirs, Vol. 32, pp. 1–9. 0435-4052/06/

Havíř, J., & Špaček, P. (2004). Recent tectonic activity and orientations of the principal stresses in the Jeseníky region. *Geolines, 17*, 38–39.

Hirschmann, G., Duyster, J., Harms, U., Kontny, A., Lapp, M., de Wall, H., & Zulauf, G. (1997). The KTB superdeep borehole: Petrography and structure of a 9-km-deep crustal section. *Geologische Rundschau, 6*, 3–14.

Jacobs, J., Hejl, E., Van den haute, P., & Wagner, G. A. (1993). A preliminary tectono-thermal model for the KTB deduced from apatite, zircon and sphene fission-track analysis. *KTB Report, 93*(2), 125–128.

Mukherjee, S. (2023). Introduction to structural geology and tectonics field guidebook—volume 2. In S. Mukherjee (Ed.), *Structural geology and tectonics field guidebook—volume 2* (pp. xi–xiv). Springer Nature Switzerland AG. ISBN 978-3-031-19575-4.

Nocquet, J.-M., Calais, E., Altamimi, Z., Sillard, P., & Boucher, C. (2001). Intraplate deformation in western Europe deduced from an analysis of the International Terrestrial Reference Frame 1997 (ITRF97) velocity field. *Journal of Geophysical Research, 106*(B6), 11239–11257.

Novakova, L. (2010). Detailed brittle tectonic analysis of the limestones in the quarries near Vápenná village. *Acta Geodynamica et Geomaterialia, 7*(2) (158), 1–8.

Novakova, L. (2014). Evolution of paleostress fields and brittle deformation in Hronov-Porici Fault Zone, Bohemian Massif. *Studia Geophysica et Geodeatica, 58*, 269–288. Springer.

Novakova, L., (2015). Tectonic phase separation applied to the Sudetic Marginal Fault Zone (NE part of the Czech Republic). *Journal of Mountain Science, 12*(2), 251–267. Springer. https://doi.org/10.1007/s11629-014-3297-5

Peterek, A., & Schröder, B. (2010). Geomorphologic evolution of the cuesta landscapes around the Northern Franconian Alb review and synthesis. *Zeitschriftfür Geomorphologie, 54*, 305–345.

Schenk, V., Schenková, Z., & Pospíšil, L. (1989). Fault system dynamics and seismicactivity—examples from the Bohemian Massif and the Western Carpathians. *Geophysical Transactions, 35*(1–2), 101–116.

Scheck, M., Bayer, U., Otto, V., Lamarch, J. (2002). The Elbe Fault System in North Central Europe-a basement controlled zone of crustal weakness. *Tectonophysics, 360*, 281–299.

Siebel, W., Höhndorf, A., Wendt, I. (1995). Origin of late Variscan granitoids from NE Bavaria, Germany, exemplified by REE and Nd isotope systematics. *Chemical Geology, 125*, 3–4, *25*, 249–270.

Stejskal, V., Štěpančíková, P., Vilímek, V. (2006). Selected geomorphological methods assessing neotectonic evolution of the seismoactive Hronov – Poříčí Fault Zone. *Geomorphologica Slovaca, 6*, 14–22.

Schulmann, K., Konopasek, J., Janousek, V., Lexa, O., Lardeaux, J.-M., Edel, J., Štípská, P., & Ulrich, S. (2009). An Andean type Palaeozoic convergence in the Bohemian Massif. *Comptes Rendus Geosciences, 341*, 266–286. https://doi.org/10.1016/j.crte.2008.12.006

Skácel, J. (2004). The Sudetic Marginal Fault between Bílávoda and Lipová Lázně. *Acta Geodynamica et Geomaterialia, 1*(3), 135, 31–33.

Skácelová, Z., Mlčoch, B., & Tasáryová, Z. (2009). Digital elevation model of the crystalline basement and Permo-Carboniferous surface (Bohemian Massif, NE part of the Czech Republic). *Acta Geodynamica et Geomaterialia, 6*(3), 155, 265–271.

Špaček, P., Sykorova, Z., Pazdirkova, J., Švancara, J., & Haviř, J. (2006). Present-day seismicityof the south-eastern Elbe Fault System (NE Bohemian Massif). *Studia Geophysica Et Geodaetica, 50*(2), 233–258.

Stejskal, V., Kašpárek, L., Kopylova, G. N., Lyubyshin, A. A., & Skalský, L. (2009). Precursory groundwater level changes in the period of activation of the weak intraplate seismic activity on the NE margin of the Bohemian Massif (Central Europe) in 2005. *StudiaGeophysica et Geodetica, 53*, 215–238.

Stejskal, V., Skalský, L., & Kašpárek, L. (2007). Results of two years' seismo-hydrological monitoring in the area of the Hronov-Porici Fault Zone, Western Sudetes. *Acta Geodynamica et Geomaterialia, 4*(4), 148, 59–76.

Štěpančíková, P. (2005). Selected analyses of the morphostructure of the NE part of the Rychlebskéhory Mts. (Czech Republic). *Acta Geodynamica et Geomaterialia, 2*(1), 137, 59–67.

Štěpančíková, P., Fischer, T., Stemberk, J., Novakova, L., & Hartvich, F. (2019). Active tectonics in the Cheb basin: Youngest documented Holocene surface faulting in central Europe? *Geomorphology, 327*, 472–488.

Štěpančíková, P., & Rowberry, M. (2008). Rock landforms that reflect differential relief development in the North-eastern sector of the Rychlebskéhory and the adjacent area of Žulovskápahorkatina (SE Sudeten Mts., Czech Republic). *Acta Geodynamica et Geomaterialia, 5*, 151, 297–321.

Tásler, R., Čadková, Z., Dvořák, J., Fediuk, F., Chaloupský, J., Jetel, J., Kaiserová-Kalibová, M., Prouza, V., Schovánková-Hrdličková, D., Středa, J., Střída, M., & Šetlík, J. (1979). Geology of the Czech part of the Intra-Sudetic Basin [in Czech]. ÚÚG, Praha, pp. 292. https://doi.org/10.1016/j.geomorph.2018.11.007

Valenta, J., Gazdova, R., & Kolinsky, P. (2011). Seismo-hydrological monitoring in thearea of the Hronov-Poríčí Fault Zone, Northern Czech Republic, Central Europe. *American Geophysical Union, Fall Meeting, 2011*, S11B-2211.

Valenta, J., Stejskal, V., & Štěpančikova, P. (2008). Tectonic pattern of the Hronov-Porici Trough as seen from pole-dipole geoelectrical measurements. *Acta Geodynamica et Geomaterialia, 5*(2), 150, 185–195.

Wauschkuhn, B., Jonckheere, R., & Ratschbacher, L. (2015). The KTB apatite fission-track profiles: Building on a firm foundation? *Geochimica Et Cosmochimica Acta, 167*, 27–62.

Wemmer, K., & Ahrendt, H. (1997). Comparative K–Ar and Rb–Sr age determinations of retrograde process-es on rocks from the KTB deep drilling project. *Geologische Rundschau, 86*, S272–S285.

Wojewoda, J. (2007). Neotectonic aspect of the Intrasudetic shear zone. *Acta Geodynamica et Geomaterialia, 4*(4), 148, 31–41.

Woldřich, J. N. (1901). Earthquake in the north-eastern Bohemia on January 10, 1901 [in Czech]. *Transactions of the Czech Academy of Sciences, series II, 10*(25), 1–33.

Ziegler, P. A., & Dèzes, P. (2007). Cenozoic uplift of Variscan Massifs in the Alpine foreland: Timing and controlling mechanisms. *Global Planetary Change, 58*, 237–269.

Zulauf, G., & Duyster, J. (1997). Supracrustal intraplate thickening of Variscan basement due to Alpine fore-land compression: Results from the superdeep well KTB (Bohemian Massif, Germany). *Tectonics, 16*, 730–743.

Geological Field Observations Along the Pandoh Syncline: The Mandi-Kataula-Bajura Section of Himachal Pradesh, NW-India

Paramjeet Singh, Pratap Chandra Sethy, Hrithik Rastogi, M. Rajanikanta Singh, A. Krishnakanta Singh, Satyajit Singh Thakur, and Saurabh Singhal

Abstract The Mandi-Kataula-Bajura section of Himachal Himalaya offers an invaluable opportunity to observe the lithology and structures of Pandoh Syncline. It comprises of (i) the Lesser Himalayan Sequence, which includes Shali and Rampur Groups, and (ii) low to medium grade metamorphic of the Lesser Himalayan Crystallines, which includes the Kulu thrust sheet of the Himachal Himalaya. The road dip-section is almost perpendicular to the major strike parallel and cross-cuts the litho-tectonic units. The excursion guide proposes a field itinerary across the meta-sedimentary sequence of LHS in the south, and through the low to medium-grade metamorphic of the synclinal structure (also known as Pandoh syncline). Based on our field observations, this field trip guide allows us to identify (i) the deformation behaviors, shear-sense indicators near major thrusts, and (ii) the grade of metamorphism of the Pandoh syncline of the Himachal Pradesh, NW-Himalaya. Our study suggests that the southern and northern limbs of the Pandoh syncline reached up to garnet-grade of metamorphism and the core part remains low-grade of metamorphism (i.e. biotite grade). In both limbs of the synclinal structure, the garnet crystal size varies from 0.5 to 5 mm along the above-mentioned section.

1 Introduction

The Himalaya mountain belt was formed in response to a continent–continent collision between Indian and Eurasian platesca. ~55 Ma ago (Jain, 2014; Mukherjee, 2013, 2015a, 2015b; Singh et al., 2020, 2021a; Yin, 2006) and offers a unique opportunity to know the geological, structural and metamorphic processes of the orogeny (Thakur, 1992).The architecture of the Himalayan orogen is expressed by several north-dipping, crustal-scale thrust zones from north to south, such as South Tibetan Detachment System (STDS), Main Central Thrust (MCT), Main Boundary Thrust

P. Singh (✉) · P. C. Sethy · H. Rastogi · M. R. Singh · A. K. Singh · S. S. Thakur · S. Singhal
Wadia Institute of Himalayan Geology, #33 Genaral Mahadev Singh Road, Dehradun 248001, India
e-mail: psinghgeol@gmail.com

S. Mukherjee (ed.), *Structural Geology and Tectonics Field Guidebook–Volume 2*, Springer Geology, https://doi.org/10.1007/978-3-031-19576-1_4

(MBT), Main Frontal Thrust (MFT) and their associate thrust systems (Banerjee et al., 2019; Bose & Mukherjee, 2019a, 2019b, 2020; Jain & Anand, 1988; Le Fort, 1986; Mahato et al., 2019; Patel et al. 2011a, 2011b, 2012, 2015; Singh, 2014; Singh & Patel, 2017, 2022; Singh et al., 2012, 2021, 2022a, 2022b; Thakur, 1987). Based on the grade of metamorphism, each of the litho-tectonic units of the Himalaya has been separated by the major thrust/faults (Fig. 1) such as the STDS separates the Tethys Himalayan Sequence (THS) in the hanging wall and Higher Himalayan Crystalline (HHC) in the footwall, the MCT divides the HHC and Lesser Himalayan Sequences (LHS), and the MBT separates the LHS in the hanging wall and the Neogene sediments of the Siwalik in footwall (Singh et al., 2021, 2022b; Yin, 2006).

It is well known fact that the Himalaya mountain belt evolution has undergone three phases of deformation: the first began with the continental collision between Indian-Eurasian plates ~55 Ma ago (Jain, 2014) and in this process the THS was thrusted over the Indian continental lithosphere (Mandal et al., 2019; Robinson & McQuarrie, 2012; Webb, 2013; Webb et al., 2011 and references therein); During the second phase, the HHC was thrust towards south along the MCT (ca. 22–25 Ma), (Patel et al., 2015; Singh & Patel, 2022; Singh et al., 2012; Valdiya, 1980). In the last phase, Compression and folding of transported crystalline rocks of HHC along the entire Himalayan orogen took place and Lesser Himalayan Duplex (LHD) structures were developed in the footwall of MCT during the middle Miocene (ca. 15–18 Ma)

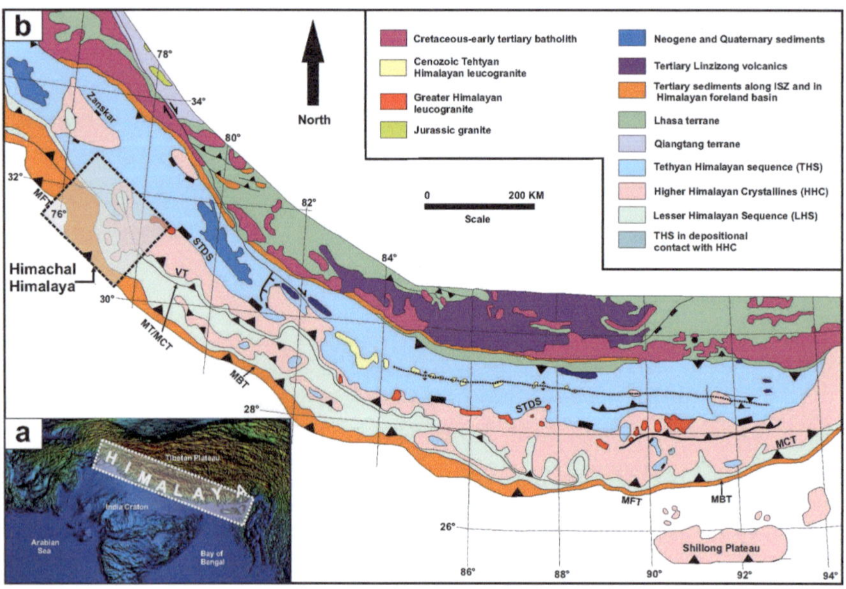

Fig. 1 **(a)** based on the GTOPO30 digital elevation model, (U.S. Geological Survey) topographic map the Himalayan mountain belt, **(b)** Simplified geological and tectonic setting map of the Himalaya (black square box show the our study area: Himachal Himalaya) MFT: Main Frontal Thrust, MBT: Main Boundary Thrust, MCT: Main Central Thrust, VT: Vaikrita Thrust; STDS: South Tibetan Detachment System (Taken after Singh & Patel, 2022)

(DeCelles et al., 2001; Robinson & Martin, 2014; Robinson et al., 2006; Singh & Patel, 2017; Srivastava & Mitra, 1994) and formation of the MBT was initiated (ca. 12–14 Ma) (Singh & Patel, 2022).

In the NW-Himalaya, the geological and tectonic settings of the central part of Himachal Pradeshare very complex especially regarding the existence of the Kulu and Jutogh thrustsheet and actual demarcation of the thrust known with the name "MCT". Some of the recent studies (Harton et al., 2014; Robyr et al., 2002, 2006; Vannay and Grasemann, 2001; Webb et al., 2011 reference therein) considered the Punjal/Chail Thrust of Thakur (1992) as "MCT". In the western part of the Himachal Himalaya, the Himalayan metamorphic belt (HMB) zone consists of inverted metamorphic isograds from kyanite-garnet grade down to biotite grade are well described and bound between the Zanskar shear zone (ZSZ) in the northern side and the MCT in south (Harton et al., 2014; Robyr et al., 2002, 2006 and reference therein). Whereas, and in the central-eastern part, metamorphic isograde (Webb et al., 2011) and retrograde pre-Himalayan metamorphism (Thakur, 2014) are well described. In the central and eastern part the STDS is considered as the northern most limit of the HMB (Jain, 2014; Jain et al., 2014; Singh et al., 2021; Thakur, 1992).

This chapter explains the geological and tectonic setup of LHS, the Pandoh syncline of the Kulu Thrust sheet. We also tried to distinguish the Pandoh syncline and its root connections with the help of their lithological, structural and metamorphic studies. Our study is a continuation of previous works from this terrain by two Indian geologists O.N. Bhargava and S.V. Srikantia, (Bhargava & Bassi, 1994; Bhargava & Srikantia, 2014; Bhargava et al., 1991, 2016, 2021; Bhargava, 1995a, 1995b; Srikantia & Bhargava, 1988, 1998) and provides detailed constraints on the geology and tectonic setup of the Himachal Himalaya.

In this field guidebook chapter (Fig. 2), we highlight the following key geological features of the Pandoh syncline along the Mandi-Kataula-Bajura section of Himachal Pradesh:

(a) Definition and characterization of the MBT, Panjal Thrust (PT) vis-à-vis MCT/Kulu Thrust.
(b) Deformation pattern, shear sense analysis and metamorphism along the Mandi-Kataula-Bajaura section.

2 How to Access the Area

The Himachal Himalaya as defines here stretches from the Tons River on the eastern side which defines the Himachal-Uttrakhand border, to the Ravi River on the western side demarcating the Himachal-Jammu & Kashmir border. It comprises 12 districts and all are well connected to other parts of India. The Mandi town is ~500 km towards west of Delhi and to reach the Mandi town, anyone may travel by air to Chandigarh airports and further road journey along the Chandigarh-Bilaspur-Sundernagar-Mandihighway. This road journey offers the beautiful sightseen through the Cenozoic

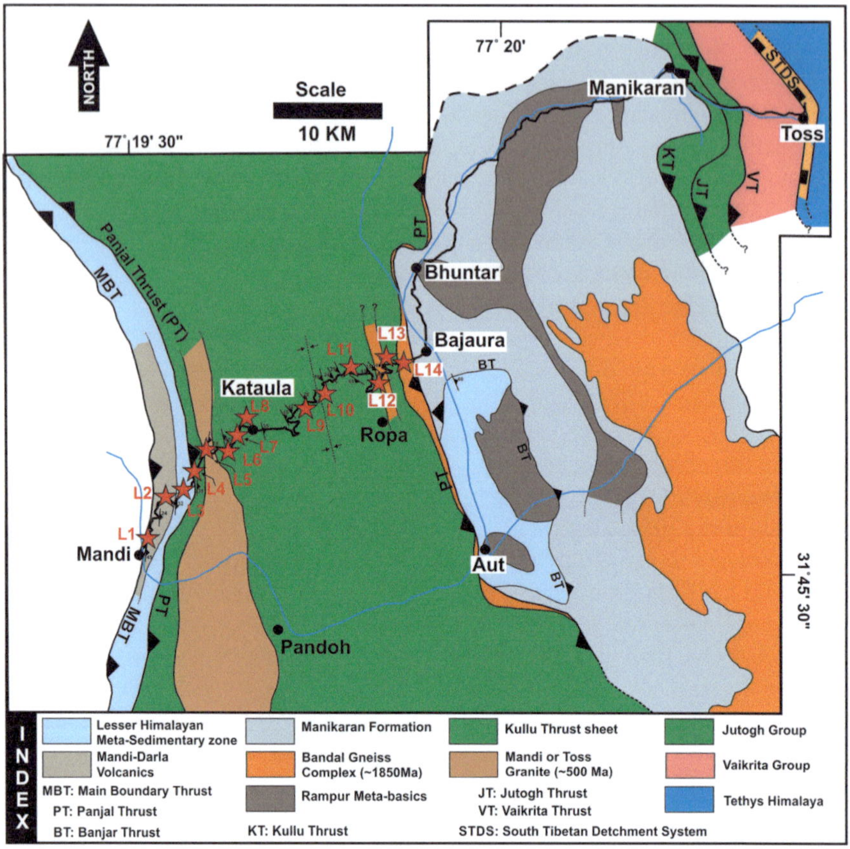

Fig. 2 Geological and Tectonic map of the study area between Mandi-Kataula-Bajaura sections of Himachal Pradesh. The star indicate our locations (Modified after: Bhargava and Bassi, 1994; Singh et al., 2021; Thakur, 1992)

sedimentary deposits of the Himalayan foothills (Siwalik rocks). Our north-westward journey along Kalka-Nalagarh-Swarghat-Mandi national highway provides a wide range of beautiful meandering and picturesque of Satluj Valley. Our destination town Mandi is situated on the bank of the Beas River. On the first day evening, we reach to Mandi town and take rest to acclimatize our bodies for the field work. Before start a geological field work, we have to complete some of the basic formalities of the H.P. Government. The formalities include permission to carry GPS, topo sheets, roadmaps, geological maps etc. as these items are not permitted in these Border States. The official permission may be got from the District Magistrate (Mandi town) to carry these items during the geological field work. In this field guide chapter, our field locations are numbered as location 1, 2, 3… etc. (see Fig. 2), starting from Mandi town. Location-wise descriptions of the rocks are explained in detail below.

3 Geology

In the NW-Himalaya, the central part of Himachal Pradesh (also see Mukherjee & Koyi, 2010) consists of almost all the litho-tectonic units of the Himalayan fold-thrust belt. Here, we review the geological and tectonics background of the Himalayan fold-thrust belt along the Mandi-Kataula-Bajaura traverse (Singh et al., 2021; Thakur, 1992; Yin, 2006). The central part of the Himachal Pradesh is divided into four tectonic zones from south to north as explained below (see Fig. 1):

(a) **Sub-Himalayan sedimentary sequence**

The youngest sedimentary sequence lies in the outermost part of the Himalayan fold-thrust belt toward south mainly consists of the fluvial molasses sediments of Paleo-Neogene time and overridden by the LHS rocks along the MBT. These sediments were deposited in the shallow-marine environments during Eocene time and were described as Subathu/Dagshai, and Kasauli Formations, consisting of limestones, shales, minor fine-grain sandstones and quartz arenite, followed by fluvial deposits as an equivalent of the Dharmashala Group (Jain et al., 2009; Srikantia & Bhargava, 1998; Singh et al., 2021). Here, in central Himachal Himalaya the Paleo-Neogene sedimentary group (Dharamshala Group) is further subdivide into two Kasauli and Dagshai Formations containing sandstones, minor shales, siltstones, and thrusted over the Siwalik rocks along the Bilaspur Thrust. Similarly, the Siwalik rock rises abruptly above the Indo-Gangetic alluvial plains along the Himalayan Frontal Thrust (HFT) between Nalagarh-Swarghat-Bilaspur sections (Srikantia & Bhargava, 1998).

(b) **Lesser Himalayan Sequence (LHS)**

In the central part of Himachal Pradesh, the main complexity is the Lesser Himalayan rocks. The LHS rocks raised a lot of controversy in terms of their existence and their similarity and variability with the rocks along strike parallel of the Himalayan mountain belt. The Proterozoic age rocks of LHS also known as the Lesser Himalayan Metasedimentary Sequence (LHMS) Zone and are autochthonous to para-autochthonous, un-fossiliferous, and low-grade metamorphic with greenschist-facies. The LHMS is mainly represented by Shali and Rampur groups. The Shali group is further subdivided into three parts the lower most is Sundernagar Formation overridden by the Shali structural belt including the Mandi-Darla volcanic and the Simla group rest at the top (Bhargava, 1982, 1995a, 1995b; Srikantia & Bhargava, 1988). The Shali Group in the area, mainly comprised of the grey slates and phyllites in the lowermost part, volcanic in the middle part and white-to-purple quartzite, grey, olive green to purple shales, slates, phyllite at top (Srikantia & Bhargava, 1988, 1998; Thakur & Choudhury, 1983; Yin, 2006).The LHMS zone is tectonically overlain by low to medium-grade crystalline rocks, also known as the Lesser Himalayan Crystalline (LHC) Zone (Bhargava & Bassi, 1994; Singh et al., 2020, 2021; Srikantia & Bhargava, 1976, 1998; Thakur, 1992) and separated by Panjal Thrust (PT). The southern part of LHC zone is commonly known as Kulu Thrustsheet and consisting the prominent feature an intruded igneous bodies (i.e. Mandi Granite body) of early Cambrian to early Permian age and outliers by the Proterozoic sequences (Purkayastha et al., 1999). This crystalline thrust sheets is bounded by PT and form a synclinal structure named as the Pandoh Syncline (Singh et al., 2021). The Kulu Thrust sheet is the southern extension of Jutogh Thrust sheet and lies in northern side of tectonic window of the Proterozoic/Paleozoic rocks (i.e. Kulu-Larji-Rampur window) (Bhargava, 2000; Bhargava & Bassi, 1994).

We observe that the Pandoh syncline and the Kulu thrust sheet are same tectonic affinities of low-grade crystalline rocks that exist in the Himachal Pradesh and ascend in tectonic order (Bhargava & Bassi, 1994). In previous literature, these thrust sheets were referred to as Chail (Thakur, 1992), Jutogh (Singh et al., 2021), and Salkhala, Kulu (Srikantia & Bhagava, 1984). The available literature of this area correlates the Kulu Thrust sheet with the Almora-Dudatoli-Bainath klippen of the Kumaun-Garhwal region in east and Salkhala Thrust-sheet in Jammu & Kashmir and eastern Pakistan of NW-Himalaya (c.f. Bhargava, 2000). This study favors the correlation and suggests that the Salkhala in west and Kulu Thrust sheet in Himachal are equivalent to the Almora-Baijnath-Lansdown klippen of the Kumaun-Garhwal region (Singh, 2014; Singh et al., 2012, 2022a, 2022b). Based on geology, metamorphism, and tectonic setup of the studied section, in this book chapter we are distinguish and correlate the Pandoh Syncline and the Kulu thrust sheet.

(c) **Higher Himalayan Crystallines (HHC)**

The core part is high-grade metamorphic and is known as the Higher Himalayan Crystalline Sequence, which occupies the highest elevation of the Himalayan range (Singh, 2014). It is characterized by a 40–50 km thick sequence of Proterozoic to Ordovician (DeCelles et al., 2000; Parrish & Hodges, 1996) crystalline rocks overriding the LHC through north-dipping south-vergent MCT along the strike parallel of the Himalaya (Fig. 1). In Himachal Pradesh, the rocks of the Jutogh group are considered as the southern extension of the HHC and consist of low to medium grade metamorphics of the biotite-chlorite schist, mica schist, gneisses, and intrusive granites of Ordovician-Cambrian ages (0.9–0.5 Ga age) (Dhiman & Singh, 2021; Miller et al., 2001; Singh et al., 2002,, 2017, 2020). In this study, we considered Jutogh and Vaikrita Thrust sheets to be part of HHC (Bhargava, 2000; Bhargava & Bassi, 1994; Bhargava & Srikantia, 2014; Singh et al., 2020, 2021, 2022b). The Jutogh Thrust sheet is underlain by the muscovite-biotite-garnet grade of the Kulu Thrust sheet and the tectonic contact between them marks the Jutogh Thrust (JT/MCT) which has been considered to be a continuation of the MCT. Further north, the Jutogh Group of rocks is overridden by the high-grade metamorphic and migmatites rocks of the Vaikrita Group. It is suggested that the Vaikrita Thrust sheet was extruded along a broad northeast dipping intracontinental ductile shear zone between coeval VT in the south and STDS in the north during the late-Oligo early Miocene (~23 Ma) (Burchfiel et al., 1992; Coleman, 1998; Dezes et al., 1999; Patel et al., 1993; Robyr et al., 2006; Webb et al., 2011).The rocks of the Vaikrita Group are bounded by the VT in the south and STDS in the north, which separates it from the THS (Bharagava et al., 1991; Patel et al., 2011a, 2011b; 2012; Singh, 2014; Thakur, 1992).

(d) **Tethyan Himalayan Sequence (THS)**

The sedimentary series of the Tethys Himalaya is exposed to the north of the Higher Himalayan Crystallines and comprises primarily of a massive pile of sedimentary rocks often comprising grey-green-slate and phyllite. The THS rocks either underwent low-grade Cenozoic metamorphism or were unmetamorphosed (Steck et al., 1993; Yin, 2006). The siliciclastic and carbonate sedimentary rocks interbedded with Palaeozoic and Mesozoic volcanic rocks and fossiliferous rocks exhibits typical platform sediments of marine facies deposited on the northern passive margin of the Indian continental plate (Baud et al., 1984; Brookfield, 1993; Critelli & Garzanti, 1994; Gaetani & Garzanti, 1991; Gansser, 1964; Garzanti, 1993, 1999; Steck et al., 1993). The TSS is commonly referred to as the Haimanta Group of rocks based on the low-grade metamorphism and sedimentary characteristics of rocks (Yin, 2006).

4 Field Excursion

4.1 Observation of Main Boundary Thrust (MBT)

Location 1: Our first day in the field starts with characterizing the contact zone of the Main Boundary Thrust (MBT) between the Mandi Volcanics and the Paleogene sediments of the Kasauli (Dharamshala) Formation along the Mandi-Jogindernagar road section (Fig. 3a). The contact zone is well exposed near Bhiuli village on the right bank of the Beas River. The outcrop was partially covered with green vegetation and the alternative bed of grey coloured sandstone and volcanic rocks were partially fractured due to the thrust zone (Fig. 3b). The bedding plane/dipping direction of the strata is clearly visible with dip amount and direction of 35–41° due east direction. This outcrop belongs to the contact zone of MBT. The hangingwall rocks are part of the Mandi-Darla volcanics of the Shali structure belt (Bhargava, 2000) and the footwall rocks belong to coarse grain sandstone (Fig. 3c) of the Kasauli Formation. The GPS location is N 31° 42′ 43″ E 76° 57′ 7″ with the elevation ~755 m and the strike almost NW–SE and dip ~45° NE.

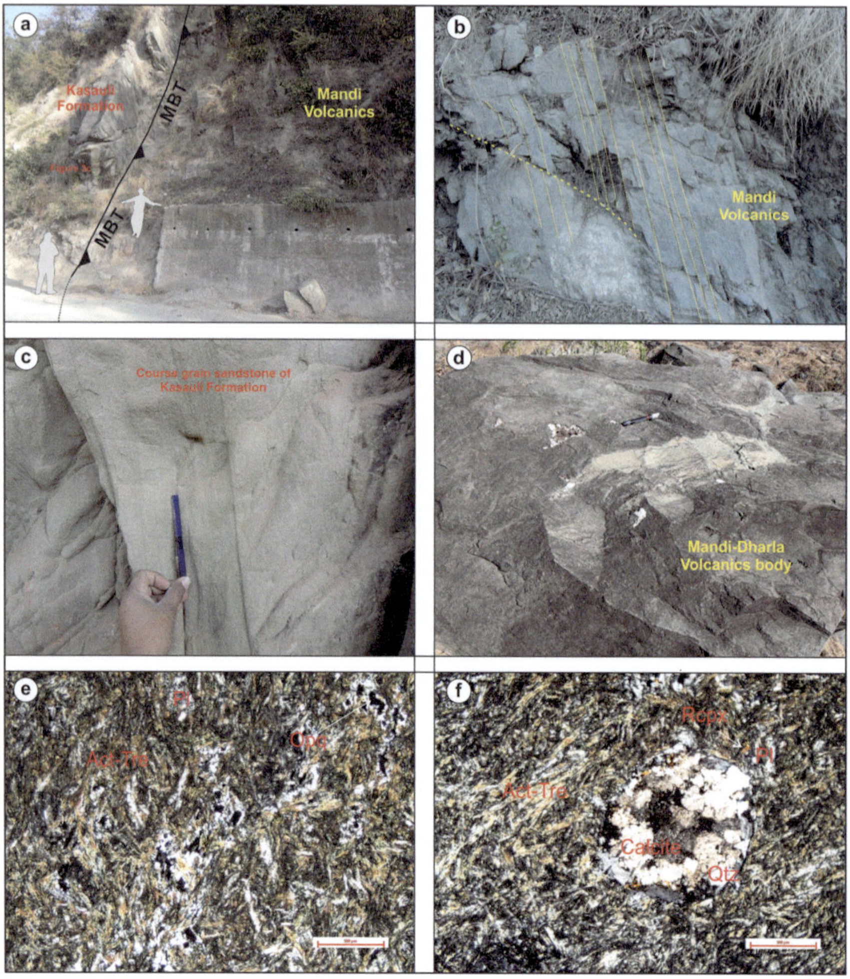

Fig. 3 Along the Mandi-Drang road section (**a**) location of Main Boundary Thrust (MBT) contact between Mandivolcanics and Kasauli Formation, (**b**) highly fractured zone of Mandi volcanic, close proximity to MBT hanging wall, (**c**) a exposure of coarse grain sand stand of the Kasauli Formation, close proximity to MBT footwall, (**d**) massive and undeformed mandi volcanic exposure, (**e- f**) thin-section of undeformed volcanic dark greenish in colour and contains amygdules

5 The Hanging Wall of MBT (Mandi-Darla Volcanic and LHS Rocks)

Location 2: After ~10 km towards the north, we turned towards the Drang salt mine and after a few kilometers, we observed a massive volcanic body known as the Mandi-Darla volcanic of Shali structure belt. This volcanic body is fractured at both edges and undeformed at the core part. The core part is fine grained, dark greenish in colour and contains amygdules (Fig. 3d). The GPS location is N 31° 43′ 55″ E 76° 58′ 41″ with the elevation ~975 m. In microscopy, the majority of these mafic volcanic rocks are composed of fine-to medium-grained green schist facies mineral assemblages with spherulitic, spinifex, and granoblastic textures. The crisscrossing sheaves of closely spaced subparallel acicular high-Mg minerals such as tremolite-actinolite and anthophyllite define the texture of spinifex (Fig. 3e). Porphyroblasts comprising (Mukherjee, 2015a, 2015b, 2020; Passchier & Trouw, 2005) anthophyllite, actinolite, tremolite, and plagioclase feldspars are embedded in a finer actinolite, plagioclase feldspar, chlorite, opaque, and quartz groundmass locally. Plagioclase feldspars that have been sericitized are common, and amagdules are filled with secondary minerals like calcite, quartz, and zeolite (Fig. 3f).

Location 3: Near Katindhi village, we observed the first exposure of the Simla group of rocks and separated from the Shali Structure Belt by Mandi Thrust (Srikantia & Bhargava, 1976). The quartzite of the Simla-Jaunsar group is white to light pinkish in color, deformed and has high specific gravity (Fig. 4a, b, c). The quartzite is a mono mineral rock and has a thin layer of dark colored mineral that appears along the foliation plain (Fig. 4b). The deformed quartzite is fractured and open to close folds are developed. Figure 4c indicates a light pinkish colored quartzite, foliation plain and kink band are developed. This type of kink band was developed near the thrust zone. The strike of the foliation plain of quartzite is nearly N–S. The dip amount and direction are 35 due east respectively. The GPS location is N 31° 45′ 32″ and E 76° 58′ 23.8″ with an elevation ~1350 m. A recent study by Bhargava et al. (2021) suggests that the Simla group is part of the undifferentiated Jaunsar group and is equivalent to the Nathuakhan and Nagthat Formations, based on detrital zircon peaks of ca. 880 Ma (Mandal et al., 2014).

6 The Hanging Wall of Panjal Thrust (Mandi-Granite and Pandoh Syncline)

Location 4: After crossing the UR River, we observed a change in the lithology and saw the highly crushed zone of chlorite + muscovite association of Chlorite schist. Here, along the UR River, we marked the Panjal Thrust (PT) or southern boundary of the Pandoh syncline. On the left bank (i.e. footwall of PT) is quartzite of the Simla-Jaunsar Group and on the right bank (i.e. hanging wall of PT is green schist facies of the chlorite + muscovite association (see Fig. 4d). Close to the PT, the rocks are sheared, foliation plain (S1) is parallel to the schistoscity (S0) and indicates top to SW shearing. The intruded quartz veins are sheared and intruded across the foliation plane in the first phase of deformation and folded in the second phase during shearing in the south (Fig. 4e). Similarly, the boudin structures of the quartz vein are also visible and indicate the second phase of deformation of the Pandoh Syncline. This location belongs to the Kulu Thrust sheet of Bhargava and Bassi (1994) that mainly contains muscovite-biotite schist, gneiss, and intrusive granite. In our earlier work,

Fig. 4 **(a -b)** Near Katindhi village, exposure of quartzite of Simla group shows the development of fold, **(c)** development of foliation and kink band, **(d)** Location and contact zone of Panjal Thrust (PT) along the UR river on Mandi-Bajaura road traverse, **(e)** outcrop exposure of garnitiferous-biotite-schist and **(f)** a thin section shows garnet poikiloblasts containing inclusion of quartz, biotite and muscovite of Pandoh syncline

we correlated it with 'Chail formation' (Singh et al., 2021). From this location, we have entered the garnet zone and the mica schist rock mainly contains garnet, biotite, muscovite, chlorite, plagioclase, epidote, quartz, and iron-oxides. The thin-section of garnet bearing biotite-schist indicates that the garnet porphyloblast (Mukherjee, 2015a, 2015b, 2020; Passchier & Trouw, 2005) is fractured due to shearing and contains inclusions of biotite (Fig. 4f).The GPS location is N 31° 46′ 33.2″ and E

76° 59′ 00″ with the elevation ~990 m. The strike of the foliation is almost NW–SE and dip 38° due east/NE.

Location 5: The first exposure of the Mandi granite body, near IIT Mandi, lies within the southern limb of the Pandoh syncline (Fig. 5a).The strike of the mylonitic gneiss is almost NW–SE and dips 39° NE along with the GPS location N 31° 46′ 37″ and E 76° 59′ 17″ and the elevation is ~1075 m. The granite body is fractured on both edges and also has randomly orientated feldspar phenocryst. Some of the feldspar phenocryst are starched up to 3–5 cm and some are rounded. The thin-section study revealed that the feldspar crystals of the granite body indicate the shearing and are wrapped with minerals like chlorite-muscovite-biotite at the edges (Fig. 5b). The feldspar crystals clearly suggest the shearing sense towards the south-west along the PT. The granite body mainly consists of quartz-feldspar-muscovite-biotite-zircon-apatite minerals. Shearing and deformation are observed on both margins and the core part remains undeformed as discussed by (Purkayastha et al., 1999; Singh et al., 2021). The grain size of feldspar augen increased toward the core part and also indicates southward direction shearing (Fig. 5b).

Location 6: On the Mandi-Bajaura road, around 3 km towards Kataula village, this location corresponds to part of the southern limb of synclinal structure, the northern contact zone of granite and mica-schist, and also observed the sheared zone along the contact. The strike of the bed is nearly NW–SE and dip 30° due East. The main rock type is mica-schist, which crumbled due to upward thrust and caused the folding (Fig. 5c). The mica-schist mainly consists of the following mineral associations: quartz, muscovite, chlorite, biotite, and a few opaque minerals. We also observed that the rock is fine-grained and dominated by muscovite + chlorite flakes. The GPS location is N 31° 50′ 07″ E 76° 59′ 15.7″ and the elevation is ~1211 m.

Location 7: Further moving towards Kataula, after ~1 km, the exposed mica-schist rocks have quartz veins intruded parallel to the bedding plain, however the quartz veins are stretched during the thrusting and indicate the top-to-SW shearing (Fig. 5d). In addition to this, the boudin structure indicates more stretching towards the SW direction (Fig. 5d, e). The GPS location is N 31° 47′ 57.8″ and E 77° 00′ 53.8″ and the elevation is ~1211 m. The dip amount direction is 42° toward NE. This location lies very close to the core part and the southern limb of the Pandoh Syncline. The amount of dip in the beds has moderately increased as compared to the previous locations. It is important to observe that the strike of the beds slightly moves towards the west and the dip amounts are also moderately high.

Location 8: The locality of this stop is IIT Mandi north campus near Kataula village. The GPS location is almost the same as the previous location, i.e. N 31° 47′ 58″ and E 77° 00′ 55″ and the elevation is ~1230 m. Here, we observed garnet bearing mica schist and the size of the garnet crystals is 3–5 mm, visible with necked eyes (Fig. 5f). The typical garnetiferous-mica-schist of the garnet zone within syncline is exposed and the garnet crystals are randomly oriented, i.e. less deformed crystals indicate that we reached close to the core part of the Pandoh syncline. In our earlier chapter, we observed the garnet within the thin-section only and were not able to trace the ~5 mm size garnet in the garnet bearing mica-schist zone along the Mandi-Pandoh section as reported by earlier workers. The orientation of the bed

Fig. 5 **(a)** Outcrop image showing the Paleozoic granite of Mandi area and inset photo taken from the thin section of same granite taken from contact zone, which indicate the shearing sense developed, **(b)** thin section of Mandi granite showing the shearing sense top-to-south along the PT, **(c -d -e)** exposure of mica-schist along the Mandi-Bajaura road section with highly crushed zone, intrusion of quartz veins and development of top-to-south shear sense, **(f)** outcrop image shows the garnetiferous-mica-schist near (IIT Mandi) Kataula village

slight rotates and the dip amount also changes from moderate to gently. A similar cross-over into the biotite zone in the core of the Pandoh syncline was also observed by Leger et al. (2013) and constrained the temperature condition to about ~480 °C.

Location 9: After moving further towards Segli village, the first exposure of highly fractured quartzite, conglomeratic stone, and black shale is observed (Fig. 6a, b). The strike of the bedding plain is almost NNW-SSE and the dip amount is gentle dipping, i.e. 20°. At this location, we mark a small outlier of the Manjir Formation, may be equivalent of the Batal Formation, the Tethyan Sequence, which was thrust during the mega thrusting event of the Himalayan orogeny in the late-Oligo early Miocene period (Singh et al., 2012, 2022a, 2022b). It is considered that a minor pebbly layer of Batal Formation is overlying at top of Pandoh Syncline and known as Manjir Formation. The most interesting thing to observe is that we did not notice any thrusting evidence along the contact zone, whereas the underling schist rock represents a weak zone. Here we follow (Bhargava & Bassi, 1994) and (Bhargava, 2000) geological maps and suggest that this location might be part of the Manjir Formation (may be equivalent of the Batal Formation of the Tethyan Sequence), but we do not confirm it. The GPS location: N 31° 48′ 5.9″ E 77° 03′ 30.5″ and elevation is ~1792 m.

Location 10: After observing the overlying part of the Manjir Formation, we moved further northwards, where mica-schist with quartzite of 20–30 cm in width exposed near the Khokhan wild life sanctuary is seen. The strike of beds NE-SW and dip amount and direction is 10° and SE direction respectively. The open folds are observed and mica schist rocks are crushed in the hinge zone (Fig. 7a). The thin-section study reveals the mineral assemblage and mainly contains quartz, chlorite, muscovite, biotite, and feldspar minerals (Fig. 7b). The biotite mineral is stretched along the foliation plains. Additionally, 50 m towards the north, another compact and massive exposure of the psammitic-mica-schist rocks that indicate the intrusion of quartz veins and the veins are stretched along the foliation plains, also formed a train of shearing sense indicators in thrust-sense (Fig. 7c). We suggest that during the first phase of deformation, the quartz veins were intruded and the second phase indicates the stretching along the foliation plains and, later on, shearing towards the south took place. [GPS location: N 31° 50′ 00″ E 77° 06′ 34″ and elevation is ~1835 m].

Location 11: On the Mandi-Bajaura road, the exposer of mica schist is continuous for up to 3 km and we stop at a new location near Rahla village Nala. We observed the development of crenulations in the mica-schist rocks (Fig. 7d). It is considered that when an early planar fabric is overprinted by a later planar fabric, crenulations occur. It might be possible that during the metamorphism of the Kulu thrust sheet (i.e. before thrusting along Kulu thrust and formation of synclinal structure), the recrystallization of mica minerals formed the crenulations. The thin-section of the same sample reveals that pinch and swell formed mainly consists of the mineral associations: quartz, chlorite, and muscovite, which are seen in microscopic studies (Fig. 7e). One of the most noticeable things is that the dip direction of the bedding plain changed to a south-west direction and we entered the northern limb of the Pandoh syncline structure. The strike of beds SEE-NWW and dip amount and direction are 45° and SW direction respectively. The GPS reading of this location is N 31° 49′ 15″ E 77° 06′ 36″ and the elevation is ~1646 m.

Location 12: After moving further north towards Bajara town, between Rahla and Ropa villages, a sharp contact of granitic gneiss with the mica-schist of the Khokhan area is seen. Here, the strike of the bed slightly rotates toward the east direction and is almost NE–SW and the dip amount is 35° SW. The augen of the granitic gneiss are stretched, deformed and form an asymmetrical boudin structure (Fig. 7f). The physiography of the area suggests that the rock cliffs of the valley are

Fig. 6 **(a -b)** Outcrop exposure of conglomeratic rocks of Manjir Formation, highly fractured quartzite and carbonaceous rock (might be equivalent to Batal Formation of Tethys Himalayan Sequence) The Manjir Formationis overlying at top of Pandoh Synclinal Structure

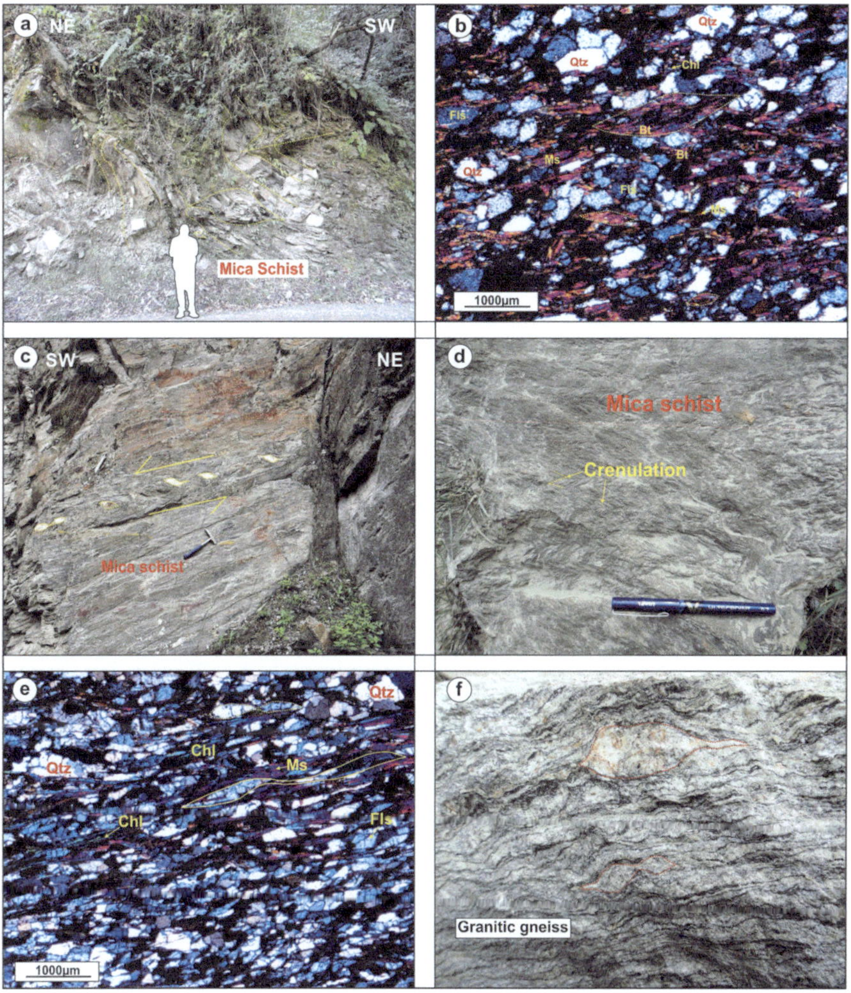

Fig. 7 **(a -b)** Show the outcrop image of mica-schist and thin section indicate the mineral assemblage quartz + chlorite + feldspar + biotite-mucovite and opque minerals, **(c)** intrusion of quartz veins along the foliation plain and development of shear-sense indicator as top-to-south, **(d -e)** another exposure of mica-schist and micro fold/crenulations are developed along the foliations and the thin section indicates the pinch and swells structures along with mineral assemblage quartz + chlorite + feldspar + biotite-mucovite and opque minerals on the Mandi-Bajaura road, near Rahla village Nala, **(f)** towards Bajara town, between Rahla and Ropa villages, a sharp contact of granitic gneiss with the mica-schist exposed. The augen of the granitic gneiss are stretched, deformed and form an asymmetrical boudin structure

very high and we are moving from top to bottom along the Mandi-Bajaura road. The granitic gneiss mainly contains the following mineral associations: quartz-feldspar-muscovite-chlorite-zircon and opaque minerals. The GPS location is N 31° 50′ 7.3″ and E 77° 07′ 58.7″ and the elevation is ~1391 m.

Location 13: Near the Ropa village, we move along the ropa Nala to observe the rock exposure and a beautiful exposure of garnet bearing mica schist is seen (Fig. 8a). The size of the garnet is approximately 3–5 mm and it is repeated in the northern limb of the Pandoh syncline. A similar type of garnet bearing mica-schist exposure was seen in the southern limb near Kataula village and along the Mandi-Aut section as well (Singh et al., 2021). On the other hand, we have observed that the quartz veins are intruded and folded (Fig. 8b, c, d). The intruded quartz veins of mica-schist rock show a normal sense of stretching and then folding, a normal-sense of shearing (Fig. 8c) and asymmetrical boudin structures. Figure 5d, e of this study reports a similar type of intruded quartz veins in mica-schist showing top-to-the-southward shearing were observed in the southern limb of the Pandoh syncline. The repetition of lithology and the same mineral assemblage indicates the Kuluthrust sheet was thrusted over the Lesser Himalayan meta-sedimentary rocks of the Simla-Jaunsar group and folded during the Himalayan orogenesis and formed the synclinal structure. The GPS location of this site is N 31° 50′ 12″ and E 77° 07′ 26″ and the elevation is ~1334 m.

Location 14: In the Kakrunala village and ~3 km before Bajaura town, we have seen a clear contact between the augen gneiss and quartzite of the Rampur group and this contact zone is known as the Panjal Thrust (Fig. 8e). The published literature from central Himachal Pradesh generally marks it as 'MCT' but we mark it as 'Panjal Thrust' or the northern boundary of the Pandoh syncline as North Panjal Thrust. The GPS location of this site is N 31° 50′ 20.1″ and E 77° 08′ 23.9″ and the elevation is ~1080 m. The thin-section prepared of augen-gneiss and revealed that the feldspar augen are rounded, stretched, and fractured. The fractured augen of feldspar clearly indicates normal-sense (see Fig. 9a, b, c) and the feldspar augen are rounded from the SW direction (Fig. 9b). The stretched feldspar augen shows the normal sense of shearing (see Fig. 9d). Further north of this location, the existence of a tectonic 'window' within the metamorphic zone extended further east–west up to the Kulu-Larji-Rampur area. The KLR tectonic window mainly consists of a thick sequence of chlorite, phyllite, and quartzite, which is considered as part of the Shali Structural belt of LHS and is tectonically overlain by the metamorphic rocks of the Kulu and Jutogh thrust sheets (Srikantia & Bhargava, 1998). It is divided into two tectonic zones such as: The lower zone is the Larji structural zone, also known as the Larji window zone of autochthonous nature. Similarly, the upper zone is the next higher structural level and is known as the Rampur window, which is para-autochthonous in nature (Singh et al., 2021).

Fig. 8 **(a)** The outcrop showing the garnet bearing mica schist in the northern limb of Pandoh syncline (near the Ropa village), **(b)** intruded quartz veins of mica-schist rock shows a stretching of quartz veins in normal sense of and, **(c -d)** folding of quatz veins and a normal-sense of shearing along the northern limb of Pandoh syncline, **(e)** contact zone of the Panjal Thrust between the augen gneiss of Pandoh synclinal structure and quartzite of the Rampur group, near Kakrunala village and ~3 km before Bajaura town

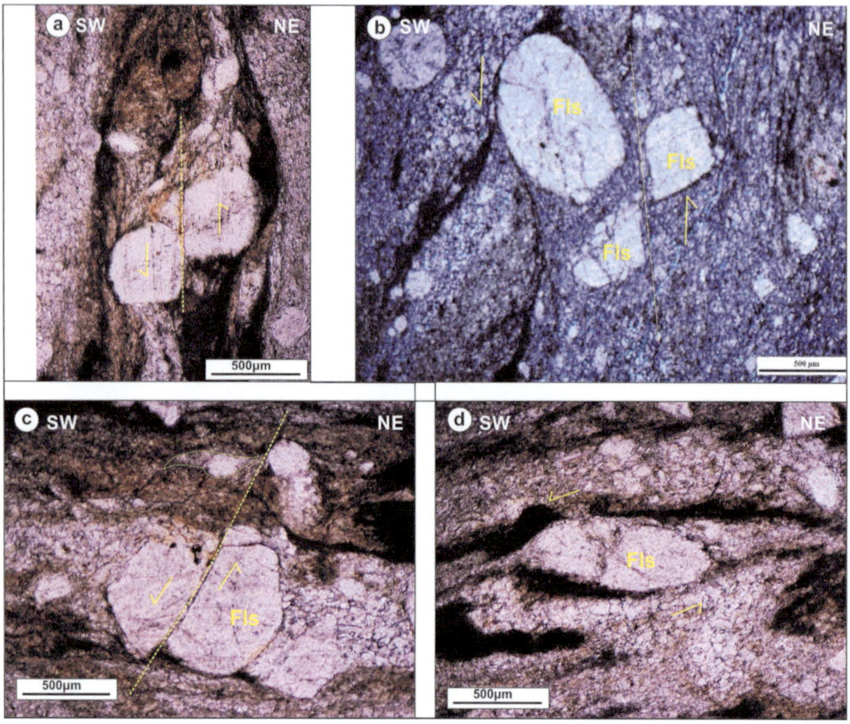

Fig. 9 (a–d) the thin-section of augen-gneiss and indicates the feldspar augen are rounded, stretched, and fractured and also indicate the normal-sense and rounded augen are rounded from the SW direction

7 Observations and Conclusion

Based on our field observations, we inferred that the portion exposed between the Mandi-Kataula-Bajaura sections belongs to the Pandoh syncline, and that the thrust sheet is equivalent to the Almora-Bajinath-Lansdown klippen of Kumaun Garhwal region of NW Himalaya. Based on the whole rock Rb–Sr data, It is suggested that three crystalline thrust sheets are exiting and the ascending tectonic order are Kulu, Jutogh and Vaikrita thrust sheets (Fig. 2) (Bhargava & Bassi, 1994). Our structural and metamorphic studies revealed that the Pandoh syncline was the southern extension of the Kulu Thrust sheet lying in the north of the tectonic window (i.e. KLR window) and also suggest that the syncline structure has undergone two different phases of metamorphism and deformation during the Himalayan orogen. We conclude that:

- The structural analyses of the Pandoh syncline (Mandi-Kataula-Bajaura section) suggest two stages of deformation: (i) the first phase was the pre-syn orogeny. During this phase, the quartz veins intruded and folded into open to close folds. (ii) The second phase suggests post-orogenic deformation, as seen in deformed or folded quartz veins and the development of recumbent/overturned folds. During this phase the shear-sense markerwas developed as a thrust sense (top-to-south) and a normal sense (top-to-north) of shearing along the Punjal Thrust.

- The metamorphic study reflects that the core part of Pandoh syncline reached up to biotite grade of metamorphism and both limbs indicate higher grade i.e. garnet-grade metamorphism. Along the Mandi-Kautaula-Bajaura section, the size of garnet grain varies from 0.5 mm to ~5 mm.
- Additionally, based on the study carried out by (Bhargava & Bassi, 1994 and Webb et al., 2011), we correlate that the Pandoh syncline is the southern extension of the Kulu Thrust sheet and both zones indicate the same lithology, deformation and metamorphism phases. However, this study is purely based on the field observation of rocks and tectonic zones.

Acknowledgements The authors are thankful to the Director, WIHG Dehradun for allowing us to carry out the field work under Institute Research Project Activity 1A and providing all the facilities and moral support (Wadia contribution no. WIHG/0189). P. Singh is thankful to Professor O.N. Bhargava (Panjab University) for his continuous encouragement, support and discussion on the Chail, Salkhala, Kulu and Jutogh thrust sheets. He is also thankful to Soumyajit Mukherjee (IIT Bombay) for accepting our proposal for field excursion-based work in the field guidebook. Mukherjee (2023) summarized this chapter.

References

Banerjee S, Bose N., Mukherjee, S. (2019). Field structural geological studies around Kurseong, Darjeeling-Sikkim Himalaya, India. In: Mukherjee S. (Ed) Tectonics and Structural Geology: Indian Context. Springer. pp. 421–440.

Baud, A. Y. M. O. N., Gaetani, M., Garzanti, E., Fois, E., Nicora, W., & Tintori, A. (1984). Geological observation in southeastern Zanskar and adjacent Lahul area (northern Himalaya). *Eclogae Geologicae Helvetiae, 77*(1), 177–197.

Bhargava, O. N. (1982). Tectonic windows of the lesser Himalaya. *Himalayan Geology, 10*, 135–155.

Bhargava, O. N. (1995a). The Bhutan Himalaya: A geological account. *Special Publication Geological Survey of India, 39*, 1–245.

Bhargava, O. N. (1995b). Stratigraphy. In Bhargava, O.N. (Ed.) The Bhutan Himalaya: A geological account. Geological Survey of India, Special Publication, 39, 173–181.

Bhargava, O. N. (2000). The Precambrian sequences in the Western Himalaya. *Special Publication Geological Survey of India, 55*, 69–84.

Bhargava, O. N., & Bassi, U. K. (1994). The crystalline thrust sheets in the Himachal. Himalaya and the age of amphibolite facies metamorphism. *Journal of the Geological Society of India, 43*, 343–352.

Bhargava, O. N., Bassi, U. K., & Sharma, R. K. (1991). The crystalline thrust sheets, age of metamorphism, evolution and mineralisation of the Himachal Himalaya. *Indian Minerals, 45*(1&2), 1–18.

Bhargava, O. N., Singh, B. P., Frank, W., & Tangri, S. K. (2021). Evolution of the lesser Himalaya in space and time. *Himalayan Geology, 42*(2), 263–289.

Bhargava, O. N., & Srikantia, S. V. (2014). Geology and age of metamorphism of the Jutogh and Vaikrita Thrust sheets. *Himachal Himalaya. Himalayan Geology, 35*(1), 1–15.

Bhargava, O. N., Thoni, M., & Miller, C. (2016). Isotopic evidence of Early Palaeozoic metamorphism in the lesser Himalaya (Jutogh Group), Himachal Pradesh, India: Its implication. *Himalayan Geology, 37*(2), 73–84.

Bose N, Mukherjee, S. (2019a). Field documentation and genesis of the back-structures from the Garhwal lesser Himalaya, Uttarakhand, India. In R. Sharma, I. M. Villa, S. Kumar (Eds) Crustal architecture and evolution of the Himalaya-Karakoram-Tibet Orogen. Geological Society of London Special Publications 481, 111–125.

Bose, N., & Mukherjee, S. (2019b). Field documentation and genesis of back-structures in ductile and brittle regimes from the foreland part of a collisional orogen: Examples from the Darjeeling-Sikkim lesser Himalaya, India. *International Journal of Earth Sciences, 108*, 1333–1350.

Bose, N., & Mukherjee, S. (2020). Estimation of deformation temperatures, flow stresses and strain rates from an intra-continental shear zone: The main boundary thrust, NW Himalaya (Uttarakhand, India). *Marine and Petroleum Geology, 112*, 104094.

Brookfield, M. E. (1993). The Himalayan passive margin from Precambrian to Cretaceous times. *Sedimentary Geology, 84*(1–4), 1–35.

Burchfiel, B. C., Zhiliang, C., Hodges, K. V., Yuping, L., Royden, L. H., Changrong, D. (1992). The South Tibetan detachment system, Himalayan orogen: Extension contemporaneous with and parallel to shortening in a collisional mountain belt (Vol. 269). Geological Society of America.

Coleman, M. E. (1998). U-Pb constraints on Oligocene-Miocene deformation and anatexis within the central Himalaya, Marsyandi Valley, Nepal. *American Journal of Science, 298*(7), 553–571.

Critelli, S., & Garzanti, E. (1994). Provenance of the lower Tertiary Murree redbeds (Hazara-Kashmir Syntaxis, Pakistan) and initial rising of the Himalayas. *Sedimentary Geology, 89*(3–4), 265–284.

DeCelles, P. G., Gehrels, G. E., Quade, J., LaReau, B., & Spurlin, M. (2000). Tectonic implications of U-Pb zircon ages of the Himalayan orogenic belt in Nepal. *Science, 288*(5465), 497–499.

DeCelles, P. G., Robinson, D. M., Quade, J., Ojha, T. P., Garzione, C. N., Copeland, P., & Upreti, B. N. (2001). Stratigraphy, structure, and tectonic evolution of the Himalayan fold-thrust belt in western Nepal. *Tectonics, 20*, 487–509.

Dezes, P. J., Vannay, J. C., Steck, A., Bussy, F., & Cosca, M. (1999). Synorogenic extension: Quantitative constraints on the age and displacement of the Zanskar shear zone (northwest Himalaya). *Geological Society of America Bulletin, 111*(3), 364–374.

Dhiman, R., & Singh, S. (2021) Neoproterozoic and Cambro-Ordovician magmatism: episodic growth and reworking of continental crust Himachal Himalaya India. *International Geology Review 63*(4) 422-436 2 https://doi.org/10.1080/00206814.2020.1716399

Gaetani, M., & Garzanti, E. (1991). Multicyclic history of the Northern India continental margin (Northwestern Himalaya) (1). *AAPG Bulletin, 75*(9), 1427–1446.

Gansser, A. (1964). Geology of the Himalayas.

Garzanti, E. (1993). Himalayan ironstones, "superplumes", and the breakup of Gondwana. *Geology, 21*(2), 105–108.

Garzanti, E. (1999). Stratigraphy and sedimentary history of the Nepal Tethys Himalaya passive margin. *Journal of Asian Earth Sciences, 17*(5–6), 805–827.

Harton, F., Lee, J., Hacker, B., Bowman-Kamaha'o, M., & Cosca, M. (2014). Himalayan gneiss dome formation in the middle crust and exhumation by normal faulting: New geochronology of Gianbul dome, northwestern India. Geological Society of America Bulletain. https://doi.org/10.1130/B31005.1

Jain, A. K. (2014). When did India–Asia collide and make the Himalaya? Current Science, 254–266.

Jain, A. K., & Anand, A. (1988). Deformational and strain patterns of an intracontinental collision ductile shear zone—an example from the Higher Garhwal Himalaya. *Journal of Structural Geology, 10*(7), 717–734.

Jain, A.K., Shreshtha, M., Seth, P., Kanyal, L., Carosi, R., Montomoli, C., Iaccarino, S., and Mukherjee, P.K. (2014). The Higher Himalayan Crystallines, Alaknanda – Dhauli Ganga Valleys, Garhwal Himalaya, India. In: (Eds.) Chiara Montomoli, Rodolfo Carosi, Rick Law, Sandeep

Singh, and Santa Man Rai, Geological field trips in the Himalaya, Karakoram and Tibet, *Journal of the Virtual Explorer, Electronic Edition,* ISSN 1441-8142, volume 47, paper 8.

Jain, A. K., Lal, N., Sulemani, B., Awasthi, A. K., Singh, S., Kumar, R., & Kumar, D. (2009). Detrital-zircon fission-track ages from the Lower Cenozoic sediments, NW Himalayan foreland basin: Clues for exhumation and denudation of the Himalaya during the India-Asia collision. *Geological Society of America Bulletin, 121*(3-4), 519-535.

Le Fort, P. (1986). Metamorphism and magmatism during the Himalayan collision. *Geological Society, London, Special Publications, 19*(1), 159–172.

Leger, R. M., Webb, A. A. G., Henry, D. J., Craig, J. A., & Dubey, P. (2013). Metamorphic field gradients across the Himachal Himalaya, northwest India: Implications for the emplacement of the Himalayan crystalline core. *Tectonics, 32*, 540–557. https://doi.org/10.1002/tect.20020

Mahato, S., Mukherjee, S., & Bose, N. (2019). Documentation of brittle structures (back shear and arc-parallel shear) from Sategal and Dhanaulti regions of the Garhwal lesser Himalaya (Uttarakhand, India). In S. Mukherjee (Ed.), *Tectonics and structural geology: Indian context* (pp. 411–424). Springer International Publishing.

Mandal, S., Robinson, D. M., Khanal, S., & Das, O. (2014). Redefining the tectono-stratigraphic and structural architecture of the Almoraklippe and the Ramgarh-Munsiari thrust sheet in NW India. *Geological Society, London, Special Publications, 412*(1), 247–269.

Mandal, S., Robinson, D. M., Kohn, M. J., Khanal, S., & Das, O. (2019). Examining the tectono-stratigraphic architecture, structural geometry, and kinematic evolution of the Himalayan fold-thrust belt, Kumaun, northwest India. *Lithosphere, 11*(4), 414–435.

Mukherjee, S. (2013). Channel flow extrusion model to constrain dynamic viscosity and Prandtl number of the higher Himalayan Shear Zone. *International Journal of Earth Sciences, 102*, 1811–1835.

Mukherjee, S. (2015a). Atlas of structural geology Vol. 1. Elsevier.

Mukherjee S. (2015b). A review on out-of-sequence deformation in the Himalaya. In S. Mukherjee, R. Carosi, P. van der Beek, B. K. Mukherjee, D. Robinson (Eds.) Tectonics of the Himalaya. Geological Society, London. Special Publications 412, 67–109.

Mukherjee, S. (2020). Atlas of structural geology, Vol. 2. Elsevier Publication, ISBN: 978-0-12-816802-8.

Mukherjee, S. (2023). Introduction to structural geology and tectonics field guidebook—volume 2. In S. Mukherjee (Ed.) *Structural geology and tectonics field guidebook—volume 2* (pp xi–xiv). Springer Nature Switzerland AG. Cham. ISBN 978-3-031-19575-4.

Mukherjee, S., & Koyi, H. A. (2010). Higher Himalayan shear zone, Sutlej section-structural geology & extrusion mechanism by various combinations of simple shear, pure shear & channel flow in shifting modes. *International Journal of Earth Sciences, 99*, 1267–1303.

Miller, C., Thöni, M., Frank, W., Grasemann, B., Klötzli, U., Guntli, P., & Draganits, E. (2001). The early Palaeozoic magmatic event in the Northwest Himalaya, India: source, tectonic setting, and age of emplacement. *Geological Magazine, 138*(3), 237-251.

Parrish, R. R., & Hodges, V. (1996). Isotopic constraints on the age and provenance of the lesser and greater Himalayan sequences, Nepalese Himalaya. *Geological Society of America Bulletin, 108*(7), 904–911.

Passchier, C.W., & Trouw, R. A. (2005). Microtectonics. New York.

Patel, R. C., Adlakha, V., Lal, N., Singh, P., & Kumar, Y. (2011a). Spatiotemporal variation in exhumation of the Crystallines in the NW-Himalaya, India: Constraints from fission track dating analysis. *Tectonophysics, 504*(1–4), 1–13.

Patel, R. C., Adlakha, V., Singh, P., Kumar, Y., & Lal, N. (2011b). Geology, structural and exhumation history of the higher Himalayan Crystallines in Kumaon Himalaya, India. *Journal of the Geological Society of India, 77*(1), 47–72.

Patel, R. C., Singh, S., Asokan, A., Manickavasagam, R. M., & Jain, A. K. (1993). Extensional tectonics in the Himalayan orogen, Zanskar, NW India. *Geological Society, London, Special Publications, 74*(1), 445–459.

Patel, R. C., Singh, P., & Lal, N. (2012). Plio-Quaternary exhumation history of the Garhwal-Kumaon Himalayas, NW-India: An analysis on low temperature thermochronological data. *Extended Abstracts Journal of Nepal Geological Society, 45*, 136–138.

Patel, R. C., Singh, P., & Lal, N. (2015). Thrusting and back-thrusting as post-emplacement kinematics of the Almora klippe: Insights from low-temperature thermochronology. *Tectonophysics, 653*, 41–51.

Purkayastha, S., Verma, P. K., Jain, A. K., Manickavasagam, R. M., & Ghosh, T. K. (1999). Growth and rotation of porphyroblasts and development of shear/extension foliation in Baggi-Mandi-Pandoh region, Western Himalaya-implication on Himalayan tectonics. *Jain and Manickavasagam Edition, Geodynamics of the NW-Himalaya, Gondwana Research Group Memoir, 6*, 125–133.

Robinson, D. M., DeCelles, P. G., & Copeland, P. (2006). Tectonic evolution of the Himalayan thrust belt in western Nepal: Implications for channel flow models. *Geological Society of America Bulletin, 118*(7–8), 865–885.

Robinson, D. M., & Martin, A. J. (2014). Reconstructing the Greater Indian margin: A balanced cross section in central Nepal focusing on the lesser Himalayan duplex. *Tectonics, 33*(11), 2143–2168.

Robinson, D. M., & McQuarrie, N. (2012). Pulsed deformation and variable slip rates within the central Himalayan thrust belt. *Lithosphere, 4*(5), 449–464.

Robyr, M., Hacker, B. R., & Mattinson, J. M. (2006). Doming in compressional orogenic settings: New geochronological constraints from the NW Himalaya. Tectonics, 25(2).

Robyr, M., Vannay, J.-C., Epard, J.-L., & Steck, A. (2002). Thrusting, extension and doming during the polyphase tectonometamorphic evolution of the high Himalayan crystalline zone in NW India. *Journal of Asian Earth Science, 21*, 221–239.

Singh, P. (2014). *Exhumation history of the Higher and lesser Himalayan crystallines in the Kumaon-Garhwal region, NW-India: As revealed from fission track thermochronology* [Unpublished Ph.D. thesis]. Department of Geophysics, Kurukshetra University Kurukshetra. https://doi.org/10.13140/RG.2.2.25869.87524

Singh, P., Ao, A., Thakur, S. S., Rana, S., Sharma, R., Singh, A. K., Singhal, S. (2021). Geology, structural, metamorphic and mineralization studies along the Mandi-Kullu-Manali-Rohtang section of Himachal Pradesh, NW-India. In S. Mukherjee (Eds.) Structural geology and tectonics field guidebook—volume 1. Springer Geology. Springer, Cham. https://doi.org/10.1007/978-3-030-60143-0_15

Singh, P., Bhakuni, S. S., Singhal, S. (2017). Tectonic implications of U-Pb (zircon) Geochronology of Chor Granitoids of the lesser Himalaya, Himachal Pradesh, NW Himalaya. AGU meeting, 11–14 Dec, 2017 held at New Orleans, USA.

Singh, P., & Patel, R. C. (2017). Post-emplacement kinematics and exhumation history of the Almora klippe of the Kumaun-Garhwal Himalaya, NW India: Revealed by fission track thermochronology. *International Journal of Earth Sciences, 106*(6), 2189–2202.

Singh, P., & Patel, R. C. (2022). Miocene development of the main boundary thrust (MBT) and Ramgarh Thrust (RT), and their influence on the exhumation of lesser Himalayan rocks of the Kumaun-Garhwal region, NW-Himalaya: Insights from fission track thermochronology. *Journal of Asian Earth Sciences, 224*, 104987. https://doi.org/10.1016/j.jseaes.2021.104987

Singh, P., Patel, R. C., & Lal, N. (2012). Plio-Plistocene in-sequence thrust propagation along the main central thrust zone (Kumaon–Garhwal Himalaya, India): New thermo-chronological data. *Tectonophysics, 574–575*, 193–203.

Singh, P., Singhal, S., & Das, A. N. (2020). U-Pb (zircon) geochronologic constraint on tectonomagmatic evolution of Chaurgranitoid complex (CGC) of Himachal Himalaya, NW India: Implications for the Neoproterozoic magmatism related to Grenvillian orogeny and assembly of the Rodinia supercontinent. *International Journal of Earth Sciences, 109*, 373–390. https://doi.org/10.1007/s00531-019-01808-5

Singh, P., Patel, R. C., & Chaudhary, S.K. (2022a) Apatite Fission Track Thermochronology: fundamentals, technique and review study from the Garhwal-Kumaun region, NW-Himalaya. *Himalayan Geology, 43* (1 A), 96-110.

Singh, P., Sethy, P. C., Singh, A. K., Singhal, S., Maurya, A. K., & Giri, S. R. (2022b). Geochemistry and U–Pb zircon geochronology of the Jutogh Thrust sheet, Himachal Pradesh, NW-Himalaya: Implications to the petrogenesis and regional tectonic setting. *Geological Journal,* 1–19. https://doi.org/10.1002/gj.4583

Singh, S., Barley, M. E., Brown, S. J., Jain, A. K., & Manickavasagam, R. M. (2002). SHRIMP U–Pb in zircon geochronology of the Chor granitoid: evidence for Neoproterozoic magmatism in the Lesser Himalayan granite belt of NW India. *Precambrian Research, 118*(3-4), 285-292.

Srikantia, S. V., & Bhargava, O. N. (1976). Geology of the Shali Belt and adjoining areas. *Memoir Geological Survey of India, 106*(1), 31–199.

Srikantia, S. V., & Bhargava, O. N. (1984). The Jutogh klippe of Shimla area of the Himachal Himalaya: Its geology and structural evolution. *Journal of Geological Society of India, 25*(4), 218–230.

Srikantia, S. V., & Bhargava, O. N. (1988). The Jutogh Group of meta-sediments of the Himachal Himalaya: Its litostraitigraphy. *Journal of Geological Society of India, 32*, 279–294.

Srikantia, S. V., & Bhargava, O. N. (1998). Geology of Himachal Pradesh. GSI Publications, 2(1), 282.

Srivastava, P., & Mitra, G. (1994). Thrust geometries and deep structure of the outer and lesser Himalaya, Kumaon and Garhwal (India): Implications for evolution of the Himalayan fold-and-thrust belt. *Tectonics, 13*(1), 89–109.

Steck, A., Spring, L., Vannay, J. C., Masson, H., Bucher, H., Stutz, E., Marchant, R., & Tieche, J. C. (1993). The tectonic evolution of the Northwestern Himalayan in eastern Ladakh and Lahul, India. In: P. J. Treloar, & M. P. Searle (Eds.) Himalayan tectonics. Geological Society Special Publication, no. 74, pp. 265–276.

Thakur, S. S. (2014). Retrograde corona texture in pre-Himalayan metamorphic mafic xenoliths, Sutlej valley, NW Himalaya: Implication on rare occurrence of high-grade rocks in the Himalaya. *Journal of Asian Earth Sciences, 88*, 41–49.

Thakur, V. C. (1987). Development of major structures across the northwestern Himalaya, India. *Tectonophysics, 135*, 1–13.

Thakur, V. C. (1992). Geology of western Himalaya. *Physics and Chemistry of the Earth, 19*, 1–355.

Thakur, V. C., & Choudhury, B. K. (1983). Deformation, metamorphism and tectonic relations of central crystallines and main central thrust in Eastern Kumaun Himalaya. In P. S. Saklani (Ed.), *Himalayan sears* (pp. 45–57). Himalayan Books.

Valdiya, K. S. (1980). *Geology of Kumaun lesser Himalaya* (p. 281). Publisher Wadia Institute of Himalayan Geology.

Webb, A. A. G. (2013). Preliminary balanced palinspastic reconstruction of Cenozoic deformation across the Himachal Himalaya (northwestern India). *Geosphere, 9*(3), 572–587.

Webb, A. A. G., Yin, A., Harrison, T. M., Célérier, J., Gehrels, G. E., Manning, C. E., & Grove, M. (2011). Cenozoic tectonic history of the Himachal Himalaya (northwestern India) and its constraints on the formation mechanism of the Himalayan orogen. *Geosphere, 7*(4), 1013–1061.

Yin, A. (2006). Cenozoic tectonic evolution of the Himalayan orogen as constrained by along strike variation of structural geometry, exhumation history, and foreland sedimentation. *Earth Science Review, 76*, 1–131.

The Rock Outcrops at Raghunathdi, SE of Ghatsila (Jharkhand, India): a Spectacular Preservation of Polyphase Folding

Srinanda Ganguly, Arpita Samanta, and Abhik Kundu

Abstract The present chapter describes and illustrates an interesting outcrop of rock from a location near Ghatsila, Jharkhand. This outcrop exposes spectacular preservation of signatures of multiple phases of folding. It is an ideal site for studying superposed fold interference patterns and relations between axes and axial surfaces of different generations of folds. Teachers of structural geology from eastern India, along with their students, are frequent visitors of this outcrop; they consider this outcrop as a museum of superposed fold interference patterns. This place is an easy reach as the Ghatsila town is well connected with major cities of India by both railway and road. We expect that this chapter will encourage teachers from all over to bring their students here in order to teach how to study and map a multiple folded terrain.

Keywords Superposed fold · Pucker axis lineation · Recumbent fold · Hook interference pattern · Raghunathdi

1 Introduction

The well-known rock outcrop of Paleo-Proterozoic Chaibasa Formation (Table 1) in the Raghunathdi (alias Sushnikalmi) area, 4.4 km to the southeast of Ghatsila town in the East Singbhum District, Jharkhand preserves very complex fold geometry formed due to multiple phases of folding. The outcrop (22°33′40.65″N, 86°30′30.69″E, Fig. 1) is overall trends NE–SW, with ~255 m^2 area. This outcrop, popularly known as the *'Tentuldanga outcrop'*, is extensively studied by structural geologists especially from eastern India (Ghosh & Sengupta, 1990, Sengupta & Ghosh, 1997). This is one of the most favourite outcrop for teachers of structural geology to train their students the styles and manifestations of superposed folding. Therefore, this outcrop has been visited by a very large number of geologists for almost last four decades (or even more). However, except in one field guide on Ghatsila and neighbouring area published by the Department of Geological Sciences, Jadavpur University, India

S. Ganguly · A. Samanta · A. Kundu (✉)
Department of Geology, Asutosh College, 92 S. P. Mukherjee Road, Kolkata, West Bengal 700026, India
e-mail: kundu.abhik@gmail.com

S. Mukherjee (ed.), *Structural Geology and Tectonics Field Guidebook–Volume 2*, Springer Geology, https://doi.org/10.1007/978-3-031-19576-1_5

(Ghosh et al., 1985) and in an atlas of structural geology (Mukherjee, 2015), this exposure has not been awarded with any mention in any geology texts so far. Explanations on the origin have been dealt with in a few research articles proposing contrasting ideas (Ghosh & Sengupta, 1990; Paul, 2014; Sengupta & Ghosh, 1997). This article presents a detailed description of the structures so that it can act as a field guide for the students of structural geology who will visit the outcrop for the first time, and can be an important testing ground of structural geology especially for the Indian students (Mukherjee, 2019a, 2019b). We intentionally submit the work to an international; publisher (Springer) for the widest publicity. Students can look at recent atlas (e.g., Mukherjee, 2013, 2014, 2021; Mukherjee et al., 2020) to get eyes trained on geological structures before visiting the field outcrop. In terms of structural diversity in the field, the place can be compared with Ambaji Gujarat (Mukherjee et al., 2020).

Singhbhum is a crucial terrain in Indian geology. Several geologists studied lithological diversity, mineral deposits, and deformation signatures in order to understand the terrane evolution (Acharyya et al., 2010; Chakraborty et al., 2015; Dunn, 1937,

Table 1 Stratigraphy of the Singhbhum Fold belt on the north of Singhbhum Shear zone and the Singbhum Shear zone (after Chakraborty et al., 2015)

Chandil Formation	↑	Rhyolite, tuff, agglomerates, shale, C-shale (Low greenschist (phyllite) metaorphites)
Dalma Formation		Basic lava and volcano-clastic sediments of continental affinity (Greenschist facies metamorphites)
Dhalbhum Formation		Sandstone, siltstone, orthoquartzite, shale, tuff of continental affinity (Greenschist facies metamorphites)
Singbhum Shear Zone melange		Chlorite quartz schist, talc/actinolite schist, magnetite–apatite rock, Cu–U ores, tourmalinite, kyanite-bearing quartzite, chloritoid schist, etc. (Greenschist to amphibolites facies metamorphites)
Soda granite (only in Singbhum Shear Zone)		Granitoids (Hydrothermally altered greenschist to amphibolites facies metamorphites)
Chaibasa Formation		Shale, sandstone, and orthoquartzite with minor basic to ultrabasic rocks of marine affinity (Greenschist to amphibolites facies metamorphites)
Dhanjhori Formation		Mainly basic and dacitic lavas, minor quartzite, shale, and basal conglomerate of continental affinity
Archaean basement		Unconformity—metasedimentary rocks, tonalite, trondhjemite and granitoid intrusions

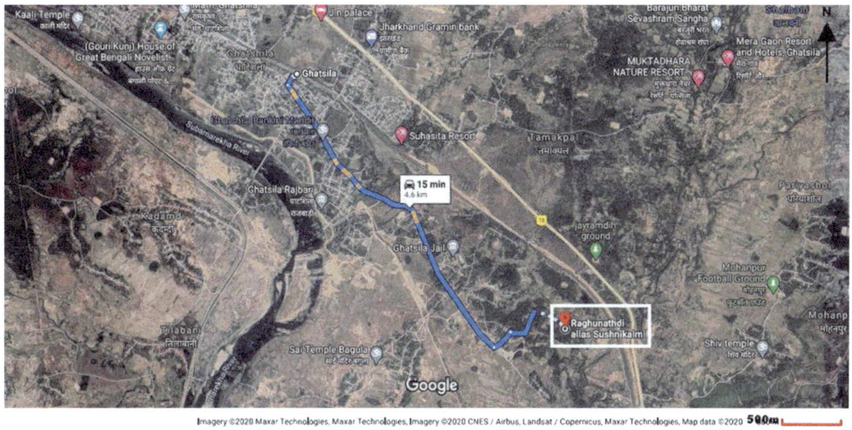

Fig. 1 Location, on Google Map, of the outcrop studied (the deep red balloon inside the white rectangle). The blue line shows the way to the spot from Ghatsila railway station

1940; Mazumder, 2005; Mazumder et al., 2012; Mukhopadhyay, 2001; Mukhopadhyay et al., 2008; Nelson et al., 2014; Roy et al., 2004; Saha, 1994; Sarkar & Saha, 1977; Sengupta and Mukhopadhyay, 2000).

The outcrop dealt in this article belongs to Chaibasa Formation of Singbhum Group and consists of mainly of mica and garnet bearing schists and micaceous quartzite intercalations in mm to dm scale (Mazumder & Alterman, 2016). A detailed description of the structures and fold interference patterns preserved in this outcrop could be a ready reference for students studying superposition of folds; the present chapter aims to provide the same. The deformation structures are preserved in an alternate sequence of pelitic meta-sediments (kyanite-muscovite schist, Paul, 2014; identifiable in the field as mica schist) and quartzite (Figs. 2, 3).

On the first look on the vertical sections two distinct series of antiform-synforms can be revealed in the mica schist—quartzite sequence of Chaibasa Formation at near Raghunathdih (22°33′40.65″N, 86°30′30.69″E, Fig. 1). In both of these folds the rock layers and the pervasive bedding-parallel schistosity plane are folded against steeply dipping distinct zonal crenulation cleavages (Figs. 4, 5 and 6). The thickness of the folded layers also varies from <1 cm to ~110 cm (Figs. 4, 5). The folds in both the sets vary widely in hinge geometry—chevron, rounded and box (Figs. 5, 7, 8). One of the axial planar cleavages (crenulation cleavages) strikes NW–SE. The wavelength and amplitude of crenulations/folds in the quartzite—schist sequence vary widely and define disharmonic fold (Fig. 4). Mullion and pucker axis lineation parallel to the fold hinges are discernable on the schistosity surface (Fig. 9a, b). The wavelength varies from <a cm up to ~800 cm and the amplitude varies from less than a cm up to ~120 cm. The other set of folds also vary in dimensions, however, the variation is not as wide of the other set. The crenulation cleavage corresponding to this set of folds strikes almost perpendicular to the other steeply dipping crenulations cleavage. This fold set is also manifested by a prominent set of pucker axis

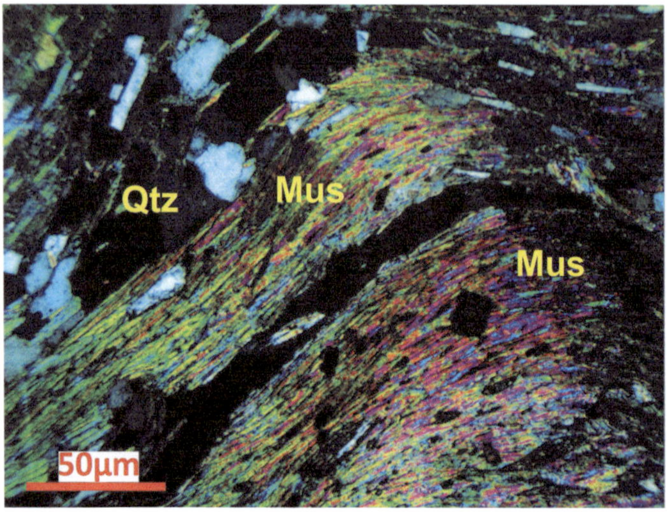

Fig. 2 Photomicrograph of muscovite schist under crossed polarised light. Qtz = quartz, Mus = muscovite (Thin section was prepared from a loose sample as we avoided use of hammer on the outcrop)

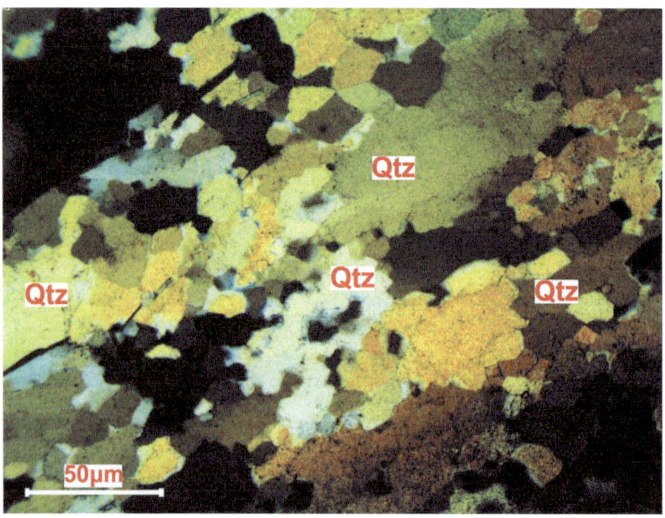

Fig. 3 Photomicrograph of quartzite. Qtz = quartz (Thin section was prepared from a loose sample)

lineation at high angle/~perpendicular to the aforementioned set of lineation (Fig. 9a, b). These two sets of antiform-synform sequences superimpose to form dome-basin type-1 interference pattern (Fig. 10) (Ramsay, 1967).

Fig. 4 Series of disharmonic antiform-synforms. Both rock layers and layer parallel schistosity are folded against subvertical axial plane/crenulations cleavage. Note variation in wavelength and amplitude of folds from layer to layer. Length of the marker is 16 cm

Fig. 5 Series of antiforms-synforms against subvertical axial plane cleavage. Note variation in shapes (broad, rounded, box shaped) of hinge zones and fold dimenstions. Length of the marker is 16 cm

Overall, the rock layers are tightly folded giving rise to isoclinal recumbent folds with varying hinge geometry (Figs. 11, 12), as seen in profile sections at high-angle to the mullion structures (Fig. 13). The pervasive schistosity is axial planar to these recumbent folds and is parallel to the limbs. These Recumbent folds are refolded against the NW–SE striking steep to subvertical crenulations cleavage, axial

Fig. 6 Crenulated pervasive schistosity and zonal crenulation cleavage (Steep to subvertical). Length of the marker is 16 cm

Fig. 7 Outcrop preserving box folds. Length of the marker is 16 cm

Fig. 8 Sharp hinged folds in folded schistosity. Length of the marker is 16 cm

(a) (b)

Fig. 9 **a** Mullion linetation (red broken line) and perpendicular pucker axis lineation (yellow broken line); **b** two sets of pucker axis lineations (broken lines of black and red colours) at high angle to each other. Length of the marker in both photographs is 16 cm

Fig. 10 Dome-basin formed by superimposition of two folds. Note presence of folded pucker axis, one perpendicular to the other. Length of the marker is 16 cm

planar to the series of disharmonic antiforms-synforms, giving rise to a type-3 hook interference patterns (Fig. 14; Ramsay, 1967) discernable on ~ ENE–WSW trending vertical sections oblique/~perpendicular to the axes of both the folds. The limbs of these refolded recumbent folds show antiform-synform trains.

Another set of hook pattern is seen on vertical section where the recumbent fold are refolded against subhorizontal younger axial planes (Fig. 15).

Close observation of the outcrop reveals cm scale folds in the <1 cm thick rock layers, which are refolded by multiple phases of folding (Figs. 16, 17). Type 1, 2 and 3 interference patterns can be identified in these smaller-scale folds.

We avoid discussing the sequence of folding events or presenting a structural map here. Earlier authors have discussed this sequence of folding (Ghosh & Sengupta, 1990; Paul, 2014). However, a careful and systemic observation of the folds and interference patterns may attract interests in any student of structural geology to prepare a structural map that should unfold the deformation history of this area. Lastly we request the readers not to hammer on the outcrop and we wish that appropriate measures will be taken to preserve and protect this spectacular outcrop that would continue fascinating the future generation students for years.

Fig. 11 Isoclinal recumbent folds (shown by sketch within the rectangular box) with sharp hinge in quartzite layers. Length of the marker is 16 cm

Fig. 12 Recumbent fold, in quartzite layers, with box shaped hinge. Length of the marker is 16 cm

Fig. 13 Subhorizontal mullions. Red line marks a mullion axis. Length of the marker is 15 cm

Fig. 14 Hook interference pattern due to superposition of antiform-synforms with subvertical axial planes on recumbent fold. Length of the marker is 16 cm

Fig. 15 Hook innterference pattern. Both older and younger axial planes are subhorizontal. Length of the marker is 16 cm

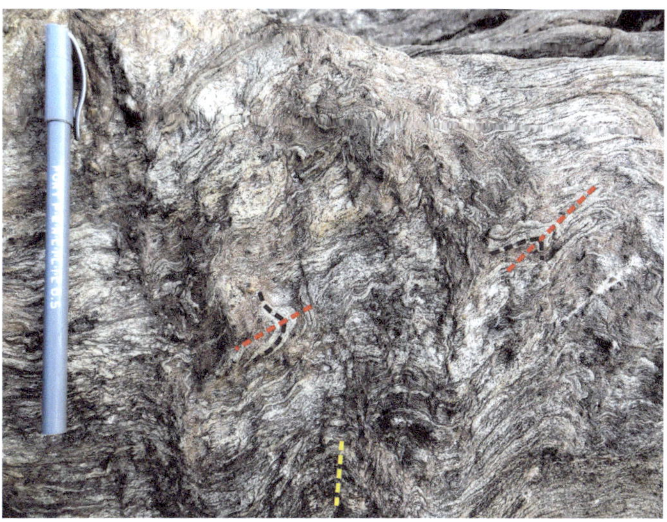

Fig. 16 A portion of the exposure showing complex fold geometry due superposition of multiple generation of folds (axial planes are marked by broken lines). Length of the marker is 15 cm

Fig. 17 A portion of the exposure showing complex fold geometry due superposition of multiple generation of folds. Length of the marker is 16 cm

Acknowledgements The authors are thankful to Dr. Biswajit Ghosh (University of Calcutta) for a critical internal review of the manuscript. Soumyajit Mukherjee (IIT Bombay) invited, edited and reviewed this article. The Springer team (Alexis Vizcaino and proof readers) are thanked for assistance. Mukherjee (2023) summarized this chapter.

References

Acharyya, S. K., Gupta, A., & Orihashi, Y. (2010). New U-Pb zircon ages from Paleo-Mesoarchean TTG gneisses of the Singhbhum Craton, eastern India. *Geochemical Journal, 44*(2), 81–88.

Chakraborty, M., Sengupta, N., Biswas, S., & Sengupta, P. (2015). Phosphate minerals as a recorder of PT-fluid regimes of metamorphic belts: Example from the Palaeoproterozoic Singhbhum Shear Zone of the East Indian shield. *International Geology Review, 57*(11–12), 1619–1632.

Dunn, J. A. (1937). The mineral deposits of eastern Singhbhum and surrounding areas. *Memoirs Geological Survey of India, 69*, 211–213.

Dunn, J. A. (1940). The stratigraphy of south Singhbhum. Office of the Geological Survey of India.

Ghosh, S. K., & Sengupta, S. (1990). Singhbhum Shear Zone: Structural transition and a kinematic model. *Proceedings of the Indian Academy of Sciences-Earth and Planetary Sciences, 99*(2), 229–247.

Ghosh, S. K., Sarkar, S. C., & Sengupta, S. (1985). A field guide for Ghatsila and neighbouring mineral-belt. Department of Geological Sciences, Jadavpur University, Kolkata, India.

Mazumder, R. (2005). Proterozoic sedimentation and volcanism in the Singhbhum crustal province, India and Their Implications. *Sedimentary Geology, 176*(1–2), 167–193.

Mazumder, R., & Alterman, W. (2016). A brief overview of Palaeoproterozoic geology of the Singhbhum crustal province, Eastern India. Palaeoproterozoic Supercontinents and Global Evolution. UNESCO IGCP 509.

Mazumder, R., Van Loon, A. J., Mallik, L., Reddy, S. M., Arima, M., Altermann, W., & De, S. (2012). Mesoarchaean–Palaeoproterozoic stratigraphic record of the Singhbhum crustal province, eastern India: a synthesis. Geological Society, London, Special Publications, *365*(1), 31–49.

Mukherjee, S. (2013). *Deformation microstructures in rocks* (pp. 1–111). Springer Geochemistry/Mineralogy. ISBN 978-3-642-25608-0.

Mukherjee, S. (2014) *Atlas of shear zone structures in Meso-scale* (pp. 1–124). Springer Geology. ISBN 978-3-319-0088-6.

Mukherjee, S. (Ed.). (2015). *Atlas of structural geology* (p. 183). Elsevier.

Mukherjee, S. (2019a). Teaching structural geology in Indian context. In S. Mukherjee (Ed.), *Teaching methodologies in structural geology and tectonics* (pp. 221–232). Springer. ISBN 978-981-13-2781-0.

Mukherjee, S. (2019b). Teaching methodologies in structural geology and tectonics: An introduction. In S. Mukherjee (Ed.), *Teaching methodologies in structural geology and tectonics* (pp. 1–251). Springer. ISBN 978-981-13-2781-0.

Mukherjee, S. (2021). *Atlas of structural geology* (Second Edn., pp. 1–260). Elsevier. ISBN: 978012816802.

Mukherjee, S. (2023). Introduction to structural geology and tectonics field guidebook—volume 2. In S. Mukherjee (Ed.), *Structural geology and tectonics field guidebook—volume 2* (pp xi–xiv). Springer. ISBN 978-3-031-19575-4.

Mukherjee, S., Bose, N., Ghosh, R., Dutta, D., Misra, A. A,, Kumar, M., Dasgupta, S., Biswas, T., Joshi, A., & Limaye, M. (2020). *Structural Geological Atlas*. Springer. ISBN: 978-981-13-9825-4.

Mukhopadhyay, D. (2001). The Archaean nucleus of Singhbhum: The present state of knowledge. *Gondwana Research, 4*(3), 307–318.

Mukhopadhyay, J., Gutzmer, J., Beukes, N. J., & Hayashi, K. I. (2008). Stratabound magnetite deposits from the eastern outcrop belt of the Archaean Iron Ore Group, Singhbhum craton, India. *Applied Earth Science, 117*(4), 175–186.

Nelson, D. R., Bhattacharya, H. N., Thern, E. R., & Altermann, W. (2014). Geochemical and ion-microprobe U-Pb zircon constraints on the Archaean evolution of Singhbhum Craton, eastern India. *Precambrian Research, 255*, 412–432.

Paul, J. (2014). Deformation pattern as indicator of structural evolution: Ghatsila-Galudih-Tentuldanga fold belt, Tentuldanga, Jharkhand, Eastern India. *Earth Science India*, Popular Issue, *VII*(IV), 1–8. www.earthscienceindia.info

Ramsay, J. G. (1967). *Folding and fracturing of rocks* (p. 568). Mc Graw Hill Book Company.

Roy, A., Sarkar, A., Jeyakumar, S., Aggrawal, S. K., Ebihara, M., & Satoh, H. (2004). Late Archaean mantle metasomatism below eastern Indian craton: Evidence from trace elements, REE geochemistry and Sr–Nd–O isotope systematics of ultramafic dykes. *Journal of Earth System Science, 113*(4), 649–665.

Saha, A. K. (1994). Crustal evolution of Singhbhum North Orissa Eastern India. *Memoirs Geological Society, India, 27*, 341.

Sarkar, S. N., & Saha, A. K. (1977). The present status of the Precambrian stratigraphy, tectonics, and geochronology of the Singhbhum-Keonjar, Mayubhanj region, eastern India. Indian Journal of Earth Sciences, 37–65.

Sengupta, S., & Ghosh, S. K. (1997). The kinematic history of the Singhbhum shear zone. *Proceedings of the Indian Academy of Sciences-Earth and Planetary Sciences, 106*(4), 185–196.

Sengupta, S., & Mukhopadhyay, P. K. (2000). Sequence of Precambrian events in the eastern Indian craton. *Geological Survey of India Special Publication, 57*, 49–56.

Spectacular Soft-Sediment Deformation Structures in Sedimentary Rock Outcrops of Damodar Valley Basin, West Bengal, India: A Field Guide

Arpita Samanta and Abhik Kundu

Abstract This chapter deals with a few outcrops of sedimentary rocks present in the Damdor River valley, West Bengal. These outcrops preserve different types of soft sedimentary deformation (SSD) structures in a fluvial set up. As preservation potential is very low for sedimentary rocks, outcrops with good quality of primary sedimentary structures and SSDs are rare for training of process-based sedimentology. Moreover, these SSDs of Mesozoic time are of aseismic origin, which are less explored and reported. This chapter presents some glimpses of SSDs formed in the Gondwanaland mainly during the early Triassic Period which considered as the most unstable interval of the Phanerozoic Eon. SSDs, such as convolute-laminations, chevron folds, overturned cross stratifications, sand volcanoes, and load casts produced by different mechanisms but possibly by the same autokinetic force are preserved in these sections. This chapter will be a helpful guide in field for recognition of different types of SSDs generated by autocyclic rearrangement processes.

1 Introduction

Sedimentary rocks preserve excellent clues for understanding the evolving Earth system. Sediments spread in wide range of environments that characterize dynamics of physical (e.g., Dasgupta & Mukherjee, 2020; Mukherjee & Kumar, 2018) as well as biological processes on the Earth's surface. Like every other domain of Earth science, the knowledge on sedimentary rocks and ideas about processes of their evolution have been changing and expanding continuously; field observation and primary data collection from the field still remain as basic building block of every study on sedimentary rocks. Any study of sedimentary rocks begins with a systematic observation of rock outcrops or drill-core samples. Sediment provenance as well as the nature of transport and depositional processes can be inferred directly

A. Samanta · A. Kundu (✉)
Department of Geology, Asutosh College, 92, S. P. Mukherjee Road, Kolkata, West Bengal 700026, India
e-mail: kundu.abhik@gmail.com

S. Mukherjee (ed.), *Structural Geology and Tectonics Field Guidebook–Volume 2*, Springer Geology, https://doi.org/10.1007/978-3-031-19576-1_6

from sedimentary structures, textures, and composition (Boggs, 1995; Stow, 2005; Tucker, 2011).

Although many sedimentary exposures spanning from Paleoarchean to Cenozoic in different part of India preserve fantastic primary sedimentary structures and pose as models for conceptual sedimentological studies, soft-sediment deformation structures (SSDS) are very little explored to decipher the contemporary basin tectonics and tectono-sedimentary processes (Bhattacharya & Bandyopadhya, 1998; Bhattacharya et al., 2016; Dasgupta, 2008; Kundu & Goswami, 2008; Kundu et al., 2011; Mazumder et al., 2006). These structures are important clues to both the depositional process as well as of natural disturbances which might have occurred during and/or immediately after deformation (Seilacher, 1984; van Loon, 2009).

In this context, outcrops of the Permian–Triassic terrestrial successions (those are rare in global scale also; Fielding et al., 2019) of the Raniganj coalfield area within the Damodar Valley Basin (outcrops at Tetulerakh Nala section, Dumdumi, Iswarda, Sunuri Nala section and near Damodar Railway station) can act as ideal field areas for studies of SSDS (Fig. 1 a, b). The locations are in the West Bardhaman district of West Bengal, India.

(a)

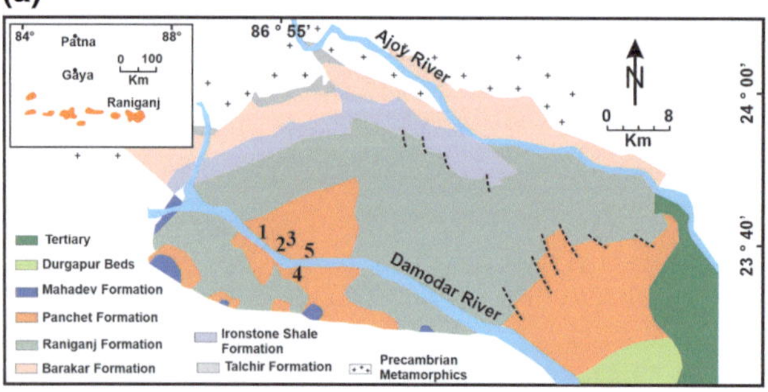

(b)

	Age	Formation
Triassic	Upper	Supra-Panchet
	Middle	
	Lower	Panchet
Permian	Upper	Raniganj/ Barren Measures
	Lower	Barakar
		Karharbari
		Talchir

Fig. 1 **a** Simplified geological map of the Raniganj Coalfield area of the Damodar Valley Basin, West Bengal, India (modified after Gupta, 2009) showing the disposition of different Formations and locations of the outcrop sections. 1. Sunuri, 2. Dumdumi, 3. Damodar Railway station, 4. Tentulerakh and 5. Iswarda; **b** Generalized stratigraphic column and depositional ages of the different lithounits of Damodar Valley Basin (simplified after Raja Rao, 1987). Section numbers are referred in field photographs (Figs. 2, 3, 4, 5, 6, 7, 8 and 9).

2 Background and Present Work

Lithology and associated primary sedimentary structures give clues about depositional environment (Boggs, 1995). The Late Permian to Early Triassic deposits of Damodar Valley Basin records accumulation of fluvial siliciclastic sediments (Ghosh, 2002). Sediments of the Late Permian Raniganj Formation consist of coal-bearing sandstone and shale, whereas the Early Triassic Panchet Formation is mainly made up of sandstone and is devoid of any coal (Ghosh, 2002). Ubiquitous primary structures are preserved in both the Raniganj and Panchet Formations (Figs. 2, 3 and 4).

The Damodar Valley Basin used to be an intra-cratonic rift basin, where the Permian rocks of the Damuda Group unconformably rest over the older Precambrian basement (Biswas, 1999; Ghosh, 2002; Mitra, 1994; Tewari & Casshyap, 1996). The Early Permian glacial deposits of the Talchir Formation make the initiation of sedimentation in Damodar Valley Basin. The Talchir Formation is conformably overlain by sandstones of the Barakar and Ironstone Shale Formations (Krishnan, 1982). The topmost unit of the Damuda Group is represented by the Late Permian Raniganj Formation (with coal), which is overlain by the Early Triassic Panchet Formation (devoid of coal). The contact between the Raniganj Formation and the

Fig. 2 Field photographs of the Raniganj Formation in section 4 (Fig. 1) showing: **a** A Laminated silty shale unit forming ~5–10 cm thick beds (Length of the marker is 28 cm); **b** Mud cracks (Length of the marker is 28 cm); **c** Climbing ripples and flaser lamination in the flood plain deposits (Length of the marker is 16 cm); **d** Palustrine/lake carbonates associated with verve in small lakes of the flood plain (cf. Alonso-Zarza, 2003) (Diameter of the marker is 6 cm)

Fig. 3 Field photographs of the Panchet Formation showing **a** Large-scale trough cross-stratification with convex upper bounding surface, characteristic of bar deposits (Length of the marker is 28 cm) (section 3 in Fig. 1); **b** Setulf structures (Sarkar et al., 2011) in the Panchet Formation (Length of the marker is 28 cm) (section 1 in Fig. 1); **c** Bone of *Lystrosaurus* in the Panchet Formation (Length of the marker is 16 cm) (section 2 in Fig. 1) and **d** Sigmoidal cross-stratification in fine to medium-grained sandstone (Length of the marker is 16 cm) (section 5 in Fig. 1)

overlying Panchet Formation is either gradational (1B) or is characterized by a local unconformity (Krishnan, 1982). The Raniganj coalfield area is surrounded on three sides by Archean rocks and faulted down on the south and west (Krishnan, 1982).

From the Gondwana sedimentary rocks (Panchet Formation) of the Damodar Valley Basin, SSDS have been reported by previous workers (Bhattacharya et al., 2016; Kundu & Goswami, 2008; Kundu et al., 2011). The Panchet Formation hosts fantastic SSDS, along with other primary structures, which are very helpful to understand the sedimentary processes particularly those responsible for deformation of the soft-sediment. Different types of SSDS have been recognized in the Panchet Formation sediments. In the following section different types of SSDS present in the area are described; followed by a brief discussion on the possible processes of formation (Allen, 1982, 1986; Allen & Banks, 1972; Moretti et al., 1999, 2014; Kundu & Goswami, 2008; Kundu et al., 2011; Owen & Moretti, 2011).

Fig. 4 Field photograph showing fine-grained shale deposits of the Raniganj Formation is overlain by relatively coarse grained sand stone deposits of the Panchet Formation, which infers changes of flow energy/velocity from relatively lower to higher (section 1 in Fig. 1)

2.1 *Overturned Cross-Stratification/Recumbently Folded Cross-Beds*

Overturned cross-stratifications are observed in coarse-grained sandstone of the Panchet Formation and consist of completely overturned cross-strata (Fig. 5). The deformation is restricted to individual sets and does not continue into the underlying or overlying strata. Liquefaction and the action of shear in current on liquefied sand bed can generate this type of overturned cross-stratifications.

2.2 *Penecontemporaneous Folds*

Numerous complex folds occur within coarse-grained sandstone and sometimes in siltstone to fine-grained sandstone of the Panchet Formation. Often the fold forms are chevrons (Fig. 6). Liquefaction and horizontal shearing can cause these kinds of folds in sediments. But the geometry of folds depends on initial foreset shape and sometimes being localized only in the sigmoidal foresets (Røe, 1987).

Fig. 5 Field photographs of discrete overturned cross-stratification in medium to coarse-grained sandstone of the Panchet Formation (section 1 in Fig. 1). Length of the marker is 28 cm

Fig. 6 Field photographs of **a** Complex folds (Length of the marker is 16 cm) and **b** Chevron folds (Length of the marker is 28 cm) preserved in the laminated fine to coarse-grained sandstone of the Panchet Formation (section 1 in Fig. 1)

Fig. 7 Convolute laminations in medium to coarse-grained sandstone of the Panchet Formation (section 1 in Fig. 1). Length of the marker (visible portion of the hammer) is 23 cm

2.3 *Convolute Laminations*

Convolute laminations occur in siltstone to coarse-grained sandstone of the Panchet Formation (Fig. 7). In some places the geometry of the convolution is more complex. Either sediments deposited in a quick condition or caused to lose strength through being partially or fully liquidize can generate convolute laminations. These are generally formed by turbidity current in normally graded clastic sediments in different depositional environments (Allen, 1977).

2.4 *Load Structures*

Load casts are abundant in the Panchet Formation and they seem to be localized at the interface between fine to medium sandstone and sometimes shale to medium sandstone (Fig. 8). Sediment with a relatively higher bulk density overlies sediment with a lower bulk density, load structures may form if the lower layer loses strength.

2.5 *Sand Volcanoes*

Sand volcanoes are observed on exposed bedding surfaces of sand stones and are usually less than 1 m in diameter in the study area. They are preserved in medium to coarse sandstones of the Panchet Formation (Fig. 9) and the most common structure in this formation. Sometimes water escape pillar structures are associated with upward deflected primary laminations. Subaqueous sediments overlain by younger deposits and causes pressure forces and results the pore water out between the grains. Rapid migration of a fluid through sediment can produce sand volcanoes.

Fig. 8 Field photographs of load cast and flame structures in medium to coarse-grained sandstone of the Panchet Formation. Note near-complete destruction of laminations in the lower part of the load casts (section 2 in Fig. 1). Length of the marker is 28 cm

Fig. 9 Field photographs of sand volcanoes in the top surface of a medium to coarse-grained sandstone of the Panchet Formation (section 1 in Fig. 1). Length of the marker is 16 cm

Acknowledgements Authors thank Asutosh College for providing the necessary facilities to execute the work. We thank Sudipta Das, Dipayan Dasgupta, Rohit Jaswal, and Nishant Chauhan for their help during field works. Soumyajit Mukherjee (IIT Bombay) invited, edited and reviewed this article. The authors are thankful to Prof. Rajat Mazumder (German University of technology in Oman, Muscat) for a critical internal review of the manuscript. Springer team (Alexis Vizcaino, Doerthe Mennecke-Buehler, Boopalan Renu and proofreaders) are thanked for assistance. Mukherjee (2023) summarized this chapter.

References

Allen, J. R. L. (1977). The possible mechanics of convolute lamination in graded sand beds. *Journal of the Geological Society, 134*(1), 19–31.

Allen, J. R. L. (1982). Sedimentary structures: Their character and physical basis. *Developments in Sedimentology, 30*, 137–177.

Allen, J. R. L. (1986). Earthquake magnitude-frequency, epicentral distance, and soft-sediment deformation in sedimentary basins. *Sedimentary Geology, 46*(1–2), 67–75.

Allen, J. R. L., & Banks, N. L. (1972). An interpretation and analysis of recumbent-folded deformed cross-bedding. *Sedimentology, 19*(3–4), 257–283.

Alonso-Zarza, A. M. (2003). Palaeoenvironmental significance of palustrine carbonates and calcretes in the geological record. *Earth-Science Reviews, 60*(3–4), 261–298.

Bhattacharya, H. N., & Bandyopadhya, S. (1998). Seismites in a Proterozoic tidal succession, Singhbhum, Bihar, India. *Sedimentary Geology, 119*, 239–252.

Bhattacharya, B., Bhattacharjee, J., Banerjee, S., Bandyopadhyay, S., & Das, R. (2016). Seismites in Permian Barakar Formation, Raniganj Basin, India: Implications on Lower Gondwana basin evolution. *Arabian Journal of Geosciences, 9*(4), 300.

Biswas, S. K. (1999). A review on the evolution of rift basins in India during Gondwana with special reference to western Indian basins and their hydrocarbon prospects. In A. Sahni, & R. S. Loyal (Eds.), *Gondwana assembly: New issues and perspectives, Proceedings of Indian National Science Academy*, Special Issue (pp. 261–283).

Boggs, S. (1995). *Principles of Sedimentology and stratigraphy*. Prentice-Hall.

Dasgupta, P. (2008). Experimental decipherment of the soft-sediment deformation observed in the upper part of the Talchir Formation (Lower Permian), Jharia Basin, India. *Sedimentary Geology, 205*(3–4), 100–110.

Dasgupta, T., Mukherjee, S. (2020). *Sediment compaction and applications in petroleum geoscience*. Springer. Series: Advances in Oil and Gas Exploration & Production. ISSN: 2509–372X.

Fielding, C. R., Frank, T. D., McLoughlin, S., Vajda, V., Mays, C., Tevyaw, A. P., Winguth, A., Winguth, C., Nicoll, R. S., Bocking, M., & Crowley, J. L. (2019). Age and pattern of the southern high-latitude continental end-Permian extinction constrained by multiproxy analysis. *Nature Communications, 10*(1), 1–12.

Ghosh, S. C. (2002). The Raniganj coal basin: An example of an Indian Gondwana rift. *Sedimentary Geology, 147*(1–2), 155–176.

Gupta, A. (2009). Ichthyofauna of the Lower Triassic Panchet Formation, Damodar valley basin, West Bengal, and its implications. *Indian Journal of Geosciences, 63*(3), 275–286.

Krishnan, M. S. (1982). *Geology of India and Burma* (6th Edn., p. 536). CBS Publishers & Distributors.

Kundu, A., & Goswami, B. (2008). A note on seismic evidences during the sedimentation of Panchet Formation, Damodar Basin, Eastern India: Banspetali Nullah Revisited. *Journal of the Geological Society of India, 72*, 400–404.

Kundu, A., Goswami, B., Eriksson, P. G., & Chakraborty, A. (2011). Palaeoseismicity in relation to basin tectonics as revealed from soft-sediment deformation structures of the Lower Triassic Panchet formation, Raniganj basin (Damodar valley), eastern India. *Journal of Earth System Science, 120*(1), 167–181.

Mazumder, R., van Loon, A. T., & Arima, M. (2006). Soft-sediment deformation structures in the Earth's oldest seismites. *Sedimentary Geology, 186*(1–2), 19–26.

Mitra, N. D. (1994). Tensile resurgence along fossil sutures: A hypothesis on the evolution of Gondwana basins of peninsular India. In *Abstracts of Proceedings, 2nd Symposium on Petroliferous basins of India* (pp. 55–62). KDMIPE Publishers.

Moretti, M., Alfaro, P., Caselles, O., & Canas, J. A. (1999). Modelling seismites with a digital shaking table. *Tectonophysics, 304*(4), 369–383.

Moretti, M., van Loon, A. T., Liu, M., & Wang, Y. (2014). Restrictions to the application of 'diagnostic' criteria for recognizing ancient seismites. *Journal of Palaeogeography, 3*(2), 162–173.

Mukherjee, S. (2023). Introduction to structural geology and tectonics field guidebook—volume 2. In S. Mukherjee (Ed.), *Structural geology and tectonics field guidebook—volume 2* (pp xi–xiv). Springer. ISBN 978-3-031-19575-4.

Mukherjee, S., & Kumar, N. (2018). A first-order model for temperature rise for uniform and differential compression of sediments in basins. *International Journal of Earth Sciences, 107*, 2999–3004.

Owen, G., & Moretti, M. (2011). Identifying triggers for liquefaction-induced soft-sediment deformation in sands. *Sedimentary Geology, 235*(3–4), 141–147.

Raja Rao, C. S. (Ed.). (1987). Coalfields of India. *Bulletin Geological Survey of India, A IV*(45) Part I, 336.

Røe, S. L. (1987). Cross-strata and bedforms of probable transitional dune to upper-stage plane-bed origin from a Late Precambrian fluvial sandstone, northern Norway. *Sedimentology, 34*(1), 89–101.

Seilacher, A. (1984). Sedimentary structures tentatively attributed to seismic events. *Marine Geology, 55*(1–2), 1–12.

Sarkar, S., Samanta, P., & Altermann, W. (2011). Setulfs, modern and ancient: Formative mechanism, preservation bias and palaeoenvironmental implications. *Sedimentary Geology, 238*(1–2), 71–78.

Stow, D. A. (2005). *Sedimentary rocks in the field: A color guide* (p. 320). Gulf Professional Publishing.

Tewari, R. C., & Casshyap, S. M. (1996). Mesozoic tectonic events including rifting in Peninsular India, and their bearing on Gondwana stratigraphy, and sedimentation. In *Proceedings, 9th International Gondwana Symposium Hyderabad, India* (pp. 865–880). Oxford and IBH Publishers.

Tucker, M. E. (2011). *Sedimentary rocks in the Field: A practical guide* (275p). Wiley-Blackwell.

Van Loon, A. J. (Tom). (2009). Soft-sediment deformation structures in siliciclastic sediments: An overview. *Geologos, 15*(1), 3–55.

Structural Geological Field Guide: Bhuj Area (Gujarat, India)

Nidhi Lohani, Soumyajit Mukherjee, Seema Singh, Aashu Pawar, and Mohamedharoon Shaikh

Abstract The Kachchh offshore being a petroliferous basin of great prospect, knowing the tectonics of the Kachchh offshore as well as the land region is of great attention for geoscientists. The present study put forward a composite picture of various features like faults, local folds associated with faults, shear bands etc. from the Kachchh land area, which is expected to (i) provide insights into the different structural aspects of the most tectonically active segment of Katrol Hill Range Fault Zone (KHRFZ) and, (ii) provide possible mechanism of present day structural set up of the region.

1 Introduction

Tectonics of the W-striking Kachchh Rift Basin (KRB), located in the state of Gujarat on the western continental margin of the Indian plate, has been a place of enormous national and international attention for the geoscientists from academia and industries (Biswas & Khattri, 2002). The KRB rifted in the Late Triassic or Early Jurassic, before India-Africa separation, causing extensional stresses and continual sedimentation until the Late Cretaceous (Biswas, 2016). As a response, the W-striking major faults were activated as normal faults, following the structural trend of the Mid-Proterozoic Delhi-Aravalli fold belt (Biswas, 2016). The rifting of the KRB then ceased during the Late Cretaceous, and the Indian plate began to drift counter-clockwise from the Mid-Jurassic Period onward (Biswas, 2016). Rifting of the KRB was followed by rift

N. Lohani · S. Singh · A. Pawar
Department of Geology, Panjab University, Chandigarh 160014, India

S. Mukherjee (✉)
Department of Earth Sciences, Indian Institute of Technology Bombay, Powai, Mumbai, Maharashtra 400076, India
e-mail: smukherjee@iitb.ac.in

M. Shaikh
Department of Geology, Faculty of Science, The M. S. University of Baroda, Vadodara, Gujarat 390002, India

S. Mukherjee (ed.), *Structural Geology and Tectonics Field Guidebook–Volume 2*, Springer Geology, https://doi.org/10.1007/978-3-031-19576-1_7

inversion since the Late Cretaceous, leading major intra-basinal faults to reactivate as reverse faults (Biswas & Khattri, 2002; Shaikh et al., 2020).

The Nagar Parkar Fault (NPF) to the north and North Kathiawar Fault (NKF) to the south serve as rift shoulders that separate the KRB from the rest of the landmass (Biswas, 2016). The tectonic framework of the KRB is governed by W-striking intra-basinal faults—Island Belt Fault (IBF), Gedi Fault (GF), Allah Bund Fault, South Wagad Fault (SWF), Kachchh Mainland Fault (KMF) (> 1200 m throw) (Padmalal et al., 2019, 2021; Shaikh et al., 2019) and Katrol Hill Fault (KHF) (~732 m maximum throw) (Maurya et al., 2021a). These major faults have formed large-scale uplifts produced by extension during rifting phase of the KRB (Late Triassic to Early Jurassic)—Island Belt Uplift (IBU), Desalpar Uplift (DU), ~60 km long and ~40 km wide Wagad Uplift (WU), ~193 km long and ~72 km wide Kachchh Mainland Uplift (KMU) (Biswas, 1993).The Northern Hill Range Fault Zone (NHRFZ) along the KMF, Katrol Hill Range Fault Zone (KHRFZ) along the KHF, Vigodi-Gugriana Khirasra-Netra Fault System (VGKNFS) and Bhuj structural low are the four structural zones that make up the KMU (Biswas, 1993; Shaikh et al., 2020). Based on recent seismicity data recorded for all the earthquakes that have occurred in the KRB, all the intra-basinal faults can be classified as neotectonically active.

There is another category of structure that can be equally important to the regional structural geology as it is to the seismicity of the KRB. These are the N-, NW-, NE-, NNW- and NNE-striking, m- to km-scale, transverse fault system with dip-/oblique-slip, which affected and segmented the W-striking major intra-basinal faults in the KRB (Biswas, 1993; Maurya et al., 2003, 2021b; Shaikh et al., 2020). Transversse faults are important since (i) prior research has reported periodic reactivation of these faults to accommodate the growing compressive stresses on W-striking faults (Maurya et al., 2003); (ii) they offset and change the structural attitude of many intra-basinal faults, therefore are an integral part of the KRB (Biswas, 1993; Shaikh et al., 2020); (iii) KRB is a petroliferous basin and transverse faults might give rise to various types of petroleum traps (Wilcox et al., 1973).

The KRB and Himalayan mountain belt are the most tectonically and seismically active regions of India, and they fall in the highest seismic risk zone-V. Previous workers have reported transverse faults from the Himalayas: for example, the Siwaliks and the Lesser Himalayas as well as the Sikkim Himalayas (Paul et al., 2015), where they have been related to the displacement of major faults and seismicity. Previous workers have reported transverse faults from various parts of the KRB: for example, Shaikh et al. (2020) documented NW-, NE- and E-striking unnamed transverse faults with dip-slip/oblique-slip in the NW-striking VGKNFS. They extend for tens of meters to a few kilometers before dying out or truncating in the deformation zone of major NW-striking faults (Shaikh et al., 2020). Maurya et al. (2003) reported transverse faults that offset two major intra-basinal faults of the KMU—the KMF and KHF. These faults have been mostly deciphered using field observations, ground-penetrating radar (GPR) surveys, landscape and drainage characteristics. Previous workers have discussed reactivation and subsequent geomorphic expressions of the transverse faults in various parts of the KRB (e.g., Maurya et al., 2003; Patidar et al.,

2007; Sohoni et al., 1999; Thakkar et al., 1999). However, very limited structural documentation from field has so far been made (Dasgupta & Mukherjee, 2019).

The primary aim of the present study is to report detailed field-based structural information near the Bhuj city. This site is ideal for carrying out structural studies because: (i) the transverse faults dissect and displace the KHF, which is one of the neotectonically active element of the region; (ii) in a recent study carried out on the basis of quantitative geomorphic parameters and field evidences the KHF, which is dissected into five segments by such transverse faults have been assigned values based on their tectonic activity (Das et al., 2019). The present area of study is one of the most active segments of the KHF.

Another aim of the present study is to carry out sedimentological studies because the selected outcrop consists of sandstone-shale sequences of Lower Cretaceous Bhuj Formation. Such sedimentological studies in conjunction with structural studies shall be helpful in knowing the overall structural disposition and deposition of sandstone units within the Formation.

2 The Study Area

2.1 Structure

The study area is located along the W-striking KHF in the central part of the KMU. Tectonically, KHRFZ lies parallel to one of the most important structural element of the KRB, the W-striking KHF. The KHF is an intra-basinal range-bounding fault that separates the KMU into two divisions– hilly topography of the Katrol Hill Range Fault Zone (KHRFZ) in the south and the low-relief rocky plain to the north comprising Late Cretaceous Bhuj Formation (Biswas, 1993; Maurya et al., 2016). The ~71 km long KHF as first described by and later on by many other researchers (Chung & Gao, 1995; Morino et al., 2008; Singh et al., 2014) is a south dipping, high-angle fault with reverse slip-sense (Biswas, 1993). It has a single, well exposed, sub-vertical fault plane, which forms a sharp boundary between the Jurassic (Jhumara) and the Cretaceous (Bhuj Formation) rocks (Biswas, 2016). The fault strikes ~ E-W but at places it is NWW-SEE due to the effect of several transverse faults that dissect it. Biswas (2016) described KHF as a post-rift fault as it has no direct relation with the syn-rift sediment accumulation. It activated after the inversion phase during Early Eocene (Maurya et al., 2016) along a primordial fault.

Based on detailed field mapping and GPR studies, Patidar et al., (2007, 2008) divided the KHF into four segments. These segments of different lengths show different degree of tectonic activity and are bounded by transverse faults that offset the KHF (Patidar et al., 2007). Out of these four segments, segment 2 and 3, which make up the central part of the KHRFZ are most tectonically active. The eastern delimiting transverse fault of segment 2 is the basis of present study (F2 in Fig. 2c). The exact coordinates of the field location are—23° 14′ 38.228″ N latitude and 69°

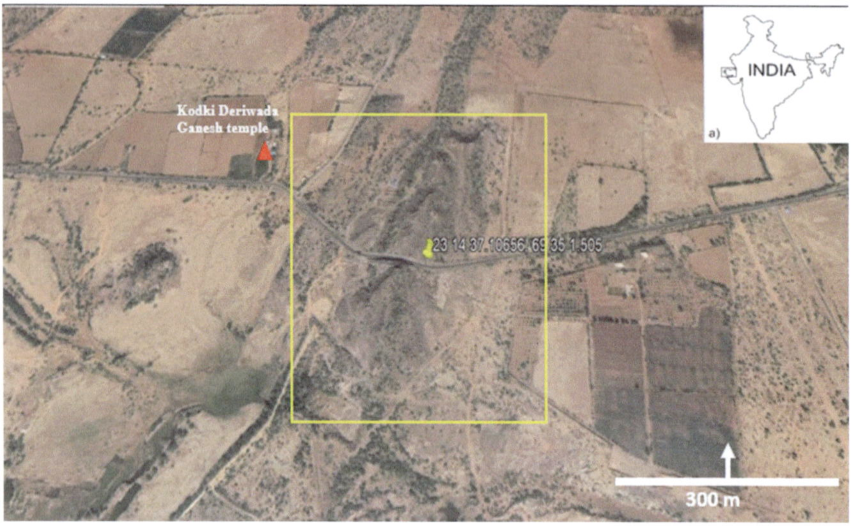

Fig. 1 Google earth image of the study area enclosed within the yellow box. Inset image *a:* study area enclosed in a rectangle on the map of India. Red triangle: KodkiDeriwada Ganesh temple

35′ 8.286″ E longitude and it lies near the Bhuj city (Fig. 1). Walid et al. (2021) has studied vein geometries in detail from our study area, but the work seems to have a limited tectonic significance.

2.2 Seismicity

The KRB is seismically very active (Zone 5, according to BIS, 2002) and is a high risk potential zone outside Himalayas in India. Inside the KRB, it is reported that the frequency of small ($M_w < 4.0$) to large ($M_w > 6.0$) earthquakes occurring near the KHRFZ is higher than the other parts of the KMU (Fig. 2c). A recent major earthquake of Mw ~6.0 recorded along the KHF was the Anjar earthquake of 21 July 1956 (Chung & Gao, 1995) (Fig. 2b). Prior studies have attributed the high seismicity and tectonically active nature of the region to the growing N-S compressive stresses due to the northward movement of Indian plate under the Eurasia. Previous researchers have reported that the E-W trending intra-basinal faults accumulate a major part of the growing compressive stress by vertical movements. In such a case the offsetting of these major faults by transverse faults becomes very significant (Maurya et al., 2016). Several workers suggest that a part of these growing stresses is transferred to these transverse faults and this might be the key reason for the present seismicity of the region (Maurya et al., 2016).

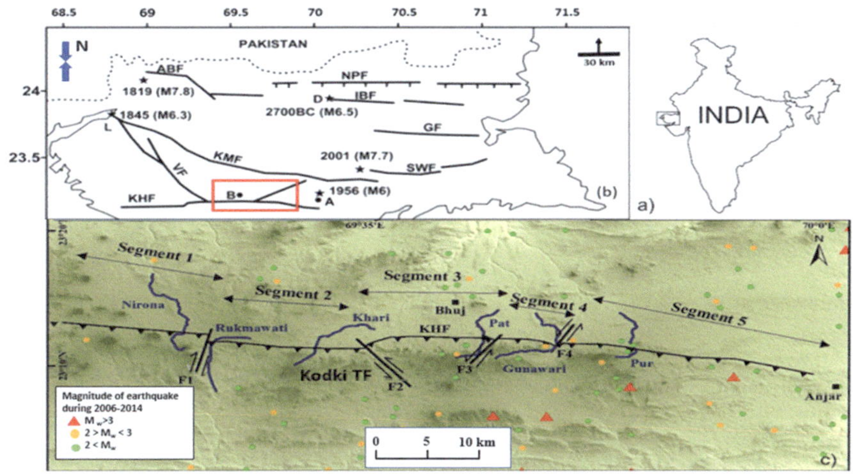

Fig. 2 **a** Map of India shows the study area in Kachchh. **b** Tectonic map of Kachchh showing major faults and associated epicenters of major earthquakes (Rastogi et al., 2014). Red box: study area. Blue arrow at top left: N-S compressive stresses acting in the region (World stress map, 2016). **c** Modified DEM of study area along with major fault segments, drainage and seismicity during the period 2006–2014 (ISR earthquake catalogue) (Das et al., 2019). The Kodki transverse fault (TF) or F2 is the area of present study

2.3 *Lithology and Stratigraphy*

The entire Mesozoic sequence except the Jhurio Formation is exposed in the Mainland Kachchh along the KHF. Stratigraphy of the Kachchh mainland has been discussed in detail by Biswas (2016) (Table 1).

The present study conforms to a field location where only the Lower Ghuneri Member of Bhuj Formation is exposed, therefore only Bhuj Formation has been discussed in the next section.

2.3.1 Bhuj Formation

The thickness of Bhuj Formation is 815 m and is completely siliciclastic, typical of a rift environment (Biswas, 1993). It deposited in the last phase of rifting of a pericratonic rift basin that originated from the disintegration of Gondwanaland in the Late Triassic (Biswas, 1977, 1987, 2005). The western exposure of Bhuj Formation is unanimously accepted to be of deltaic nature, however the nature of Bhuj sediments in the eastern part of its exposure has been debated. Biswas (1977) suggested a complete fluvial nature. On the other hand, entirely marine nature of sediments was suggested by other group of researchers (Casshyap et al., 1983; Krishna et al., 1983; Rai, 2006); Shukla & Singh, 1990.

Table 1 Updated and modified lithostratigraphic classification of the Mesozoic rocks of Kachchh

Stage	Kutch mainland group	
Tertiary	Formation	Member
Maastrichtian-Danian	Deccan traps	Basalt flows
Albian	Bhuj formation	Upper member: massive sandstones
Aptian		Ukra member: green glauconitic shale/ferruginous bands with fossil
Hauterivian to Barriasian		Ghuneri member/lower member Sandstone/shales/ferruginous bands/shales with plant fossils
Tithonian	Jhuran formation	Katesar member: massive sandstones
Kimmeridgian		Upper member: fossiliferous sandstones, shales, hard calcaereous sandstones
		Middle member: mainly shales, fossilferous with sandstone interbeds
		Lower member: sandstones/shales/arenaceous limestones with fossils
Oxfordian	Hiatus	
Callovian	Jumara formation	Dhosa Oolite member
		Gypseous shale member
		Ridge sandstone member
		Shelly shale member
Aalenian-Bathonian	Jhurio formation	Member G: thin bedded White Lst& Nod. Lst
		Member F: purple sandstones/packstones
		Member E: bedded rusty grainstone with golden oolite
		Member D: grayshales
		Member C: brick red weathering rusty grainstone with golden oolites
		Member B: grayshales
		Member A: thin bedded yellow white limestones, shales, rusty brown limestones with golden oolites Basement

The member written with bold letters constitutes the present area of study (after Biswas, 1977, 1993; Deshpande and Merh, 1980; Furisch et al., 2001; Krishna et al., 2009)

The Bhuj Formation is bounded by Deccan traps of Tertiary age (Biswas, 1977) on its upper side and it sits over an entirely marine Jhuran Formation of Jurassic age. The age of the Bhuj Formation is Lower Cretaceous (Valenginian to Santonian). It is further subdivided into three members: (i) Lower Guneri Member, (ii) Middle-Ukra Member (limited to Guneri-Ukra area) and (iii) Upper Member.

3 Field Studies

The study area comprises sandstone-shale sequences of Lower Cretaceous Bhuj Formation. Normal faults are one of the most common structural features observed in the study area. The entire study revolves around a fault-controlled ridge that has an exposed section of a major NNE-striking transverse fault, which dissects the KHF. It is a high-angle fault of normal slip-sense, which we refer to as Kodki Fault in this study. Figures 3 and 4 depict the fault exposures on the Bhuj-Kodki road.

3.1 *Small-Scale Faults*

The classic Kodki normal fault has already been reported by several authors (Kundu and Thakkar, 2011; Thakkar et al., 2017).Numerous small-scale normal faults were also observed near the main fault in the Bhuj Formation, which are reported here (Figs. 5, 6, and 7).

Although less common in the study area, few reverse faults in the vicinity of classic transverse fault of normal nature were also noted.

According to Anderson's theory of faulting a conjugate set of faults is X-shaped wherein the maximum compressive stress axis (σ_1) makes an angle of ~30° with the faults. Here, the maximum compressive stress axis bisects the acute angle, and the minimum compressive stress axis (σ_3) bisects the obtuse angle whereas the intermediate stress axis (σ_2) lies parallel to the line of intersection. Figure 8 shows an example of Andersonian conjugate strike-slip faults.

Fig. 3 Vertical section of the Kodki Fault in the Early Cretaceous Bhuj sandstone, depicting agraben structure. Location: 23°14′ 38.228″ N, 69°35′ 8.286″ E

Fig. 4 Vertical section of a normal fault in sandstone-shale sequence of Lower Bhuj succession. S. Mukherjee as scale. Location: 23°14′ 38.228″ N, 69°35′ 8.286″ E

Fig. 5 Uninterpreted (a) and interpreted (a′) images of vertical section of small-scale normal fault in Early Cretaceous sandstone of Bhuj Formation. Marker pen (13 cm) is shown as scale. Location: Road left section from E of Kodki road fault, near Bhuj city, Gujarat, India. Coordinates: 23° 14′ 38.228″ N, 69° 35′ 8.286″ E

Fig. 6 Uniform set of small-scale normal faults as observed in sub-vertical section of sandstone sequence of Early Cretaceous Lower Bhuj succession visible clearly due to the displacement of yellow marker layer on the Kodki road, near Bhuj city, Gujarat, India. Hand is used as marker (15 cm). Location: 23° 14′ 32.6461″ N, 69° 34′ 57.28681″ E

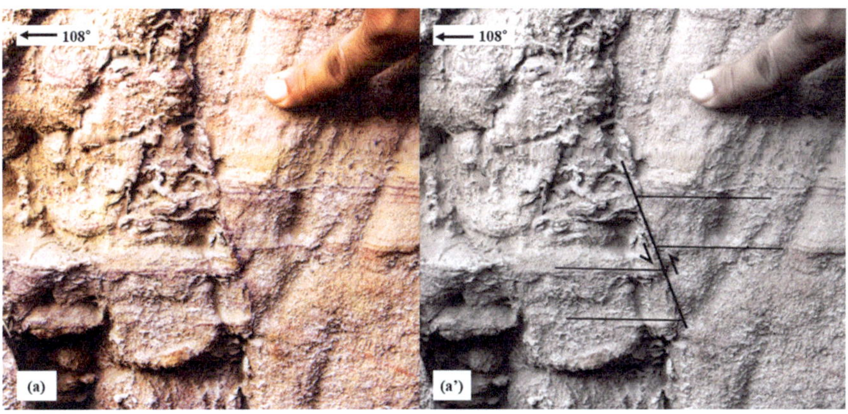

Fig. 7 Uninterpreted (a) and interpreted (a′) images of vertical section of high-angle reverse fault in Early Cretaceous sandstone of Lower Bhuj succession. Finger (6 cm) as marker. Location: 23° 14′ 38.228″ N, 69° 35′ 8.286″ E

Fig. 8 Mesoscopic conjugate faults within sandstone sequence of early Cretaceous Lower Bhuj succession. The fault plane strikes nearly N-S, and dip steeply ~70°. Slip is greater on the fault towards left of the section. S. Mukherjee is as scale. Kodki road fault, near Bhuj city, 23° 14′ 38.228″ N, 69° 35′ 8.286″ E

Sigmoidal Structures

Numerous sigmoid structures were noted in the field. The shape and inclination of P planes with respect to Y planes have been used to deduce the slip direction conclusively (Figs. 9, 10, 11, 12, 13, and 14).

Fault drag is described as the curved markers adjoining a fault (e.g. Kearey, 1993). Fault drag can be of two types—Normal drag and Reverse drag. Normal drag can be described as the convex deflection of markers in the direction of slip. It is more commonly encountered in the field. Whereas, a reverse drag can be described as the concave deflection of markers in the direction of slip (review in Mukherjee, 2014).

Normal fault with normal drag is a very common case and widely seen in the field.

Reverse faulted normal dragged rock units are widely known and reported profusely in literature. Natural examples of this type have been reported earlier by Passchier (2001), he referred them as 's-Type flanking folds' (Fig. 15).

Reverse fault with reverse drag, a rare variety of reverse fault, was documented from a single place (Figs. 16 and 17).

Fault Gouge

Refracted fault planes (e.g., Bose et al., 2020) are an important structure observed in the area. Most of the faults cut through the Lower Cretaceous sandstone shale sequence of Bhuj Formation. Due to heterogeneous lithology when these faults pass

Fig. 9 Vertical section of the Bhuj-Kodki Road Fault in sandstone-shale sequence of Early Cretaceous Bhuj Formation, depicting a normal slip (length of exposed fault plane is ~4.6 m) (210° Strike, 55° Dip, 300° Dip direction) and four sigmoidal structures at the top. The sigmoid structure marked by red box (close-up in inset) shows Top-to-225° sheared P-planes. The Y-planes that bound the curved P-planes are non-parallel. The other inset shows stereo plot of the normal fault plane. S. Mukherjee (1.74 m) is shown as scale. Location: Road right section from E of the Kodki Road Fault, near Bhuj city, Gujarat, India. Location: 23° 14′ 38.228″ N, 69° 35′ 8.286″ E

Fig. 10 Uninterpreted (a) and interpreted (a′) images of vertical section of normal fault in marker layer with sigmoidal structures in Early Cretaceous sandstone of Bhuj Formation. The P-planes are almost of equal curvature confined within parallel Y-planes which is not marked. Part of finger (6 cm) as scale. Location: Road section from E of Kodki road fault, near Bhuj city, Gujarat, India. Coordinates: 23° 14′ 38.228″ N, 69° 35′ 8.286″ E

Fig. 11 Uninterpreted (a) and interpreted (a′) images of vertical section of brittle shear Y- and P-planes observed in a vertical section in Early Cretaceous sandstone of Bhuj Formation. Top-to-30° sheared P-planes are bounded by not so clear Y-planes indicate reverse faulting. Finger (8 cm) is used as marker. Location: Kodki road fault, near Bhuj city, Gujarat, India. Coordinates: 23° 14′ 38.228″ N, 69° 35′ 8.286″ E

Fig. 12 Vertical section of the Bhuj-Kodki Road Fault in Cretaceous sandstone-shale succession of Bhuj Formation, depicting a normal fault (attitude: 210° strike, 55° dip, 300° dip direction). The red box (close-up in other boxes—Un-interpreted (a) and interpreted (a′)) shows reverse faulting within the normal fault. The top-to-30° sheared P-planes are bounded by parallel Y-planes. Finger (6 cm) as scale. The other inset shows stereo plot of the normal fault plane. S. Mukherjee (1.74 m) as scale. Location: Road section from E of the Kodki Road, near Bhuj city, Gujarat. Coordinates: 23° 14′ 38.228″ N, 69° 35′ 8.286″ E

Fig. 13 Uninterpreted (a) and interpreted (a′) images of vertical section of normal fault in marker layer with normal drag in Early Cretaceous sandstone of Bhuj Formation. Low angles favor normal drag (Grasemann et al., 2005). The low angle mentioned here is the acute angle measured from the fault to the undeformed central marker (anticlockwise angles are positive). Part of finger (6 cm) is shown as scale. Location: Road left section from E of Kodki road fault, near Bhuj city, Gujarat, India. Coordinates: 23° 14′ 38.228″ N, 69° 35′ 8.286″ E

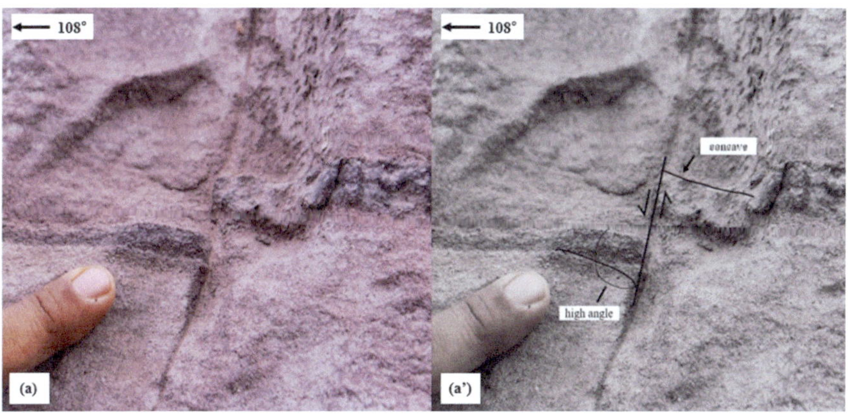

Fig. 14 Uninterpreted (a) and interpreted (a′) images of normal fault in marker layer with reverse drag in vertical section of Early Cretaceous sandstone of Bhuj Formation. Part of finger (3 cm) as scale. Location: Road section from E of Kodki road fault, near Bhuj city, Gujarat, India. Coordinates: 23° 14′ 38.228″ N, 69° 35′ 8.286″ E

through the shale units, they get refracted. Lithological heterogeneity causes fault planes to get refracted (Figs. 18 and 19).

Fault related landforms are crucial as they give indications regarding tectonic activity. Fault scarp is one such landform. Prior workers have given various definitions, in the present work the term ‘fault scarp’ is referred to as the topographic

Fig. 15 Uninterpreted (a) and interpreted (a′) images in vertical section of high angle reverse fault in marker layer with normal drag in Early Cretaceous sandstone of Bhuj Formation. Part of finger (5 cm) as scale. Location: Road section from E of Kodki road fault, near Bhuj city, Gujarat, India. 23° 14′ 38.228″ N, 69° 35′ 8.286″ E

Fig. 16 Uninterpreted (a) and interpreted (a′) images of reverse fault with reverse drag in vertical section of Early Cretaceous sandstone- shale sequence of Bhuj Formation. Part of finger (2 cm) as scale. Location: Road left section from E of Kodki road fault, near Bhuj city, Gujarat, India. Coordinates: 23° 14′ 38.228″ N, 69° 35′ 8.286″ E

expression of faulting which is associated with displacement caused by movement along the fault. It coincides with the fault plane and carries kinematic indicators of tectonic activity. Few characteristics of fault scarps are: (i) they are primarily related with episodic movements which can be due to any type of tectonic displacements like earthquakes and faults of various types, including strike-slip faults; (ii) very prone to erosion because they form due to a sudden uplift along the fault and therefore are

Fig. 17 Vertical section of Network of faults in sandstone shale sequence of Lower Bhuj succession. No scale is used. Location: Kodki road fault, near Bhuj city, Gujarat, India. Coordinates: 23° 14′ 38.228″ N, 69° 35′ 8.286″ E

Fig. 18 Vertical section of a thick white colored fault gouge (attitude: dip amount 70°, dip direction 113°, strike 23°) in Early Cretaceous sandstone of Lower Bhuj succession. Soumyajit Mukherjee (174 cm) is shown as scale. Location: Kodki road fault, near Bhuj city, Gujarat, India. 23° 14′ 38.228″ N, 69° 35′ 8.286″ E

Fig. 19 Subvertical section showing refraction of fault plane due to marker layer in Early Cretaceous sandstone of Bhuj Formation causing decrease in dip angle. Part of finger (3 cm) is shown as scale. Location: Road left section from E of the Kodki Road, near Bhuj city, Gujarat. Coordinates: 23° 14′ 38.228″ N, 69° 35′ 8.286″ E

mineralized and differently colored; (iii) they can be few centimeters to many meters in height. Several segmented fault scarps of variable height ranging from 15 cm up to few meters are present in the study area. Mostly they are NNE-SSW and NW–SE trending and gives indication of dip slip movement. Figure 20 shows one prominent exposure of fault scarp.

The studied fault scarps are comparable with 'fault-fins' structures described by Davis (1990) in Navajo sandstone of Colorado Plateau region in Southern Utah. Like fault-fin structures these fault scarps also give bladed appearance, occur as aligned elements that project upward from a common structural trace and are probably related to deformation band shear zones of tectonic origin.

Deformation bands are primary deformation elements in faults, which are a result of strain localization in highly porous rocks, usually noted in sandstones of high porosity (Fossen et al., 2007; Shaikh et al., 2020). Deformation bands differ from extensional fractures formed in low-porosity rocks as they: (i) increase cohesion and decrease permeability in contrast to extensional fractures, which tends to decrease cohesion and increase permeability; (ii) are associated with strain hardening whereas extensional fractures are associated with strain softening; (iii) are thicker and show low-displacement as compared to extensional fractures. Numerous deformation bands were noted in the study area. They are present as individual bands and in clusters as zones of bands, also reported within slip surfaces (faulted deformation bands). Their occurrence in the study area is important as: (i) prior workers have reported their association with major faults which is the case here as well (Shaikh

Fig. 20 A prominent exposure of a fault with unknown slip-sense(190° strike, 70° Dip and 215° Dip direction) in Lower Cretaceous sandstone of Bhuj Formation. Soumyajit Mukherjee (174 cm) is shown as scale. Location: Kodki road fault, near Bhuj city, Gujarat, India. Coordinates: 23° 14′ 38.228″ N, 69 °35′ 8.286″ E

et al., 2020). The density of deformation bands decreases on both sides as we move away from the Kodki transverse fault; (ii) they result in decreased permeability and increased cohesion and therefore, they act as a barrier to fluid flow (e.g., Pittman, 1981; Sample et al., 2006); (iii) they provide crucial information about their formation (e.g. Aydin & Johnson, 1978; Johnson, 1995) and progression (e.g. Schultz & Siddharthan, 2005; Wong et al. 2004) of faults in porous sandstones (Figs. 21, 22, and 23).

Another interesting feature noted in the area is the large sized honeycomb or boxwork sandstone boulders (Fig. 24). Numerous such boulders were noted in the study area around the fault-controlled ridge. A network of cm-scale deformation band shear zones that intersect each other can be seen. Honeycomb/boxwork structure are formed when the sandstone between these deformation structures weathers out.

Mineralised Fault Plane

Mineralized and differently colored surfaces are generally characteristic of fault planes. Almost all the fault planes were mineralized in the study area. Figure 25 presents an example.

Igneous Intrusion

Maurya et al. (2003) suggested that the formation of transverse faults that offset major E-W trending intra-basinal faults is contemporaneous with the dyke emplacement i.e. it is related with the Deccan trap volcanism in an extensional stress regime.

Fig. 21 Horizontal section of a cluster of deformation bands trending N-S in Early Cretaceous sandstone of Bhuj Formation. Pen (15 cm) is shown as scale. The deformation bands follow the trend of the Kodki transverse fault which we are studying. Location: Kodki road, near Bhuj. 23° 14′ 38.228″ N, 69° 35′ 8.286″ E

Fig. 22 Subhorizontal view of an almost E-W trending deformation band in weathered sandstone. The band is sinistrally faulted along a N-S trending transverse fault. Pen (15 cm) is used as scale. Location: Kodki road, near Bhuj. 23° 14′ 38.228″ N, 69° 35′ 8.286″ E

He suggested this on the basis of close association of intrusive dykes along transverse faults. One such association was observed in the field as well. A dolerite dyke was noted, which follows the trend of Kodki transverse fault. Figure 27 shows the spheroidal weathering of the dyke (Fig. 26).

Fig. 23 Sub-horizontal section of fractured deformation band in weathered Bhuj sandstone. Coin as scale. Location: Kodki road fault, near Bhuj. 23° 14′ 38.228″ N, 69° 35′ 8.28″ E

Fig. 24 Close-up of a large boulder showing the winnowed boxwork character of the deformation band shear zone in Lower Cretaceous Bhuj sandstone. The Kodki transverse fault ridge is flanked by such structures on its sides. Nidhi Lohani (163 cm) as scale. Location: Kodki road fault, near Bhuj city, Gujarat, India. Coordinates: 23° 14′ 38.228″ N, 69° 35′ 8.286″ E

Fig. 25 A sub-vertical, mineralized and striated fault plane (attitude: 210° strike, 55° dip, 300° Dip direction) in Early Cretaceous sandstone of Bhuj Formation. Lineations indicate dip-slip movement. Finger (6 cm) is shown as scale. Location: Road right section from E of the Kodki Road, near Bhuj city, Gujarat. Coordinates: 23° 14′ 38.228″ N, 69° 35′ 8.286″ E

Fig. 26 A vertical section of almost NE-SW trending dolerite dyke (attitude: 220° strike, 70° dip, 310° dip direction) of about 5–6 m width cutting through the thinly laminated sandstone-shale sequence of Early Cretaceous Lower Bhuj succession. The inset shows stereoplot of the dyke. No scale is used. Location: Kodki road fault, near Bhuj city, Gujarat, India. Coordinates: 23° 14′ 38.228″ N, 69° 35′ 8.286″ E

Fig. 27 Subvertical section of a spheroidally weathered dolerite dyke. Rounded features developed. S. Mukherjee (174 cm) is shown as scale. Location: Kodki road, near Bhuj city, Gujarat, India. Coordinates: 23° 14′ 38.228″ N, 69° 35′ 8.286″ E

Folds

Although KHRFZ is reported to have many fold-related structures, not many are noted in the area. Figure 28 shows a drape folding formed due to sagging of a shale layer due to overlying sandstone layers.

Fig. 28 Uninterpreted (a) and interpreted (a′) images of Dip-slip normal fault and fault-related folding in vertical section of thinly laminated sandstone-shale sequence of Early Cretaceous Lower Bhuj succession. S. Mukherjee (174 cm) is shown as scale. Location: Kodki road, near Bhuj city, Gujarat, India. Coordinates: 23° 14′ 32.6461″ N, 69° 34′ 57.28681″ E

4 Conclusions

Meso-scale normal faults strikingNW-SE to NE-SW is the main observation on the Kodki road. Such faultsarealso accompanied by different drag patterns, local-scale reverse faults, network of faults and conjugate faults, sigmoid shear planes, shear bands etc. Paleostress study from this location can be an important input into the tectonics of the KRB.

Acknowledgements Alexis Vizcaino and the proofreading team (Springer). CPDA grant (IIT Bombay) supported SM. NL, DM and AP worked as self-financed candidates. Mukherjee (2023) summarized this chapter.

References

Aydin, A., & Johnson, A. M. (1978). Development of faults as zones of deformation bands and as slip surfaces in sandstones. *Pure and Applied Geophysics, 116*, 931–942.

Biswas, S. K. (1977). Mesozoic Stratigraphy of Kutch, Gujarat. *The Quarterly Journal of the Geological, Mining, and Metallurgical Society of India, 49*(3,4), 1–52.

Biswas, S. K. (1987). Regional tectonic framework, structure and evolution of the Western marginal basins of India. *Tectonophysics, 135*, 307–327.

Biswas, S. K. (1993). *Geology of Kachchh.* K. D. Malaviya Institute of Petroleum Exploration

Biswas, S. K. (2016). *Tectonic framework, structure and tectonic evolution of Kutch Basin, Western India* (No. 6, pp. 129–150). Special Publication of the Geological Society of India

Biswas, S. K. (2005). Are view of structure and tectonics of Kutchbasin, Western India, with special reference to earthquakes. *Current Science, 88*, 1592–1600.

Biswas, S. K., & Khattri, K. N. (2002). A geological study of earthquakes in Kutch, Gujarat, India. *Geological Society of India, 60*(2), 131–142.

Bose, N., Dutta, D., & Mukherjee, S. (2020). Refraction of micro-fractures due to shear-induced mechanical stratigraphy in a low-grade meta-sedimentary rock. *Journal of Structural Geology, 133*, 103995.

Casshyap, S. M., Dev, P., Tewari, R. C., & Raghuvanshi, A. K. S. (1983). Ichno fossils from Bhuj formation (Cretaceous) as palaeoenvironmental parameters. *Current Science, 52*(2), 73–74.

Chung, W. P., & Gao, H. (1995). Source parameters of Anjar earthquake of July, 1956, India, and its seismotectonic implications for the Kutch rift basin. *Tectonophysics, 242*, 281–292.

Das, A., Prizomwala, S. P., Solanki, T., Chauhan, G., Thakkar, M. G., & Bhatt, N. (2019). Relative assessment of tectonic activity along the seismically active Katrol Hill Fault, Kachchh, Western India. *Journal Geological Society of India, 94*, 179–187.

Dasgupta, S., & Mukherjee, S. (2019). Remote sensing in lineament identification: Examples from Western India. *Problems and Solutions in Structural Geology and Tectonics, 5*, 205–221.

Deshpnde, S. V., & Merh, S. S. (1980). Mesozoic sedimentary model of Wagad hills, Kutch, Western India. *Journal of Geological Society of India, 21*, 75–83.

Fossen, H., Schultz, R. A., Shipton, Z. K., & Mair, K. (2007). Deformation bands in sandstone: A review. *Journal of the Geological Society, London, 164*(2007), 1–15.

Fursich, F. T., Pandey, D. K., Callomon, J. H., Jaitly, A. K., & Singh, I. B. (2001). Marker beds in the Jurassic of the Kachchh Basin, Western India: Their depositional environment and sequence stratigraphic significance. *Journal of the Palaentological Society of India, 46*, 173–198.

Grasemann, B., Martel, S., Passchier, C. (2005). Reverse and normal drag along a fault. *Journal of Structural Geology, 27*, 999–1010.

Johnson, A. M. (1995). Orientations of faults determined by premonitory shear zones. *Tectonophysics, 247*, 161–238.

Krishna, J., Pandey, B., Ojha, J. R., & Pathak, D. (2009). Reappraisal of the age framework, correlation, environment and nomenclature of Kachchh Mesozoic lithostratigraphic units in Wagad. *Journal of Scientific Research, Banaras Hindu University, Varanasi, 53*, 1–20.

Krishna, J., Sigh, I. B., Howard, J. D., & Jafar, S. A. (1983). Implications of new data on Mesozoic rocks of Kachchh, Western India. *Nature, 305*, 790–792.

Kundu, H., & Thakkar, M. (2011). Elemental concentration of U, Th and K in tectonically active regions of Khari River basin, Kachchh, Western India. *Journal of Radioanalytical and Nuclear Chemistry, 290*. https://doi.org/10.1007/s10967-011-1200-1

Maurya, D. M., Chowksey, V., Patidar, A. K., & Chammyal, L. S. (2016). A review and new data on neotectonic evolution of active faults in the Kachchh Basin, Western India: legacy of post-Deccan Trap tectonic inversion. *Geological Society, London, Special Publications, 445*. https://doi.org/10.1144/SP445.7

Maurya, D. M., Shaikh, M. A., & Mukherjee, S. (2021b). Comment on "Structural attributes and paleostress analysis of quaternary landforms along the Vigodi Fault (VF) in Western Kachchh region. *Quaternary International*. https://doi.org/10.1016/j.quaint.2021.04.029

Maurya, D. M., Tiwari, P., Shaikh, M., Patidar, A. K., Vanik, N., Padmalal, A., & Chamyal, L. S. (2021a). Late Quaternary drainage reorganization assisted by surface faulting: The example of the Katrol Hill Fault zone, Kachchh, western India. *Earth Surface Processes and Landforms*. https://doi.org/10.1002/esp.5097

Maurya, D. M., Thakkar, M. G., & Chamyal, L. S. (2003). Implications of transverse fault system on tectonic evolution of Mainland Kachchh, western India. *Current Science, 85*, 661–667.

Morino, M., Malik, J. N., Mishra, P., Bhuiyan, C., & Kaneko, F. (2008). Active fault traces along Bhuj Fault and Katrol Hill Fault, and trenching survey at Wandhay, Kachchh, Gujarat, India. *Journal of Earth System Science, 117*, 181–188.

Mukherjee, S. (2023). Introduction to structural geology and tectonics field guidebook—volume 2. In S. Mukherjee (Ed.), *Structural geology and tectonics field guidebook—volume 2* (pp. xi–xiv). Springer Nature Switzerland AG. ISBN: 978-3-031-19575-4.

Mukherjee, S. (2014). Review of flanking structures in meso- and micro-scales. *Geological Magazine, 151*, 957–974.

Padmalal, A., Khonde, N., Maurya, D. M., Shaikh, M., Kumar, A., Vanik, N., & Chamyal, L. S. (2019). Geomorphic characteristics and morphologic dating of the Allah Bund Fault scarp, great Rann of Kachchh, western India. In *Tectonics and structural geology: Indian context* (pp. 55–74). Springer.

Padmalal, A., Maurya, D. M., Vanik, N. P., Shaikh, M. A., Tiwari, P., & Chamyal, L. S. (2021). Impact of long term uplift on stream networks in tectonically active Northern Hill Range, Kachchh palaeo-rift basin, western India. *Journal of Mountain Science, 18*(6), 1609–1629. https://doi.org/10.1007/s11629-020-6420-9

Passchier, C. W. (2001). Flanking structures. *Journal of Structural Geology, 23*, 951–962.

Patidar, A. K., Maurya, D. M., Thakkar, M. G., & Chamyal, L. S. (2007). Fluvial geomorphology and neotectonic activity based on field and GPR data, Katrol hill range, Kachchh, western India. *Quaternary International, 159*, 74–92.

Patidar, A. K., Maurya, D. M., Thakkar, M. G., & Chamyal, L. S. (2008). Evidence of neotectonic reactivation of the Katrol Hill Fault during late quaternary and its GPR characterization. *Current Science, 94*, 338–346.

Paul, H., Mitra, S., Bhattacharya, S. N., & Suresh, G. (2015). Active transverse faulting within underthrust Indian crust beneath the Sikkim Himalaya. *Geophysical Journal International, 201*(2), 1072–1083.

Pittman, E. D. (1981). Effect of fault-related granulation on porosity and permeability of quartz sandstones, Simpson Group (Ordovician) Oklahoma. *AAPG Bulletin, 65*, 2381–2387.

Rai, J. (2006). Discovery of nannofossils in a plant bed of the Bhuj Member Kutch and its significance. *Current Science, 91*(4), 519–526.

Rastogi, B., Mandal, P., & Biswas, S. (2014). Seismogenesis of earthquakes occurring in the ancient rift basin of Kachchh, Western India. In P. Talwani (Ed.), *Intraplate earthquakes* (pp. 126–161). Cambridge University Press. https://doi.org/10.1017/CBO9781139628921.007

Sample, J. C., Woods, S., Bender, E., & Loveall, M. (2006). Relationship between deformation bands and petroleum migration in an exhumed reservoir rock, Los Angeles Basin, California, USA. *Geofluids, 6*, 105–112.

Schultz, R. A., & Siddharthan, R. (2005). A general framework for the occurrence and faulting of deformation bands in porous granular rocks. *Tectonophysics, 411*, 1–18.

Shaikh, M. A., Maurya, D. M., Vanik, N. P., Padmalal, A., & Chamyal, L. S. (2019). Uplift induced structurally controlled landscape development: example from fault bounded Jumara and Jara domes in Northern Hill Range, Kachchh, Western India. *Geosciences Journal*, 1–19. https://doi.org/10.1007/s12303-018-0061-9

Shaikh, M. A., Maurya, D. M., Mukherjee, S., Vanik, N. P., Padmalal, A., & Chamyal, L. S. (2020). Tectonic evolution of the intra-uplift Vigodi-Gugriana-Khirasra-Netra fault system in the seismically active Kachchh rift basin, India: Implications for the western continental margin of the Indian plate. *Journal of Structural Geology, 140*, 104124. https://doi.org/10.1016/j.jsg.2020.104124

Shukla, U. K., & Singh, I. B. (1990). Facies analysis of Bhuj sandstone (Lower Cretaceous) Bhuj area, Kachchh. *Journal of Paleontological Society of India, 35*, 189–196.

Singh, A. P., Roy, I. G., Kumar, S., & Kayal, J. R. (2014). Seismic source characteristics in Kachchh and Saurashtra regions of Western India: b- value and fractal dimension mapping of aftershock sequences. Natural Hazard. https://doi.org/10.1007/s11069-013-1005-3

Sohoni, P. S., Malik, J. N., Merh, S. S., & Karanth, R. V. (1999). Active tectonic astride Katril Hill Zone, Kachchh, Western India. *Journal of the Geological Society of India, 53*, 579–586.

Thakkar, M. G., et al. (2017). *Journal of Geology Geophysics, 6*:3(Suppl). https://doi.org/10.4172/2381-8719-C1-008

Thakkar, M. G., Maurya, D. M., Raj, R., & Chamyal, L. S. (1999). Quaternary tectonic history and terrain evolution of the area around Bhuj, Mainland Kachchh, western India. *Journal Geological Society of India, 53*, 601–610.

Walid, M., Mukherjee, S., & Dasgupta, S. (2021). Vein geometry (Bhuj, Gujarat, India). In S. Mukherjee (Ed.), *Structural geology and tectonics field guidebook—volume 1* (pp. 707–714). Springer Nature Switzerland AG.

Wilcox, R. E., Harding, T. P., & Seely, D. R., (1973) Basic wrench tectonics. *AAPG Bulletin, 57*, 74–76.

Wong, T.-f., David, C., & Menéndez, B. (2004). Mechanical compaction. In Y. Guéguen, & Boutéca, M. (Eds.), *Mechanics of fluid-saturated rocks* (pp. 55–114). Elsevier.

Structural and Sedimentary Field Studies in Angul District, Odisha, India

Subhajit Sinha, Ananya Ghosh, and Somraj Mishra

Abstract The Angul District of Odisha exposes excellent and accessible outcrops of Archean-Proterozoic metamorphics and migmatites, near-horizontal Gondwana Supergroup of sedimentary succession and the overlying Quaternary deposits with an unconformable relationship between the Archean-Proterozoic and the Gondwanas, but that is not exposed. The Archean-Proterozoic rocks are the oldest and form the basement in the region. Stomatic structures formed by alternating quartzo-feldspathic leucosome bands and the dark mafic melanosome layers are conspicuous. The granularity is high with garnets reaching nearly 1.2 cm diameter. Folding, fracturing, faulting and mafic intrusives are also observed at many places. The country rock in most part is Khondalite associated with charnockites and migmatites. Superposed folds indicate several generations of folding. The khaki shales and greyish white sandstones of Talchir Formation and the greyish sandstones of Barakar Formation constitute most of the sedimentary exposures of the Gondwana along with exposures coal seams and carbonaceous shales. Various sedimentary structures are observed in the Barakar sandstones and lithological analysis indicates their fluviatile origin with varied energy fluctuations. The Quaternary deposits are fluviatile in origin and contain boulder beds and sand-mud lithology. Paleocurrent analysis from the boulders and cross stratifications indicates flow direction sub-parallel to the present day channel.

Keywords Diastropic structure · Orissa · Granulite · Strain analysis · Sedimentary structures

---

S. Sinha (✉) · S. Mishra
Ballygunge Science College, University of Calcutta, 35, Ballygunge Circular Road, Kolkata, West Bengal 700019, India
e-mail: ssgeol@caluniv.ac.in

A. Ghosh
Department of Earth Sciences, Indian Institute of Technology Bombay, Powai, Mumbai, Maharashtra 400076, India

S. Mukherjee (ed.), *Structural Geology and Tectonics Field Guidebook–Volume 2*, Springer Geology, https://doi.org/10.1007/978-3-031-19576-1_8

1 Introduction

A geological fieldwork is carried out in and around Angul district (Odisha, India) for structural and sedimentological studies. The district is situated nearly at the central point of the state. It was formerly a part of the Dhenkanal district but later for administrative convenience, it was upgraded to a district status. The district has an area of approximately 6232 km^2. The district shares its border with Koenjhar towards NE, Dhenkanal in the east, Cuttack and Nayagarh in S-SE, Boudh in S-SW, Sambalpur and Deogarh in the west and Sundargarh towards north (Fig. 1). The district is very rich in mineral resources like coal, kyanite, China clay, graphite and many semi-precious stones.

2 Study Area-Angul District

2.1 Connectivity and Accessibility

The city is well connected by road and rail network. The district is well connected by roadways to nearly many of the important districts as it lies beside the National Highway No. 42. The district headquarter is situated in the capital city of Bhubaneswar which is at a distance of nearly 170 km. Angul Railway station forms an important railhead of the East Coast Railways on the Bhubaneswar-Talchir-Sambalpur railway track. All types of passenger and goods train especially carrying coal forms a very busy railway station. The Mahanadi and Brahmani Rivers forms the main waterways of the district and navigation is possible for nearly seven months from September to March to carry different goods like bamboos, timbers and other commodities. All the study areas are accessible from Amalapada, about 4 km from Angul Railway Station. One can reach Amalapada via Jajpur by bus from Cuttack and it takes approximately 4 h to reach the destination.

2.2 Geographic Location and Topography

Angul district is confined between 20° 31′ N and 21° 41′ N latitude and 84° 15′ E and 85° 23′ E longitude and is located nearly at the center of the state. The topography consists of ridges, hillocks plains and gentle undulations with altitude ranging between 570 and 1200 m. The name Angul originated from combining Anugola, (Anu-the last Khond Tribal Chief who was removed after a battle-gol). Two major river drains through Angul; the Mahanadi in its southern boundary and the Brahmani River in its northern region. These two rivers have many tributaries, some are perennial and some are torrent streams that had curved the landscape uniquely.

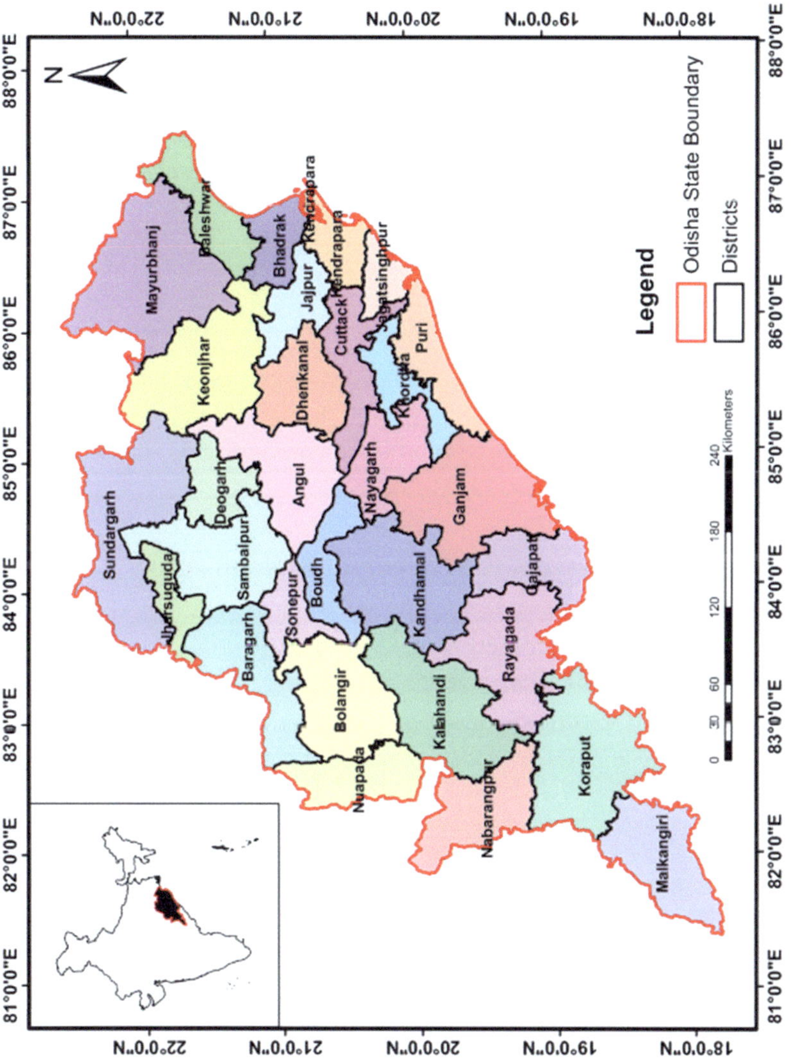

Fig. 1 Map of Odisha showing the location of Angul district and the districts sharing their boundaries

2.3 Geology

Overall the geomorphology of the region is a continuation of various structural units of the Eastern Ghat Supergroup and the Older Metamorphic Group along with the faults and lineaments. Some of these lineaments manifest as depressions and are filled with recent sediments. Ridges trend ~NW–SE and the same trend is followed by the coal-bearing Talchir Gondwana sedimentary basin.

The major part of the region of Angul district is occupied by the Khondalites along with charnockites and migmatites. The Talchir sedimentary basin is NW–SE elongated and occupies a large portion of the district towards the north and the basin abuts in the exposure of Archean Banded Iron Formations (Fig. 2) (Table 1).

3 Objectives & Approaches

This chapter deals with structural and sedimentological studies. The area comes within the Survey of India toposheet number 73.Here: (i) major lithotypes, (ii) documentation of linear and planar structures, (iii) field relationship of various structures, and (iv) sedimentological analysis of a part of Gondwana Supergroup are presented.

4 Lithological Analysis of the Basement Metamorphic Rocks

4.1 Litho-textural Observations

Exposures are located at different places around 11 m towards north of the Angul railway station. The country rock is made up of a mixed metamorphic terrain consisting of Khondalites, Migmatites and Charcnockites and various gneisses that had probably undergone upper ambhibolite-granulite facies conditions of metamorphism. Extensive migmatization is evident from its stromatic structure along deformation indicated by the presence of ptygmatic folds. The rock is greasy in appearance and contains alternate light quartzo-feldspathic and dark pyroxene and mafic minerals bands. Garnets of various sizes reaching up to 1.2 cm are copious. The thickness of the light and dark colour bands varies both laterally and vertically (Figs. 3, 4, 5, 6 and 7). Overall study area is a part of Eastern Ghat Granulite Terrain, where the large area of this mobile belt is occupied by migmatite complex which consists of banded gneisses, garnetiferous biotite gneiss and garnet bearing quartzo-feldspathic gneisses. The migmatite has probably formed by anatectic melting of the country rock in situ or by injection of granite melt into the metamorphosed country rock. Some patchy unidentified lithologies are observed nearby the shear and fracture planes might have formed by metasomatism during various metamorphic events.

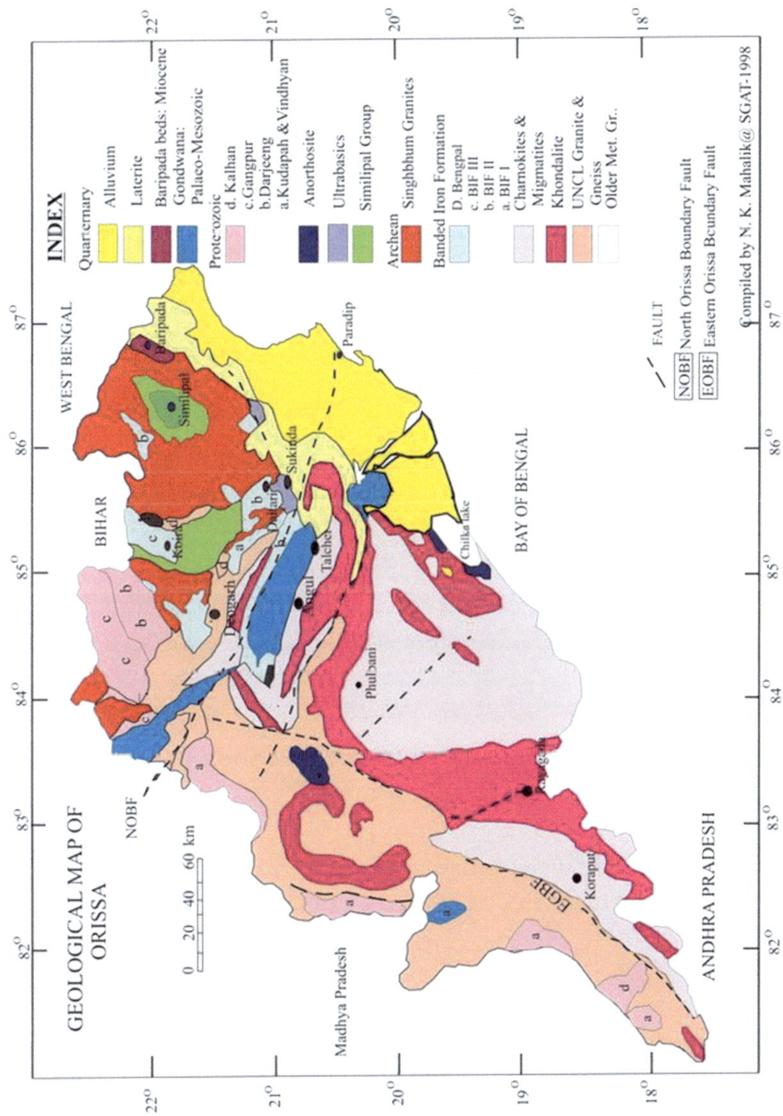

Fig. 2 Regional geological map of Odisha showing Angul district in the center (after Mahalik, 1998)

Table 1 The overall geological setup and mineralogy at a glimpse of Odisha district

Sl No	Age	Geological units	Sectors and broad litho-association
1	Phanerozoic	Cenozoic (quaternary formation and small patches of tertiary)	Eastern Coastal plain: - Alluvial sediments, interstratified quaternary ash beds and low level laterites. Residual hills and tors exposing Precambrian rock types occur in this coastal plain. Some sectors of deltaic fans extending into off-shore region have prospects of oil and gas
		Mesozoic (Gondwana Supergroup)	Central Odisha: - In the valley of Brahmani, Mahanadi and Ibriverand some of their tributaries. The sedimentary rock formations, the lower stratigraphic column have coal deposits
2	Precambrian	Proterozoic	Western Odisha: - Platformal sedimentary formations and associated limestone deposits resting on Archean gneisses and supracrustals Northwestern Odisha: - Meta-sedimentary of low or medium metamorphic grade of Gangpur Group having manganese, limestone and Pb–Zn deposits Granulites belt: - Lithopackage comprising of Khondalites, Charnockite, Migmatite, Anorthosite, and alkaline rocks having associated deposits of manganese, graphite and occurrence of gemstones

(continued)

Various sets of fractures and joints are observed that generally trend NS and EW. Some displacements are also observed at places.

4.2 *Khondalite Suite of Rocks*

The Khondalite consists of quartz-k-feldspar-garnet-silimanite-graphite bearing schists and gneisses. These are metapelites with high alumina content metamorphosed to granulite facies. Inter bands of quartzite, containing garnet and silimanite occur within Khondalite. Regional metamorphism of Eastern Ghat belt belongs to

Table 1 (continued)

Sl No	Age	Geological units	Sectors and broad litho-association
3		Archean	Northern Odisha (North Odisha Craton): a. Supracrustal belt including Iron Ore Supergroup having deposits of iron, manganese, gold and base metals b. Gniess, granites and migmatites including Singbhum, Bonai and Mayurbhanj plutons c. Mafics and ultramafic intrusives, some with deposits of chromites, Ti-V magnetite and PGE especially of Sukinda and Baula Western Odisha (Bastar Craton): - Gniess, granites and migmatites including some Sn-Ta-Nb bearing granite pegmatites and the Supracrustal assemblages

Modified after Manjrekar et al. (2006) and Senapati and Behera (2015)

Fig. 3 Outcrop of the metamorphic rock showing the light coloured leucosome and the dark coloured melanosome. Folding is conspicuous

Fig. 4 Contact between coarse and fine grained lithology

Granulite facies metamorphism of intermediate P/T facies series. In Khondalites and migmatites, index minerals such as sillimanite indicate high temperatures.

4.3 Tectonic Classification of Garnet Porphyroblasts

The garnet porphyroblasts in the study area shows three types of tectonic orientation which are classified as pre-, syn- and post-kinematic garnet porphyroblasts.

(a) *Pre-kinematic porphyroblasts* (Fig. 8): The pre-kinematic porphyroblasts were clearly recognizable by cracked and broken grains, occurrence of pressure shadow around the porphyroblast. Foliation planes do not cut through the mineral grain.
(b) *Syn-kinematic porphyroblasts* (Fig. 9): The syn-kinematic porphyroblasts were identified through the continuous schistosity along with the wrapped up pressure shadow zone of different shapes.

(c) *Post-kinematic porphyroblasts* (Fig. 10): These porphyroblasts were identified as being parallel to the foliation, randomly oriented without any pressure shadow and cut across an earlier foliation (Table 2).

Fig. 5 Large size garnet are scattered and found to be embedded within felsic-mafic bands giving rise to porphyroblastic texture

From the analysis of the garnet porphyroblasts and the nature of the pressure shadow zone, it can be inferred that the garnets are syn- to post-kinematic.

4.4 *Analysis of Superposed Folding*

The area has undergone extensive folding along with several other types of deformations. Such an exposure is observed north of Angul railway station near the Hanuman Temple.

(a) *Folds*: In general, the exposure consisted of disharmonic, ptygmatic folding. Refolded initial folds of Type-2 and Type-3 were observed.

Initial fold: fold axes $= f_1$; normal to axial plane $= c_1$; direction of heterogeneous shear displacement $= d_1$. Superposing fold: fold axes $= b_2$; normal to axial plane $= c_2$; direction of heterogeneous shear displacement $= a_2$. Angles between f_1 and $b_2 = \alpha$; between c_1 and $a_2 = \beta$; between f_1 and $c_2 = \gamma$; between c_1 and $c_2 = \delta$ (Figs. 11 and 12).

Type 2 refold ($\alpha = 90°$, β and $\gamma = 0°$): The initial axial plane and the initial fold axis are deformed. If the refold structure is progressively unroofed perpendicular

Fig. 6 Garnet porphyry and nesting are observed both in the leucosome and the melanosome. Note the size of garnets that reaches a maximum of 0.8 cm. As these porphyries are near symmetric, they do not reveal any shear sense (e.g., Mukherjee, 2017)

to b_2, interference patterns with circular forms, rounded triangular, crescent shapes and typical dome-crescent-mushroom patterns characterize the sections (Ramsay & Huber, 1987). However, oblique sections, especially if the refold structures deviate from the end-member orientation, show a great variability of complex interference patterns (Thiessen, 1986).

Type 3 refold ($\gamma = 90°$, α, $\beta = 0°$): The initial axial plane is deformed but the initial fold hinges are not bent by superposing folding. Cross sections parallel to the fold axes (f_1 and b_2) will not develop a complex interference pattern but show parallel, straight lines. However, cross sections normal to the fold axes will show complex convergent divergent or hook shaped interference patterns (Thiessen, 1986) (Fig. 13).

Fig. 7 Leucosomes and melanosomes together formed stromatic structure (leucosome-melanosome inter-layering) as observed in the metamporphic basement rocks near Angul railway station

(b) *Fault*: A fault is documented towards the eastern end crossing the railway line towards the nearby village. The trend of the fault is along ENE-WSW. The slip along the fault plane is dextral, where the SE block has moved northerly with respect to the NE block (Fig. 14).

(c) *Dyke*: An amphibolite dyke crosscutting the country rock has been observed. The NE-SW trending dyke is 7–8 cm wide (Fig. 15).

(d) *Fractures*: Two sets of conjugate fractures exist, traversing across the outcrop. One of the fracture planes trend N–S while the other follow NE–SW trend. The other set of fracture cuts the former planes almost orthogonally.

(e) *Shearing*: A syn-tectonic shear along N–S direction was crosscut by a later developed shear sense along E–W direction, indicating two different phases of deformation. The slip sense is dextral for both the shear phases (Fig. 16).

(f) *S-C fabric*: S-C fabrics exist at a vertical rock outcrop within a leucosome layer with variable thickness. The sense of shear in the S-fabric is observed. The attitude of the schistosity plane in this area is 250°/48° SW and the pitch of lineation observed on the foliation plane is 70° from west (Fig. 17).

Fig. 8 Pre-kinematic garnet porphyroblast that do not cut the foliation plane

4.5 *Strain Analysis*

Strain analysis was carried out from augen structures as they are good shear sense indicators. The porphyroblasts are sometimes garnet and sometimes quartzo-feldspathic. At places auges are observed where the quartzo-feldspathic augen is having a garnet grain in the core. Augens are phi-type porphyroblasts that show symmetric tailed geometries with respect to regional foliation. The asymmetric types are sigma-types which show stepping geometry of the tails and the delta-types that have their tails thinner and strongly curved across the median line. From the symmetric and asymmetric augen structures, strain analysis was carried out using the Fry method and the Rf-phi method. The data was generated from outcrops present in the location called Gridco. Precautions of using such method have been discussed recently in Bose et al. (2018).

(a) *Fry method*: This method is used for 2-D strain analysis also known as centre-to-centre method. In a 1 m by 1 m cross-sectional area the centers of the augen structures were plotted in a tracing paper (selecting an appropriate scale). In another tracing paper a reference origin was taken and by placing the origin on augencentre, other augencentres were marked. Then the origin was moved to another centre and again the centres of other augens were marked. This procedure was repeated until the area of interest had been covered. Finally, a

Fig. 9 The syn-kinematic garnet porphyroblast where the schistosity is continuous through the grain

strain ellipse was obtained defied by the points clouding around the origin. The aspect ratio of the final strain ellipse was measured as X/Y to be 1.52 (Fig. 18).

(b) *RF-phi method*: The method handles initially non-spherical markers. For analyzing strain by this method, the aspect ratio of the augens and phi (angle between their long axis and a reference line) was measured and the data are listed in Table 4 (Figs. 19 and 20).

5 Gondwana and Quaternary Sedimentary Successions

Angul region contains extensive sedimentary succession of both Gondwana Supergroup and the Quaternary. Coal industry had clustered in nearly all over the area. The Gondwana succession in the district is a part of the Damuda Group consisting of Karharbari, Barakar and Barren Measures Formations and it overlies the Talchir Formation (Table 5).

Fig. 10 These post-kinematic porphyroblasts are warped by the foliation containing a pressure shadow zone

Table 2 Information on the basement rock

Sub-location	Leucosome and melanosome bands	Occurrence of garnet	Average size of Porphyroblasts (cm)	Relative time relation between growth of blasts and tectonics
Sub-location 1	Indiscrete	In felsic bands	1–2	Pre-tectonic porphyroblasts
Sub-location 2	Indiscrete	In felsic bands	0.5–1	Post-tectonic porphyroblasts
Sub-location 3	Discrete	In felsic bands	0.3–0.5	Post-tectonic porphyroblasts
Sub-location 4	Indiscrete	In mafic part	0.5–1.5	Could not be determined
Sub-location 5	Discrete	In felsic part	0.5–2	Syn-tectonic porphyroblasts

Table 3 Attitude of the structural attributes of the second generation folds

Sub-locations	Type of folds	Attitude of axial plane	Attitude of limb 1	Attitude of limb 2
Northerly exposed part	Asymmetric tight folds	46° towards 70°	40° towards 270°	60° towards 270°
Same as above	Asymmetric tight folds	44° towards 270°	40° towards 270°	30° towards 270°
Same as above	Asymmetric folds	51° towards 270°	42° towards 250°	70° towards 240°
Southerly exposed part	Disharmonic asymmetric tight folds	45° towards 220°	50° towards 250°	70° towards 220°
Same as above	Tight isoclinal folds	50° towards 220°	40° towards 250°	40° towards 250°
Same as above	Tight isoclinal folds	48° towards 300°	40° towards 270°	40° towards 270°
Centrally exposed part	Asymmetric tight folds	90° along N–S	76° towards 270°	70° towards 270°

The first generation axial plane is folded and revealing the attributes of the second generation folds measured from the exposure present around the Hanuman temple

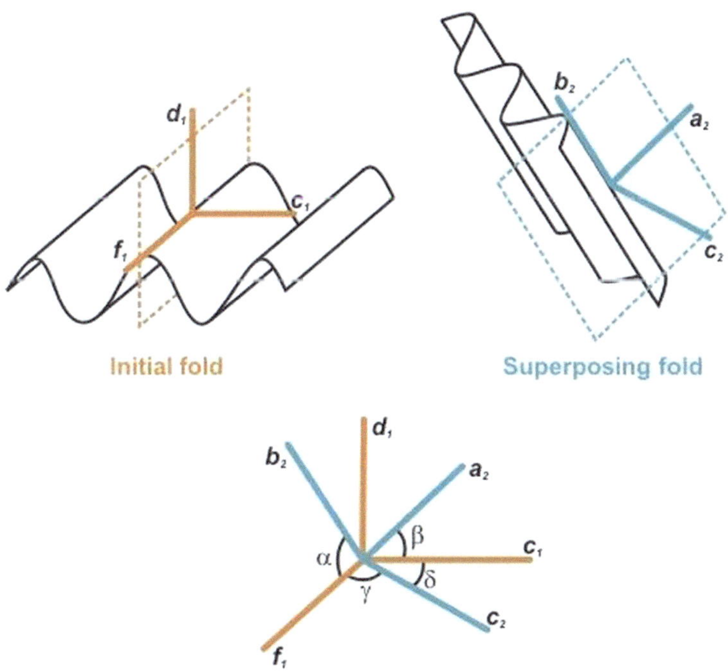

Fig. 11 Schematic description of superposed folding in relation to the kinematic axes and their relative spatial orientations (after Ramsay, 1967 and Thiessen & Means, 1980)

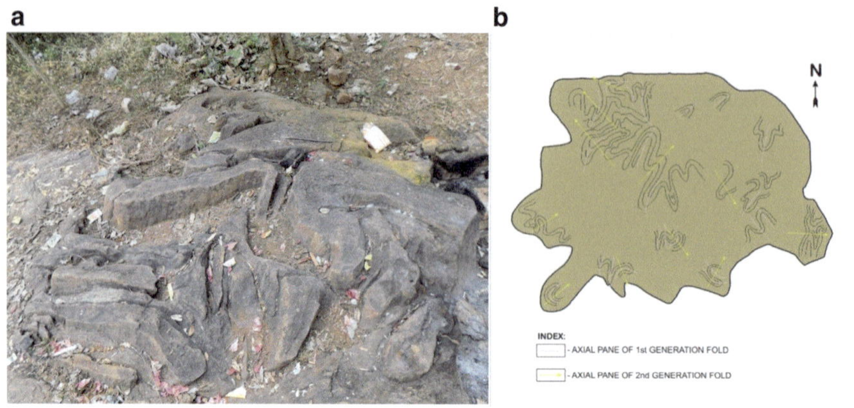

Fig. 12 **a** (left) and **b** (right). Outcrop of the superposed folds (2 and 3) in Khondalites exposed in the temple premises. The same has been traced out tentatively showing the axial planes of different generations

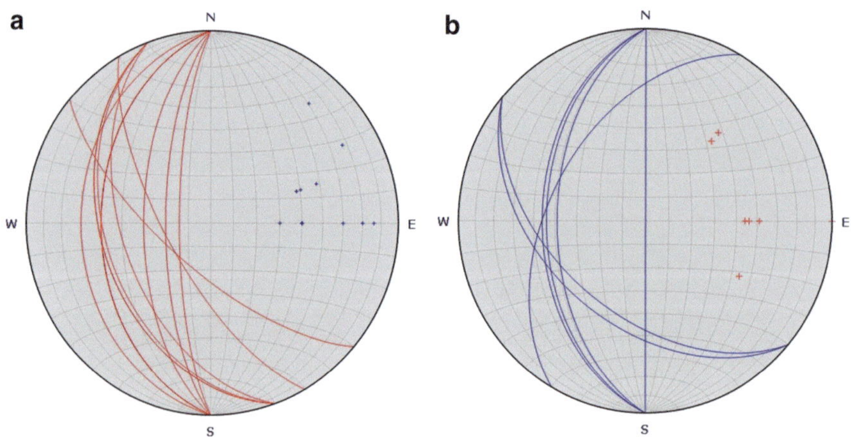

Fig. 13 **a, b** Stereographic projections, great circles and poles, of the fold data from Table 3 of the axial planes (left) and the limbs (right)

5.1 Sedimentary Sequence

The study was carried out mainly on the exposures of Talchir shale observed along the roadside while moving towards northwest from Angul town towards Nisha village (Fig. 21). Talchir shale was encountered with some sandstones that contain various sedimentary structures (Fig. 22a, b). Moving northwest again, boulders are encountered whose origin and stratigraphic position is controversial although many workers

Fig. 14 Fault zone where the right block (NE) is displaced towards south with respect to the left (SW) bock

prefer to put it in Damuda Group itself. Further northwest, Barakar Formation underlies the boulder unit is overlain, which can be accessed at SingharaJhor and its tributaries. Mainly sandstone and shale are exposed and contains sedimentary structures mostly are different types cross beddings and parallel laminations.

The Boulder Unit near niche village is clast supported mostly, at places matrix supported. There are faceted pebbles observed beside the road sections. Boulders are mostly of quartzites but a careful study also reveals the presence of gneissic and other rocks. Sedimentary structures are not discernable.

The Barakar Formation is exposed in SingharaJhor and GhurudiaNala. A lithologic description of the sedimentary rocks of the study area in brief is as follows:

(a) *Colour*: Sandstones are greyish white to light brown and the shales are black to dark grey to light brown and at places reddish to mottled clay is present, may be fire clay. Dark grey shales contain coal streaks and are carbonaceous shales (Fig. 23).
(b) *Texture*: The rocks showed an overall clastic texture. The sandstone beds comprise of fine- to coarse-grained sand-sized particles. The shale beds comprise of silt and clay. The conglomerate beds have pebble-sized clasts.
(c) *Mineralogy*: Fine to coarse-grained, quartz and feldspar with minor proportions of mafic minerals like biotite.

Fig. 15 Amphibolite dykes cross-cutting the stromatic country rock

(d) *Structure*: Different sedimentary structures are observed in the sandstones. Finer the grain size, smaller are the structures. The varied range of sedimentary structures depicts variation in energy conditions during their deposition. The sandstone beds contain planar laminations depicting a lower energy condition during deposition (Fig. 24). Small-scale trough cross stratification was identified near bounding surfaces of the individual sandstone beds. Large-scale trough cross stratification seen in the sandstone layers depicts a higher energy of deposition (Fig. 25). Mud-sand intercalation indicates fluctuating energy conditions (Fig. 26). Scour marks indicate high stream surges and erosive turbulence (Fig. 27). The shale layers contain coal streaks and fissility planes. Conglomerate bed contained faceted and striated imbricated pebbles indicating a glacio-fluvial deposit.

Coal is found extensively throughout the Barakar Formation in the area. A panoramic view of the Open Cast Coal mine near Ghurudianala (a tributary of SingharaJhor) is given below (Fig. 28). Several plant fossils are found in the dumped region of the mine. There are also coal seams exposed in a tributary channel of the nala towards south.

A small and very concise description of fluvial architecture is presented here. The Barakar sandstone-shale section is exposed in the Ghurudianala. Such sections are also exposed in SingharaJhor where sandstones are mostly greyish and contain large

Fig. 16 Shearing as observed in the outcrop, representing a dextral sense (bar measures 10 cm)

potholes. One can also go upstream and on the left bank where large trough cross stratifications are exposed. On the right bank, sporadic sections are present.

The fluviatile medium to coarse grained sandstones with various cross stratifications and horizontal laminations along with erosional surfaces sometimes containing mud clasts indicative of multi-storey sheet sandstone bodies and coalescing resemble longitudinal channel bars (Fig. 29). These bars may have formed due to migration within the channel belt and at the same time aggraded. The channel belt might have been large enough and migration of the active channel caused superposition of channel bars, and sometimes superposed on its own floodplain. Abandonment of part of floodplain at times of its migration is indicated by the presence of poor draining and anoxic condition that gave rise to the dark grey and carbonaceous shales.

5.2 *Quaternary Glimpse*

Quaternary deposits overlie in an unconformable relationship with the Gondwana Supergroup. The Quaternary deposits is observed towards the right side of the road moving northwest near the bridge over the SingharaJhor. It consists of alternate gravel and sand deposits (Fig. 30). Gravels are imbricated and it indicates a flow direction roughly parallel to the present day river course flowing nearby. High energy surges

Fig. 17 S-C fabric as observed in the outcrop showing the sense of shear

Fig. 18 Plot of centre-to-centre method or Fry method of 2-D strain analysis. The X and Y axes represent the distance and direction where the points are plotted by translating the reference centre at each case. The aspect ratio of the ellipse is 1.52. Orientation of the instantaneous stretching axis of the strain ellipse trends E–W; hence the maximum compression direction acted along N–S

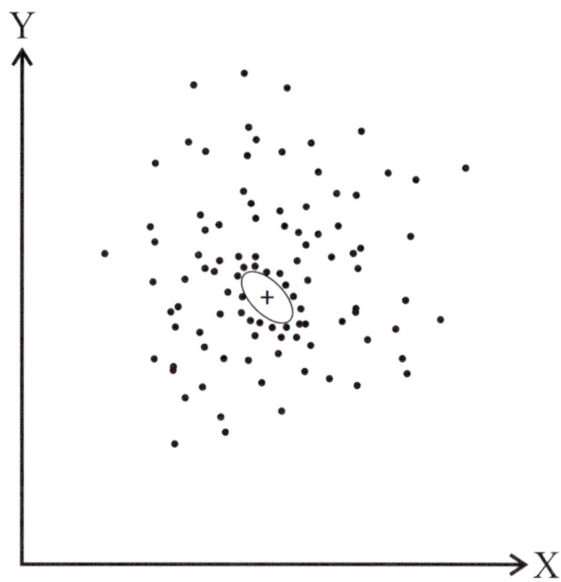

Table 4 Measurements made from augens from an outcrop present on the left side of the main road while traversing from Angul to Talchir

Sl No	Long axis (cm)	Short axis (cm)	Phi (°)	Aspect ratio
1	6	2.5	– 75	2.4
2	2.5	1.5	– 70	1.6
3	7	2.5	– 70	2.8
4	3.5	1.5	– 55	2.3
5	2	1	– 60	2
6	1.6	1.6	90	1
7	2.6	1.3	32	2
8	2.9	0.6	– 15	4.8
9	3.5	1.4	60	2.5
10	4.5	1.5	80	3
11	4	1.6	90	2.5
12	7.5	2	90	3.75
13	3	1.6	75	1.88
14	3	2	75	1.5
15	1.8	1.4	65	1.29
16	2	0.5	– 55	4
17	4	1.8	– 30	2.22
18	3.5	1	– 60	3.5
19	2.2	1	– 48	2.2
20	5	1.8	– 40	2.78
21	3.6	1.1	90	3.27
22	4	1.8	– 65	2.22
23	3.6	1.2	– 20	3
24	2	1.1	– 53	1.82
25	2	1	– 70	2

are indicated by the gravels and large gravels measuring 15–20 cm indicate high competence of flow. Faint structures like cross bedding and parallel laminations are observed at places and in most cases the exposures are masked by mud. Mottling is also a conspicuous feature in these deposits.

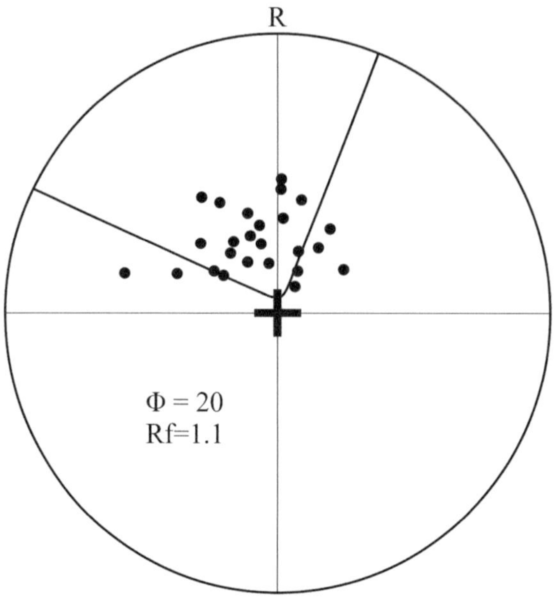

Fig. 19 Plot of Rf versus phi data of Table 4 on a hyperbolic net diagram which is generally used for strain analysis. When plotted on this net, the aspect ratios and the orientations of the augen grains define a hyperbola asymptotic to the direction of stress. The best fit aspect ratio value is 1.1 and the angle is 20°

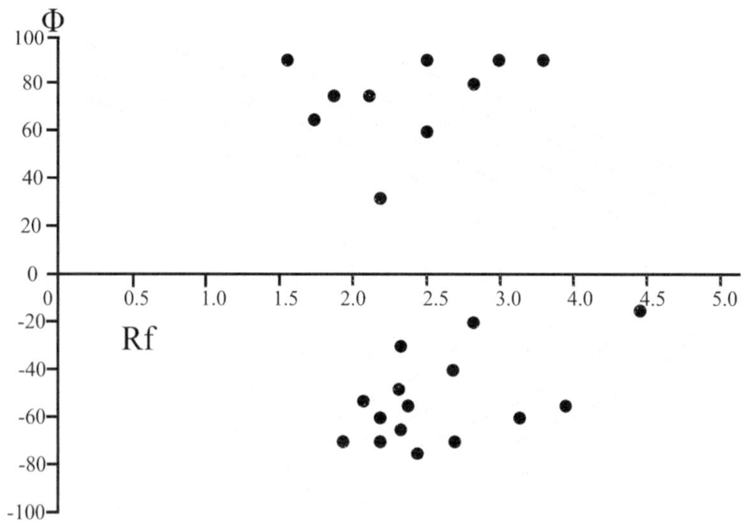

Fig. 20 Rf/phi plots of augens from Table 4. Also displayed in the plots are the estimated strains which is merely representative and far from accuracy

Table 5 Succession of the Gondwana and its associated sedimentary succession of Angul District (after Goswami et al., 2006; Hota et al., 2011)

Age	Group	Formation	Thickness (m)	Lithology
Recent	X	X	~2–30	Soil and weathered mantle
Lower Permian	Gondwana Supergroup	Barakar formation	315 +	Medium to coarse grained greyish felspathic sandstone, grey to dark grey shale and coal seams
		Karharbari formation	160 +	Pale brownish yellow coloured massive medium to coarse grained sandstone containing clasts of Talchir shale and coal seams
Upper Carboniferous to Lower Permian		Talchir formation	102 +	Diamictite, sandstone, needle shale, turbidite, rhythmites and varves
Unconformity				
Archaean to Proterozoic		PreCambrian metamorphics		Granites, gneisses and associated supracrustals

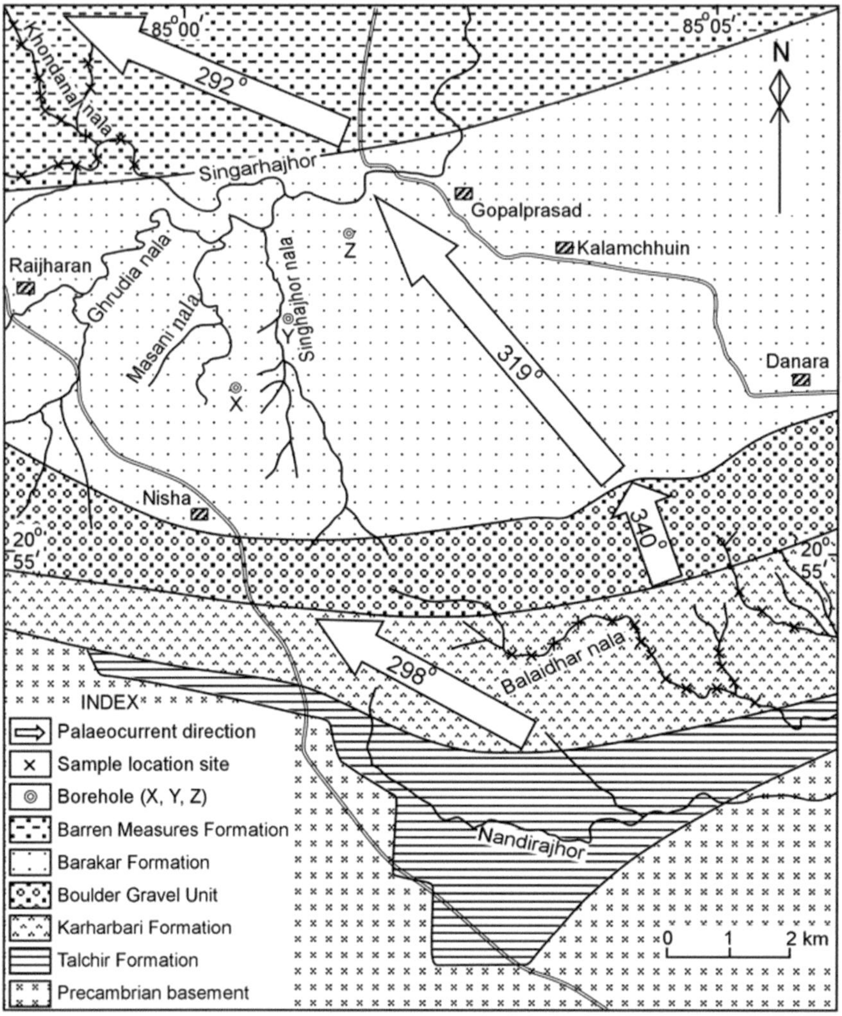

Fig. 21 Geological map of the study area showing Nandira and SingharaJhor and its tributaries from which sedimentological studies were carried out. The mean palaeocurrent directions are towards north to northwest (Hota et al., 2011)

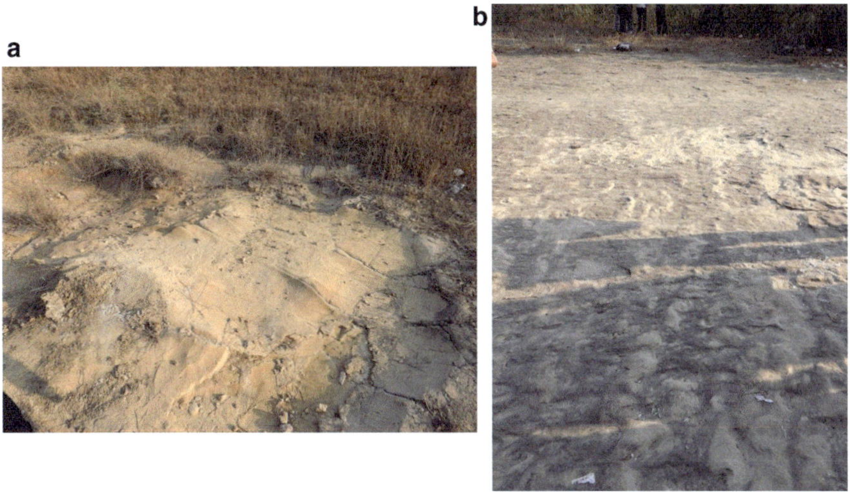

Fig. 22 **a** (left) and **b** (right). Khaki coloured Talchir shale exposed at NandiraJhor. The shales shows cracks and faint ripples at places (left). The sandy part of the shale shows prominent ripples

Fig. 23 Brown and dark grey carbonaceous shaley unit showing sharp contact observed at GhurudiaNala

Fig. 24 Horizontal to near horizontal laminations observed in fine grained sandstones

Fig. 25 Trough cross-stratification medium grained sandstone bed of Barakar unit exposed GhurudiaNala

Fig. 26 Sand-mud couplet and its intercalations indicate fluctuating energy conditions

Fig. 27 Scour in coarse grained sandstones of GhurudiaNala indicating erosive turbulence. Note the contact between medium and coarse grain sandstone towards the left side of the photograph

Fig. 28 HingulaOpen Cast Project (OCP) nearby the Ghurudianala. Plant fossils are found in abundance in the dumps seen in the foreground of the photograph

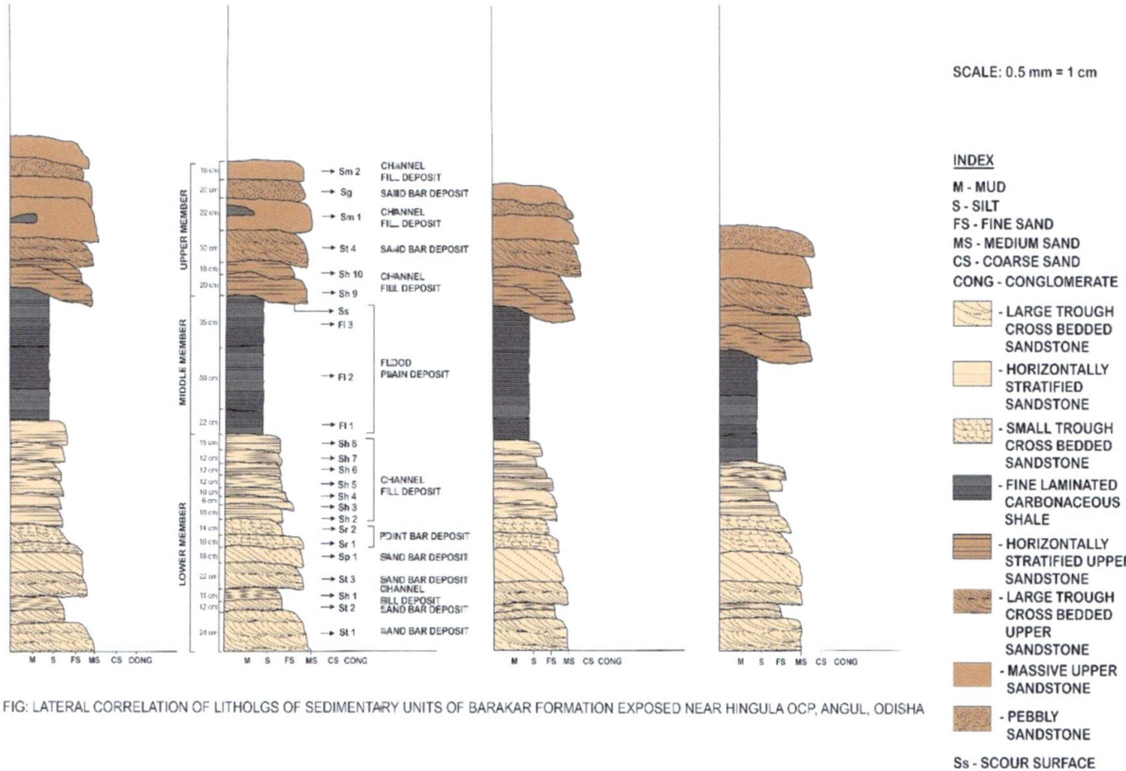

Fig. 29 Litholog prepared from the exposure of Barakar Formation exposed at Ghurudianala laterally at four locations of the same body. The diagram represents the stratigraphic architecture of the sandstone-shale units, the sedimentary structures, scours and erosional surfaces

a

b

Fig. 30 **a** (above) and **b** (below). Sectional view of the Quaternary deposits and the panaromic photo of the same showing imbricated gravel marking unconformity above Barakar Formation

Acknowledgements SSG expresses gratitude to the geology students of the University of Calcutta, 2017-2019. Dr. Subhronil Mondal and Prof. Ishita Das provided scientific input. University of Calcutta provided financial support for fieldworks. Soumyajit Mukherjee invited, edited and reviewed this chapter. Mukherjee (2023) summarized this chapter.

References

Bose, N., Dutta, D., & Mukherjee, S. (2018). Role of grain-size in phyllonitisation: Insights from mineralogy, microstructures, strain analyses and numerical modeling. *Journal of Structural Geology, 112*, 39–52.

Goswami, S., Das, M., & Guru, B. C. (2006). Permian biodiversity of Mahanadi Master Basin, Orissa, India and their environmental countenance. *Acta Palaeobotanica, 46*(2), 101–118.

Hota, R. N., Das, B. K., Sahoo, M., & Maejima, W. (2011). Provenance variability during Damuda sedimentation in the Talchir Gondwana Basin, India—A statistical assessment. *International Journal of Geosciences, 2*, 120–137.

Mahalik, N. K. (1987). Geology of the rocks lying between Gangpur Group and Iron Ore Group of horse shoe syncline in north Orissa. *Indian Journal of Earth Science, 14*, 73–84.

Mahalik, N. K. (1998). Precambrians. In *Geology and mineral resources of Orissa* (pp. 43-81). 2nd Edn., SGAT Publ.

Manjrekar, V. D., Choudhury, V., & Gautam, K. V. V. S. (2006). Coal. In N. K. Mahalik (Ed.), *Geology and mineral resources of Orissa* (pp. 205–226). Society of Geoscientists and Allied Technologist.

Mukherjee, S. (2023). Introduction to structural geology and tectonics field guidebook—volume 2. In: S. Mukherjee (Ed.), *Structural geology and tectonics field guidebook—volume 2* (pp. xi–xiv). Springer Nature Switzerland AG. ISBN 978-3-031-19575-4.

Mukherjee, S. (2017). Review on symmetric structures in ductile shear zones. *International Journal of Earth Sciences, 106*, 1453–1468.

Ramsay, J. G., Huber, M. I. (1987). *The techniques of modern structural geology, vol. 2: Folds and fractures* (p. 391). Pergamon Press.

Ramsay, J. G. (1967). *Folding and fracturing of rocks* (p. 568). McGraw-Hill.

Senapati, A., & Behera, P. (2015). Stratigraphic control of petrography and chemical composition of the lower Gondwana Coals, Ib-Valley coalfield, Odisha, India. *Journal of Geoscience and Environment Protection, 03*(04), 56–66.

Thiessen, R. (1986). Two-dimensional refold interference patterns. *Journal of Structural Geology, 8*, 563–573.

Thiessen, R. L., & Means, W. D. (1980). Classification of fold interference patterns: A re-examination. *Journal of Structural Geology, 2*, 311–316.

New Structural Geological Input from the Barmer Basin, Rajasthan (India)

Mohit Kumar Puniya, Ashish Kumar Kaushik, Soumyajit Mukherjee, Swagato Dasgupta, Nihar Ranjan Kar, Mery Biswas, and Ratna Choudhary

Abstract This study is focused on field structures especially on different brittle structures such as P- and Y-planes, normal faults and reverse faults in the eastern, western and the northern parts of the Barmer basin, Rajasthan (India). Besides aforesaid structures, basalt dykes are mapped in the northern and in the eastern parts of the basin. Reverse faults are first time documented with top-to-NNW shear in the Malani Igneous Suite (MIS). Shear sense indicator are preserved in sandstones with top-to-NE slip. The eastern part of the Barmer basin was affected by NE–SW extension and N–S compression. The western part of the basin witnessed oblique-slip faulting with NE–SW trend and can be identified in terms of brecciated zones cemented in calcareous material at several locations. The enigmatic N–S compression in the extensional Barmer basin might be the far-field effect of the India-Eurasia collision, and deserves more studies.

Keywords Barmer basin · P- and Y-planes · Brittle structures · Fault · Extension

---

M. K. Puniya (✉)
National Geotechnical Facility, Survey of India, Dehradun, Uttarakhand 248001, India
e-mail: puniyamohit@gmail.com

A. K. Kaushik
Department of Geology, Kurukshetra University, Kuruksetra, Haryana 136119, India

S. Mukherjee · S. Dasgupta
Department of Earth Sciences, Indian Institute of Technology Bombay Powai, Mumbai, Maharashtra 400076, India

N. R. Kar
Center for Earth, Ocean and Atmospheric Sciences (CEOAS), University of Hyderabad, Gachibowli, Hyderabad 500046, India

M. Biswas
Department of Geography, Presidency University, Kolkata, West Bengal 700073, India

R. Choudhary
Banasthali University, Jaipur, Rajasthan 302001, India

S. Mukherjee (ed.), *Structural Geology and Tectonics Field Guidebook–Volume 2*, Springer Geology, https://doi.org/10.1007/978-3-031-19576-1_9

1 Introduction

The Barmer/Barmer—Sanchor basin is located in western Rajasthan, with an area of ~11000 km^2. This basin is ~200 km long and 40 km wide with estimated depth of 6 km. The basin is surrounded by the Bikaner–Nagaur basin in NE and Jaisalmer basin in N/NNW (Fig. 1). It is the smallest basin in Rajasthan compared to Bikaner–Nagaur basin (70000 km^2) and Jaislmer Basin (45000 km^2). Recently Biswas et al. (2022a, 2022b) provided detail geomorphologic works from Barmer and Jaisalmer,

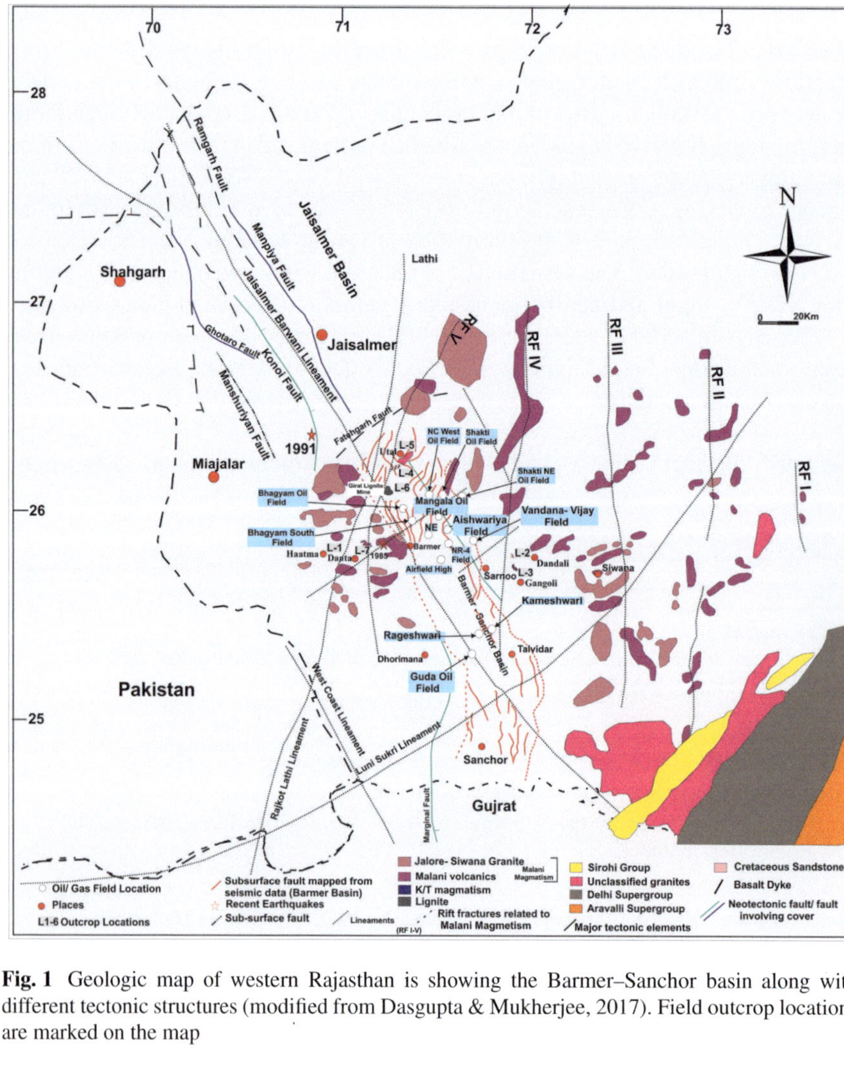

Fig. 1 Geologic map of western Rajasthan is showing the Barmer–Sanchor basin along with different tectonic structures (modified from Dasgupta & Mukherjee, 2017). Field outcrop locations are marked on the map

respectively, basins from western Rajasthan. Kar et al. (2022) has produced detail organic geochemical results from a lignite mine from Barmer. Haldar et al. (2022) has presented detail architecture of Kiradu temple located in Barmer area.

Barmer basin formed a graben structure due to the breakup of Indian craton in Late Cretaceous (Dutta, 1983; Mathur & Kumar, 2003; Dolson et al., 2015; Dasgupta & Mukherjee, 2017). This area is poorly understood in terms of surface geology due to the vast sand cover (Das Gupta, 1974; Sisodia & Singh, 2000). Dasgupta and Mukherjee (2017) added some new surface structural data in the Barmer basin to understand the geology of the area. Detail surface geological data are still lacking in the region and researchers depend more on the geophysical data.

This work presents new field structural data from the Barmer basin and enhances the knowledge of the rift formation.

2 Geology

Precambrian Malani Igneous Suite (MIS) constitutes the basement of the Barmer basin (Pareek, 1981; Bladon et al., 2015; Compton, 2015; Dasgupta & Mukherjee, 2017). Lathi Formation of Jurassic age rests over the MIS in the northern portion of Barmer basin near the Fatehgarh ridge (Fig. 1). Fatehgarh ridge is well exposed on the Barmer-Jaisalmer highway near Location-4 (Utal village). The Mesozoic Ghaggar Hakra Formation is well exposed near Sarnoo hills (Sisodia & Singh, 2000; Dolson et al., 2015; Dasgupta & Mukherjee, 2017). There are numerous basalt dykes intrusions (Figs. 1, 4, & 5) in northern and eastern portions of the basin. Some authors correlate these intrusions to the Deccan volcanism (Chandrasekaran et al., 1990; Basu et al., 1993; Sen et al., 2012; Dasgupta & Mukherjee, 2017).

Previous researchers marked these igneous activities in the eastern part of the basin but this study mapped new locations of the dykes in northern part of the basin as well (Fig. 1). Barmer and Fatehgarh Formations are the source of Bhagyam, NC west, Shakti, Mangla, Aishwarya, Vandana, NR-4, Kameshwari, Rageshwari and Guda oil fields (Farrimond et al., 2015; Lobo et al., 2015; Shiju et al., 2008) (Fig. 1).

The basin is situated at the western portion of the Aravalli range. This is the shelf-part of the Paleo-Tethys of the Gondwana period (review in Dasgupta & Mukherjee, 2017). Southern part of Barmer basin rifted into two parts: first NW–SE extension due to the east west Gondwana breakup in Mesozoic age and second NE–SW extension due to the Seychelles microcontinent breakup from India (Sharma, 2007; Collier et al., 2008; Mishra, 2011; Torsvik et al., 2013; Bladon et al., 2014, 2015; Dasgupta & Mukherjee, 2017). Devikot-Fatehgarh fault marks the northern boundary of Barmer basin, which formed due to the Himalayan collision during Late Tertiary (Compton, 2009; Mukherjee & Koyi, 2010a, 2010b; van Hinsbergen et al., 2012; Mukherjee, 2013, 2015; Mukherjee et al., 2013, 2015; Kelly et al., 2014; Dasgupta & Mukherjee, 2017).

Fig. 2 Field photographs of location-2 along the Luni river Dandali village. **a** conjugate strike slip faults, movement can be depicted along the quartz veins (along fault plane F1 dipping 65° towards NW-292° and F2 fault dips at 68° towards 350°) in basalt, observed on horizontal plane; **b** asymmetric boudin in sandstone shows slip towards 105°, observed on horizontal plane; **c** sub-vertical fault striation shows down dip movement on a plane that dips 40° towards NE (040°); **d** a gently dipping (25°) conformable contact between sandstone and underlying basalt at the right bank of Luni River near Dandali village (Fig. 1)

3 Fieldwork and Discussions

The Barmer basin can be divided into eastern domain (Barmer to Siwana), western domain (between Barmer and Miajalar) and northern domain (Barmer to Fatehgarh Fault) (Fig. 1).The basaltic eastern domain belongs to MIS (Fig. 2a), sandstone of Fatehgarh Formation (Fig. 2b, c and d) and Eranpura granite or unclassified granite (Fig. 3). This domain is studied in Luni river section and Dandali village in the eastern part of Barmer city. The faults and shear sense indicator of the eastern domain are dominated by reverse faulting with top-to-NNW slip (Fig. 2). Basalt of MIS shows conjugate strike slip faulting with sinistral and dextral movements (Fig. 2a). Asymmetric boudins in sandstone show top-to-NE slip (Fig. 2b). Fault plane trend in sandstone is NW–SE: and display up-dip movement (Fig. 2c). The contact between underlying MIS to the overlying Fatehgarh Formation sandstone is conformable and almost parallel to each other with NW–SE trend, dips gently ~10–15° towards SW (Fig. 2d).The granite hills near Gangoli village (Fig. 1) in eastern Barmer basin are intruded by basalt dykes with NE–SW trend (Fig. 3a,b). Some fault planes are also observed in the granite with strike-slip nature (with NW 315° trend). Granite of this area is non foliated, grey in colour and has porphyroclasts of feldspar. This study shows that eastern domain of Barmer basin has witnesses of two stress regimes: NE–SW extension and N–S compression.

The basalt in western domain is highly deformed by brittle fracture and is foliated locally (Fig. 4). The general foliation trend of basalt is ~N–S with moderately dip (38°) towardsN 275°.

Oblique faults are dominated in this area with NE trend (plunge-080°, trend-064°) (Fig. 4a and b). At one location near the Daanta bus stop (Fig. 1) strike-slip fault is observed with trend NW–SE, which dip steeply (81°) and show dextral slip (Fig. 5). Fault breccia and calcareous materials are observed along these fault planes (Fig. 4c). This is new field structural data from the western margin of the Barmer basin.

Here, one more episode of deformation is observed in the form of P and Y brittle shear planes. The N–S trending faults are common in western domain. Near Haatma and Kiradu temple, N–S trending faults (P-planes) terminate against the NW–SE trending faults (Y-planes) (Fig. 6a, b). They show top to NW slip in the foliated basalt of MIS (Fig. 6a, b).

Fig. 3 Field photographs of location 3, near Dandali and Gangoli villages: **a** Google Earth image of the location shows the exposures of porphyritic grey granite and later intruded basalt dykes; **b** exposure of the basalt dyke into the greyish colour granite. The dyke trends 060–240°

Fig. 4 Field photographs of location-1 near the 32 km milestone of Barmer adjacent to the Haatma village (Fig. 1): **a**, **b** slickensides are observed on NE–SW trending fault surfaces; **a** fault surface with slickensides step-like elevations (rake 40°) shows slip towards 087° (strike 080°; dip 67° and dip direction 170°); **b** fault surface show oblique movement with 086° trend and 34° plunge; **c** basalt exposure show calcareous intrusion. Part of pen (~11cm) as scale. **d** Basalt shows well developed foliation with strike 185°, dip 38° and dip direction 275°. Pen (~12cm) shown as scale

The northern domain of the study area lies between Barmer and Fatehgarh fault and is well exposed on Barmer–Jaisalmer highway (National Highway 70). There is conglomerate bed identified below the Fathehgarh Formation sandstone (Fig. 7a). This conglomerate bed represents unconformable contact between MIS and Fathehgarh Formation in the region. The Fathegarh Formation shows E–W trend with gently dip of 24° towards south (Fig. 7b). Fatehgarh sandstone is intruded by later igneous activity in terms of basalt dykes (Fig. 8a). The trend of the dykes is NE–SW and intersects the strike of sandstone at an acute angle (Fig. 8a,b). The basalt dykes show chilled margin and parallel or longitudinal joint with NE trend. Transverse joints are developed in the dykes with 320° trend (Fig. 8c). Basalt dykes show sigmoid brittle planes or the P-planes that are associated also with tangential Y-planes (Fig. 8d). The Y-plane trends 235° SW (80° dip towards 145°) and P-plane trends 325°NW (80° dip towards 055°).

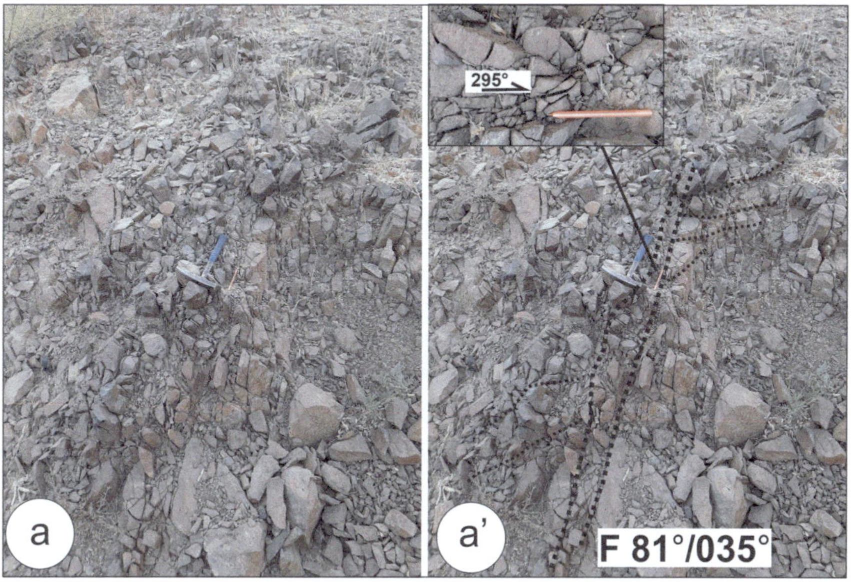

Fig. 5 Field photographs of location 7, near Danta bus stand, 1.5 km before of Haatma village towards Barmer: original (**a**) and with interpretation (**a'**) exposure of basalt shows dextral fault with clearly visible curved fracture and curved surfaces on sub horizontal surface. The trend of the fault is 125° (SE) with 81° dip amount and dip direction of 035° (NE). Geological hammer (30 cm) as scale

4 Conclusions

The studied area can be divided into three domains based on the structural data.

- Eastern domain of Barmer basin: This place has witnesses of two stress regimes, NE–SW extension and N–S compression.
- The basalt in the western domain is highly deformed by brittle fractures and at some locations it is foliated. Oblique-slip faults are dominated in this area; attitude of slip direction: 080° plunge, 064° trend. Infrequent strike-slip faults (trend NW–SE with dip amount 81°) occur alongwith breccia and calcareous cements. This is new data from the western margin of the Barmer basin.
- In the northern margin, conglomerate bed has been identified below the Fathehgarh Formation sandstone. This bed represents unconformable contact between MIS and Fathehgarh Formation. Fatehgarh sandstone is intruded be later igneous activity in the form of NE–SW trending basalt dykes.

Fig. 6 Field photographs of location-7 near Danta bus stand 1.5 km before of Haatma village towards Barmer; **a** exposure shows sinistral brittle shear within basalt, observed on a horizontal plane. The Y-plane trends NW (dip amount 70° and dipping towards NE) and P-planes trend NE (75° dip towards SE); **b** two brittle shear intersecting the basalt (one along brunton compass and second along pen 12 cm long). The Y-plane trends NW (300°) (dip amount 70° and dipping towards NE) and P-planes trend NNW (dip amount 85° and dipping towards E). Brunton compass (18 cm) as scale

Fig. 7 Field snaps of location-4 near Utal villages near wind mill no-NGB0054U124: **a** a conglomerate bed below sandstone with 1–2 m thickness. Pen (~12 cm) as a scale. **b** Sandstone beds with NE–SW trend and 24° dip towards SE (170°)

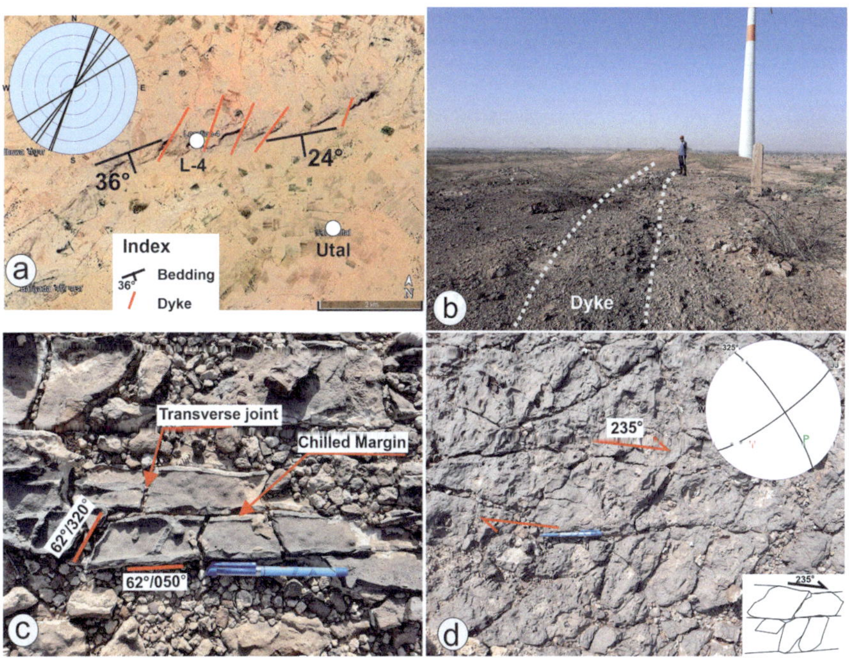

Fig. 8 Field photographs of location-4 near Utal villages: **a** Google Earth image of the location-4 shows the exposures of sandstone and later intruded basalt dykes. Bedding attitude and dyke orientation (NE–SW trend) shown; **b** trend of a dyke in front of the wind mill no-NGB0054U124 is 020–200°, thickness of the dyke is varies from 0.5 to 1.5 m; **c** transverse fractures (dip/dip direction: 62°/320°) and longitudinal joints (dip/dip direction: 62°/050°) with chilled margin. **d** Dextral brittle shear with Y- and P-planes within basalt dyke, observed on horizontal plane. The Y-plane trends 235° SW (80° dip towards 145°) and P-plane trends 325°NW (80° dip towards 055°). Pen (~11cm) as a scale

Acknowledgements The work was funded bya seed project (grant number: RD/0120-PSUCE19-001) awarded to SM by the agency: Center of Excellence in Oil, Gas and Energy (CoE-OGE, IIT Bombay). Alexis Vizcatrone and the proofreading team (Springer) are thanked for assistance. Paramita Haldar (BITS Pilani, Goa) assisted the fieldwork. Mukherjee (2023) summarized this chapter.

References

Basu, A. R., Renne, P. R., Dasgupta, D. K., Teichmann, F., & Poreda, R. J. (1993). Early and late alkali igneous pulses and a high-3He plume origin for the Deccan Flood Basalts. *Science, 261*, 902–906.

Biswas, M., Gogoi, M. P., Mondal, B., Sivasankar, T., Mukherjee, S., & Dasgupta, S. (2022a). Geomorphic assessment of active tectonics in Jaisalmer basin (western Rajasthan, India). Geocarto International. https://doi.org/10.1080/10106049.2022.2066726

Biswas, M., Puniya, M. K., Gogoi, M. P., Dasgupta, S., Mukherjee, S., & Kar, N. R. (2022b). Morphotectonic analysis of petroliferous Barmer rift basin (Rajasthan, India). *Journal of Earth System Science, 131*, 140.

Bladon, A. J., Burley, S. D., Clarke, S. M., & Beaumont, H. (2014). Geology and regional significance of the Sarnoo Hills, eastern rift margin of Barmer Basin, NW India. *Basin Research, 27*, 636–655.

Bladon, A. J., Clarke, S. M., & Burley, S. D. (2015). Complex rift geometries resulting from inheritance of pre-existing structures: Insights and regional implications from the Barmer Basin rift. *Journal of Structural Geology, 71*, 136–154.

Chandrasekaran, V., Srivastava, R. K., & Chawade, M. P. (1990). Geochemistry of the alkaline rocks of Sarnu-Dandali area, Barmer district, Rajasthan, India. *Journal of the Geological Society of India, 36*, 365–382.

Collier, J. S., Sansom, V., Ishizuka, O., Taylor, R. N., Minshull, T. N., & Whitmarsh, R. B. (2008). Age of Seychelles-India break-up. *Earth Planetary Science Letter, 272*, 264–277.

Compton, P. M. (2009). The geology of the Barmer Basin, Rajasthan, India, and the origins of its major oil reservoir, the Fatehgarh Formation. *Petroleum Geoscience, 15*, 117–130.

Compton, P. M. (2015). The geology of the Barmer Basin, Rajasthan, India, and the origins of its major oil reservoir, the Fatehgarh Formation. *Petroleum Geoscience, 15*, 117–130.

Das Gupta, S. K. (1974). Stratigraphy of western Rajasthan shelf. In *Proceedings IV Indian Colloqium, Micropalaeontology and Stratigraphy, Dehradun, India.*

Dasgupta, S., & Mukherjee, S. (2017). Brittle shear tectonics in a narrow continental rift: asymmetric nonvolcanicBarmer Basin (Rajasthan, India). *Journal of Geology, 125*, 561–591.

Dasgupta, S., & Mukherjee, S. (2019). Remote sensing in lineament identification: Examples from western India. In A. Billi, A. Fagereng (Eds.), *Problems and solutions in structural geology and tectonics. developments in structural geology and tectonics book series*, Vol. 5 (Series Editor: Mukherjee S, pp. 205–221). Elsevier. ISBN: 9780128140482.

Dolson, J., Burley, S. D., Sunder, V. R., Kothari, V., NaiduB, W. N. P., Farrimond, P., Taylor, A., Direen, N., & Ananthakrishnan, B. (2015). The discovery of the Barmer Basin, Rajasthan, India, and its petroleum geology. *American Association of Petroleum Geology Bulletin, 99*, 433–465.

Dutta, D. K. (1983). Geological evolution and hydrocarbon prospects of Rajasthan basins. In L. L. Bhandari (Ed.), Petroliferous basins of India. *Petroleum Asia Journal, VI*, 93–100.

Farrimond, P., Naidu, B. S., Burley, S. D., DolsonJ, W. N., & Kothari, V. (2015). Geochemical characterization of oils and their source rocks in the Barmer Basin, Rajasthan, India. *Petroleum Geoscience, 21*, 301–321.

Haldar, P., Puniya, M. K., Biswas, M., Kar, N. R., Mukherjee, S., & Choudhary, R. (2022). Architecture and structures of Kiradu temple (Barmer region, Rajasthan, India). In *Structural geology and tectonics field guidebook— volume 2*. Springer Nature Switzerland AG.

Kar, N. K., Mani, D., Mukherjee, S., Dasgupta, S., Puniya, M. K., Kaushik, A. K., Biswas, M., & Babu, E. V. S. S. K. (2022). Source rock properties and kerogen decomposition kinetics of Eocene shales from petroliferous Barmer basin, western Rajasthan, India. *Journal of Natural Gas Science and Engineering, 100*, 104497.

Kelly, M. J., Najman, Y., Mishra, P., Copley, A., & Clarke, S. (2014). The potential record of far-field effects of the India-Asia collision: Barmer Basin, Rajasthan, India. In C. Montomoli, S. Iaccarino, C. Groppo, P. Mosca, F. Rolfo, R. Carosi (Eds.), *Himalaya-Karakoram-Tibet Workshop, 29th (Lucca, Italy, 2014), Proceedings of Journal of Himalayan Earth Sciences Special* (pp. 80–81).

Lobo, M., Kolay, J., Sinha, P., Varghese, R., Lang, C., Kant, R., & Doodraj, S. (2015). Managing multidimensional constraints to drill ERD wells in Rajasthan with high directional difficulty index (DDI). Oil & Gas India Conference and Exhibition. Mumbai, Society of Petroleum Engineers. Accessed December 12, 2016. https://doi.org/10.2118/178073-MS.

Mathur, S. C., & Kumar, S. (2003). Sedimentation in the Barmer Basin. *Journal of Geological Society India, 61*, 368–369.

Mishra, D. C. (2011). Gravity and magnetic methods for geological studies: principles, integrated explorations and plate tectonics. Hyderabad, BS (pp. 672–675).

Mukherjee, S. (2013). Channel flow extrusion model to constrain dynamic viscosity and Prandtl number of the Higher Himalayan Shear Zone. *International Journal of Earth Science., 102*, 1811–1835.

Mukherjee, S. (2015). A review on out-of-sequence deformation in the Himalaya. In S. Mukherjee, R. Carosi, P. van der Beek, B. K. Mukherjee, D. Robinson (Eds.) *Tectonics of the Himalaya. Geological Society of London Special Publication*, Vol. 412 (pp. 67–109).

Mukherjee, S. (2023). Introduction to structural geology and tectonics field guidebook—volume 2. In Mukherjee, S. (Ed.) *Structural geology and tectonics field guidebook—volume 2* (pp xi–xiv). Springer Nature Switzerland AG. ISBN 978-3-031-19575-4.

Mukherjee, S., Carosi, R., van der Beek, P. A., Mukherjee, B. K., & Robinson, D. M. (2015). Tectonics of the Himalaya: an introduction. In S. Mukherjee, R. Carosi, P. van der Beek, B. K. Mukherjee, D. Robinson (Eds.), *Tectonics of the Himalaya*. Geological Society London Special publication

Mukherjee, S., & Koyi, H. A. (2010a). Higher Himalayan Shear Zone, Sutlej Section: Structural geology and extrusion mechanism by various combinations of simple shear, pure shear and channel flow in shifting modes. *International Journal of Earth Science, 99*, 1267–1303.

Mukherjee, S., & Koyi, H. A. (2010b). Higher Himalayan Shear Zone, Zanskar Indian Himalaya: Microstructural studies and extrusion mechanism by a combination of simple shear and channel flow. *International Journal of Earth Science, 99*, 1083–1110.

Mukherjee, S., Mukherjee, B., & Thiede, R. (2013). Geosciences of the Himalaya-Karakoram-Tibet Orogen. *International Journal of Earth Science, 102*, 1757–1758.

Pareek, H. S. (1981). Petrochemistry and petrogenesis of the Malani igneous suite, India. *Geological Society of American Bulletin, 92*(2), 206–273.

Puniya, M. K., Kaushik, A. K., Mukherjee, S., Kar, N. R., Biswas, M., & Choudhury, R. (2022). Structural geology and stability issue of the Giral lignite mine, Rajasthan, India. In: S. Mukherjee (Ed.), *Structural geology and tectonics field guidebook—volume 2*. Springer Nature Switzerland AG.

Sen, A., Pande, K., Hegner, E., Sharma, K. K., DayalAM, S. H. C., & Mistry, H. (2012). Deccan volcanism in Rajasthan: 40Ar-39Ar geochronology and geochemistry of the Tavidar volcanic suite. *Journal of Asian Earth Science, 59*, 127–140.

Sharma, K. K. (2007). K-T magmatism and basin tectonism in western Rajasthan, India: results from extensional tectonics and not from Reunion plume activity. In G. R. Foulger, D. M. Jurdy (Eds.), Plates, plumes, and planetary processes. Geological Society of America Special Paper, Vol. 430 (pp. 775–784).

Shiju, J., Bowyer, G., & Micenko, M. (2008). Mangala Field high density 3D seismic. *Biennial International Conference and Exhibition on Petroleum Geophysics, 7th. Hyderabad, Society of Petroleum Geophysicists.* https://www.spgindia.org/2008/607

Sisodia, M. S., & Singh, U. K. (2000). Depositional environment and hydrocarbon prospects of the Barmer Basin, Rajasthan, India. *Nafta (zagreb), 51*, 309–326.

Torsvik, T. H., Amundsen, H., Hartz, E. H., Corfu, F., KusznirN, GainaC., Doubrovine, P. V., Steinberger, B., Ashwal, L. D., & Jamtveit, B. (2013). A Precambrian microcontinent in the Indian Ocean. *Nature Geoscience, 6*, 223–227.

van Hinsbergen, D. J. J., Lippert, P. C., Dupont-Nivet, G., McQuarrie, N., Doubrovine, P. V., Spakman, W., & Torsvik, T. H. (2012). Greater India Basin hypothesis and a two-stage Cenozoic collision between India and Asia. *Proceedings of the National Academy of Sciences of the United States of America, 109*, 7659–7664.

Structural Geology and Stability Issue of the Giral Lignite Mine, Rajasthan, India

Mohit Kumar Puniya, Ashish Kumar Kaushik, Soumyajit Mukherjee, Nihar Ranjan Kar, Mery Biswas, and Ratna Choudhary

Abstract This field study is focused on normal faults and tension cracks of the Giral lignite mine, Barmer basin Rajasthan (India). Besides normal faults, lignite cleat and bedding data have been collected. Cleats and beddings are perpendicular to each other and filled with salt at the northern portion of the mine. Two normal faults are first time documented in the Giral mine, those can be related to initiation of strike-slip movement between India and Madagascar and main Barmer rifting. F1 fault group has trend NE–SW and F2 trend in approximately E–W. Due to the persence of these faults, northern slope of the mine shows slope failure. This slope failure can be seen on the plan view in the form of multiple stages of tension cracks. Fractured block on the slope show horizontal and vertical rotation towards N112° with 10–30° inclination. To find out the detail solution of the mine slope failure, more detail study is required.

Keywords Giral mine · Lignite · Fault · Barmer basin · Tension cracks

---

M. K. Puniya (✉)
National Geotechnical Facility, Survey of India, Dehradun, Uttarakhand 248001, India
e-mail: puniyamohit@gmail.com

A. K. Kaushik
Department of Geology, Kurukshetra University, Kuruksetra, Haryana 136119, India

S. Mukherjee
Department of Earth Sciences, Indian Institute of Technology Bombay Powai, Mumbai, Maharashtra 400076, India

N. R. Kar
Center for Earth, Ocean and Atmospheric Sciences (CEOAS), University of Hyderabad, Gachibowli, Hyderabad 500046, India

M. Biswas
Department of Geography, Presidency University, Kolkata, West Bengal 700073, India

R. Choudhary
Banasthali University, Jaipur, Rajasthan 302001, India

S. Mukherjee (ed.), *Structural Geology and Tectonics Field Guidebook–Volume 2*, Springer Geology, https://doi.org/10.1007/978-3-031-19576-1_10

1 Introduction

Lignite deposits in India occur mainly in the Tertiary sediments of the peninsular shield. These deposits are situated in Tamil Nadu, Puducherry, Kerala, Gujarat, Rajasthan and Jammu and Kashmir.

The Giral lignite mine is situated ~43 km away towards which direction to be stated from the Barmer city (Barmer district, western Rajasthan, India) (Fig. 1). It is an open cast mine and started operational since 1994. Besides Giral mine, there

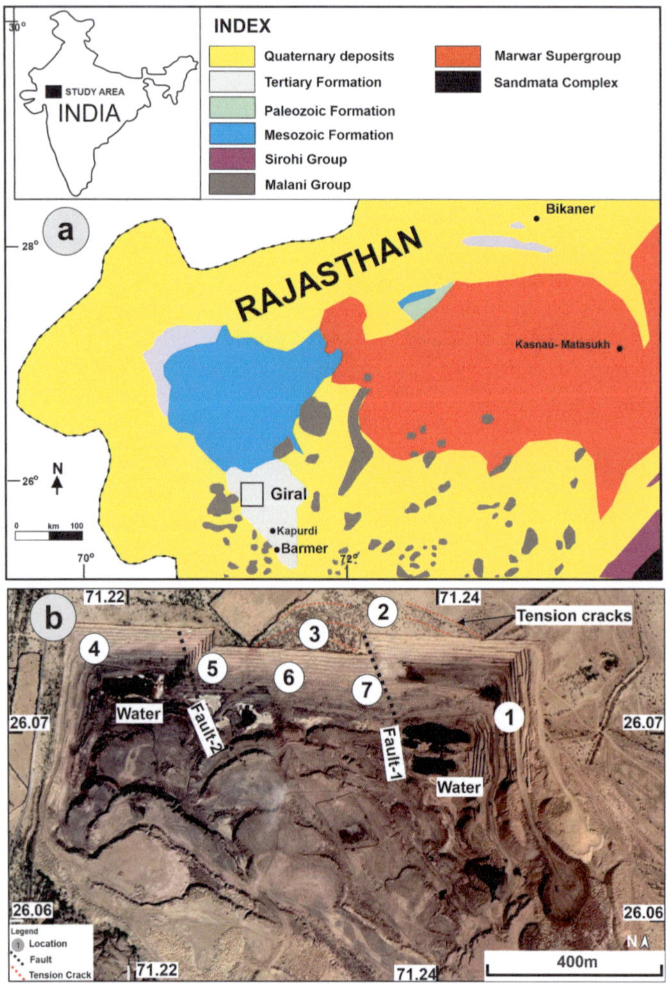

Fig. 1 Maps of the area; a Location and geological map of the area (after Roy & Jakhar, 2002); b Google Earth image showing locations of data collection sites and faults in the Giral lignite mine, Barmer (Rajasthan)

are other lignite mines too in Rajasthan- Matasukh and Kasnau (Nagaur), Sonari (Barmer) etc. Open cast mines are more cost-effective than the underground mines. The life of opencast mines depend on the steepness and stability of the slope.

Dasgupta and Mukherjee (2017) compiled the stratigraphy of the Barmer basin in their repository file. Puniya et al. (2022) summarized the geology and tectonics of Barmer basin. Besides, the geomorphological information about the Barmer (and the Jaisalmer) basin can be found at Biswas et al. (2022a, 2022b). Kar et al. (2022) presented the organic geochemical studies from a lignite mine in Barmer. The key information relevan to this artilce is as follows. The Barmer Basin is situated at the western part of Rajasthan, India, most of the part of this region is covered by the Thar Desert (Sisodia & Singh, 2000) (Fig. 1). The age of the rocks in Barmer basin ranges from Upper Proterozoic to Quaternary. Malani Igneous Suite (MIS) constitutes the base of the Barmer Basin, MIS overlain by Birmania Formation (Hughes et al., 2015; Sharma, 2007). The age of Randha and Birmania Fm is not clear. It is assumed to be of Paleozoic age (Hughes et al., 2015). Lower Jurassic Lathi Formation is exposed in the northern most portion of the Barmer basin and it overlains the Birmania Formation. Lower Jurassic Lathi Formation is exposed in the northern portion of the Barmer basin and is separated by the Fathegarh Fault from Fathegarh Formation of Maastrichtian to Early Paleocene age (Compton, 2009). Fathegarh Formation overlain by Dharvi Dungar Formation (Compton, 2009; Dolson et al., 2015; Kumar et al., 2020). Giral lignite mine and also the Sonari mine belong to the Dharvi Dungar Formation (Kumar et al., 2020). Lignite alternating with clay occurs in Giral. Singh et al. (2016) have done petrological studies of Grail mine and state that lignite of Grail mine composed of huminite group macerals, mainly telohuminite and detrohuminite, however liptinite and inertinite group macerals occur in subordinate amounts. Rajak et al. (2018) done the geochemical analysis of the Grail mine lignite and conclude that elements like Cu, Cd, Co, Ni, Zn, Pb, Na, and K occur in high concentration, while Mg and Ca have their concentrations lower than World Clarke average.

The stability of the opencast mine slopes depend on the the slope angle, rock type, joints orientations, faults, shear zones, groundwater, precipitation, earthquake and excavation method (Kumar et al., 2017; Tang et al., 2015; Xue et al., 2018). The failure of mine slope adversely affect the mining operation, and can lead to losses of human lives and degradation of agriculture land/environment.

2 Field Findings and Discussions

The Giral mine is destabilized locally by two normal faults those trend NW–SE (F-1 and F-2) and can be observed at the northern portion of the mine (Figs. 1b and 2). These faults have same trend as rift fracture in the Barmer Basin (Dasgupta and Mukherjee, 2017). These faults are new report in this study. In this mine alternate layers of lignite and shale are observed (Fig. 3b).Bedding of the lignite and shale dips

Fig. 2 Expression of fault on north face of the Giral lignite mine trending NE-SW. S. Mukherjee (170 cm) as marker

25° towards N107° direction. Cleats are near-perpendicular to the bedding planes and dipping towards N335° with dip amount 80° (Fig. 3a, a′). Few N035° trending sub-vertical joints almost orthogonal to the cleat sharply terminate against the shale bands (Fig. 3c).

As observed in a sub-vertical section inside the mine, F-1 and F-2 form a graben structure in the mine area (Fig. 4). F-1 dips64° towards N195°, and F-2 dips 58° towards N350° (Fig. 4).

The mine has been excavated ~120 m deep from the initial ground level. So the stability of the bench and berms are highly controlled by the sets of normal faults in the mine area. Parallel sets of normal faults (parallel to F1 dip direction ranges N160–195°) intersect the beds of lignite and shale at location-6 (Fig. 5). Total 14 such normal fault planes are observed within 1–2 m long outcrop at the berm and they dip 31–47° towards N160–172° (Fig. 5a′). Lignite deposit in this area shows ductile deformation before the brittle deformation in the form of normal faults. This ductile deformation is preserved in the form of asymmetric folding and later superposed by brittle deformation (Fig. 5a, a′).Asymmetric folds show top-to-N011° (NE) shear and terminate against the normal fault with N191° dip direction at location-7 (Fig. 6). Secondary mineralization of salt along the brittle planes is common in the northern portion of the mine (Fig. 5 and 6).

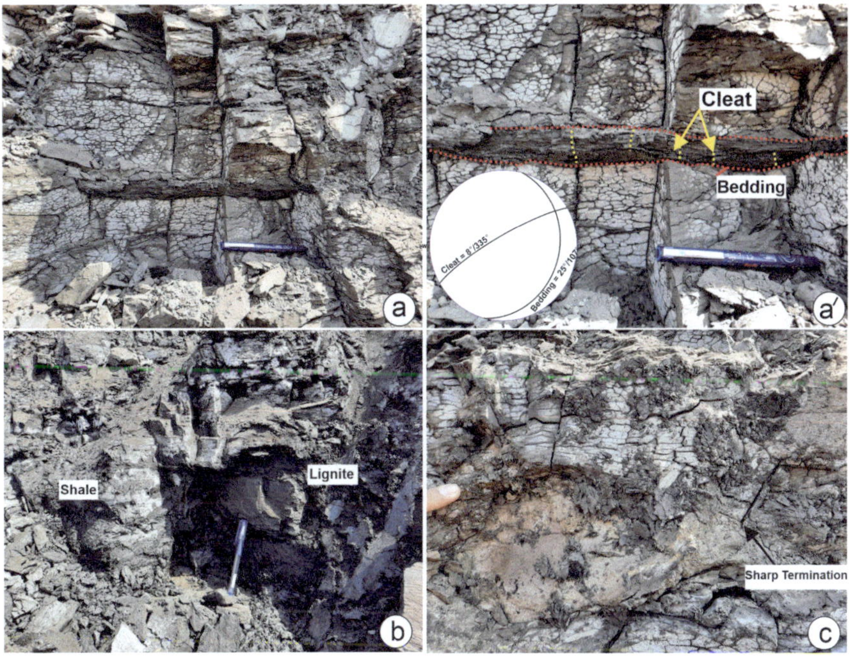

Fig. 3 Field photographs of Giral mine at location-1; **a** cleat and bedding structures structure original **a′** and with interpretation, bedding planes are well observed dipping towards N107° and inclined at an angle 25°; cleats are near-perpendicular to the bedding. Cleats strike N075° and dip towards N335°; **b** alternate bands of shale and lignite, pen (12 cm) as scale; **c** sharp termination of vertical joint in lignite against the shale with trend N035°, part of finger (2 cm) as scale

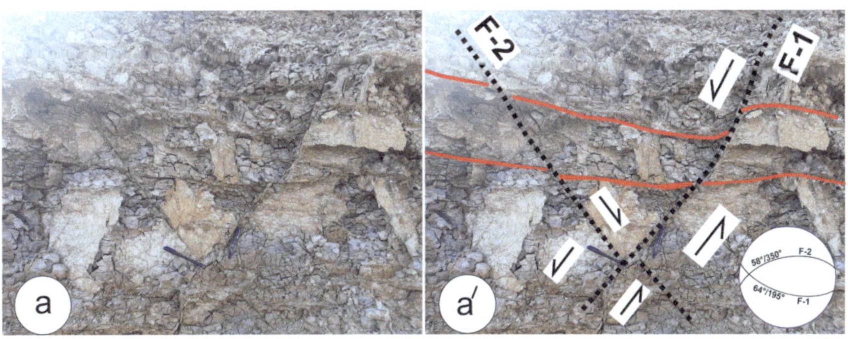

Fig. 4 Field photographs of Giral lignite mine show graben structure on a vertical plane, at location-5. **a** uninterpreted, and **a′** interpreted image., Fault-1 dips 58° towards N350°. Fault-2 dips 64° towards S015°. Brittle planes filled up by secondarily precipitated salt in the eastern portion of the mine at location-1. Pen (12 cm) as scale

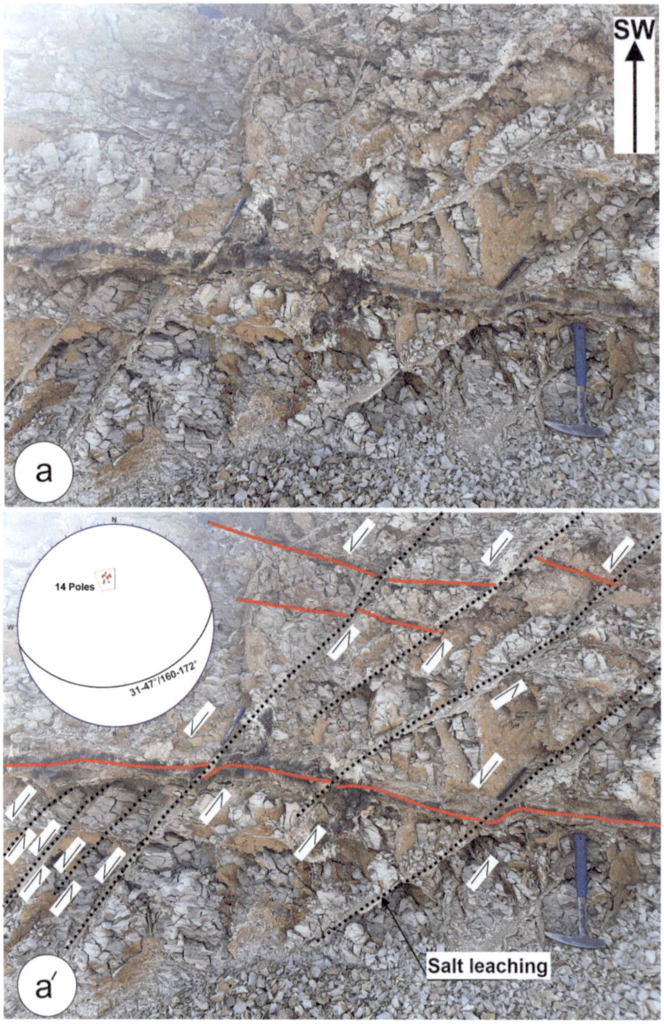

Fig. 5 Field photographs of Giral lignite mine area shows a vertical section with parallel normal faults at location-6; (**a**) and with interpretation (**a′**), the 6–9 cm thick lignite seam faulted by parallel normal faults dip 31–47° towards N160–170° at the eastern portion of the mine at location-2. Hammer (30 cm) as scale

Fig. 6 Field photographs of Giral lignite mine show on a vertical plane parallel normal faulting with 'z' shape fold at location-7 (**a**) original and with interpretation (**a′**), a 'z' shape fold shows top to N011° shear direction in folded red clay band. A steeply-dipping (70°) normal fault with dip direction N191° cuts the early deformed beds. Secondary precipitation of salt along joints is present. The joints dip 75–80° towards N190°. Geological hammer (30 cm) as scale

At the footwall of the faults (Fault 1) asymmetric boudins show a top-to-South slip (Fig. 7) at location-4. A few "random" normal faults are also observed, dipping 58° towards N085° (Fig. 8).

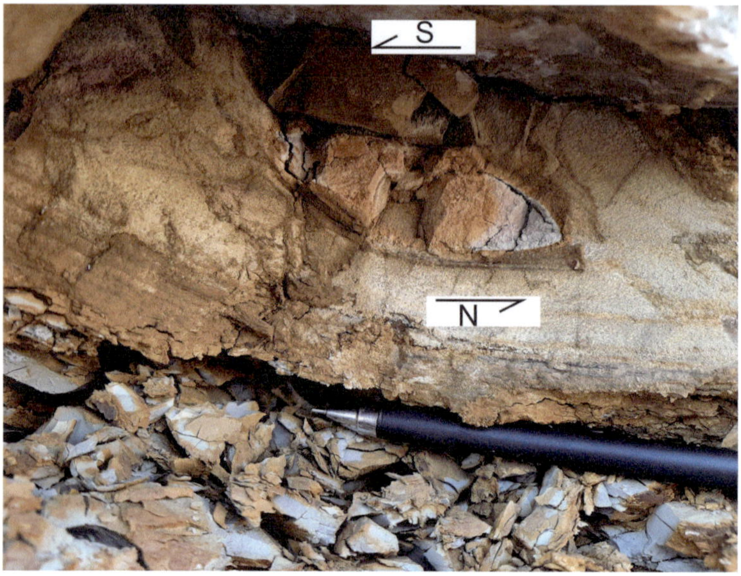

Fig. 7 A rootless clast of a boudin with top-to-S slip, location-4, observed on a vertical exposure. Part of pen (8 cm) as scale

Faulting, lignite mining and leaching of groundwater at the northern portion of mine promote ground failure. The failure is evident in the form of tension cracks at the northern portion of the mine as observed on the ground surface, i.e., on the plan view (Fig. 9). These cracks are sub-circular and show multiple stages of failure.

Fractured blocks show horizontal and vertical movement. Horizontal left-lateral slip along N295 to 308°isdeciphered with the help of curved fractures (Fig. 10a, c). Vertical and rotational movement are preserved. Difference in inclination of the adjacent fractured blocks indicates their rotation (Figs. 10b and 11). The inclination of rotated planes varies from 10 to 30° towards N115° (Fig. 10b). Vertical displacement varies from 10 cm up to 0.7 m at different levels (Fig. 11a, b). From north towards the western crown of the mine, detached block's orientation change from N112 to N210° (Fig. 11). This is due to the change in the direction of slope. Besides structural data, angle of the berm is also responsible for the mine slope stability. In this mine the angle of berm is varies from 78 to 85°. The angle of berm is very steep at northern portion and elevation difference between ground and mine deepest part is > 100 m. The width of bench also varies from place to place. At some places, it is 5 m and at elsewhere 8 m.

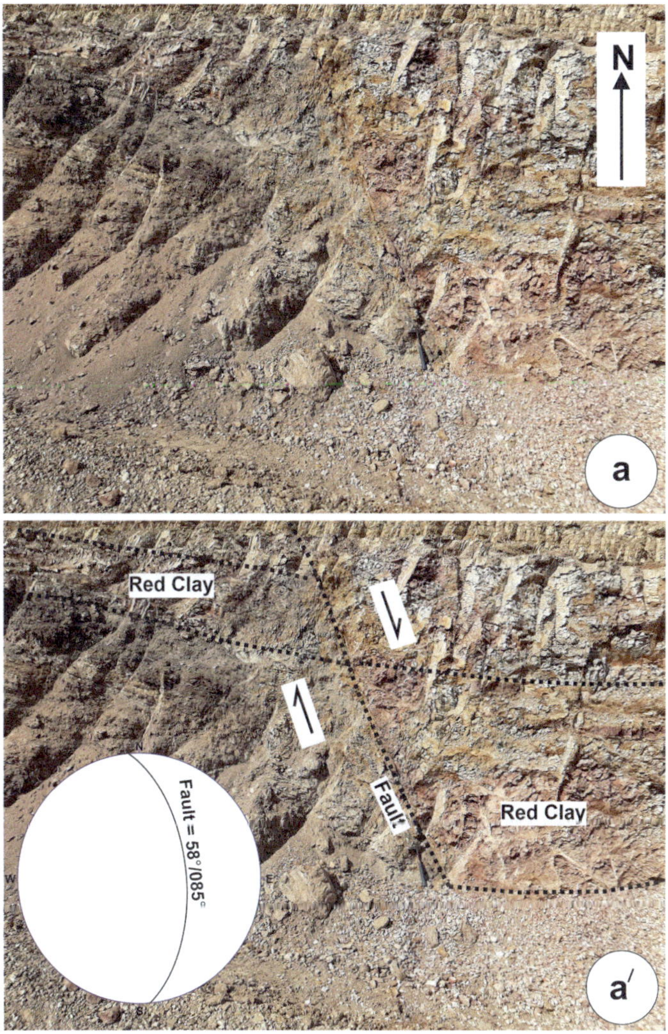

Fig. 8 Field photographs of Giral lignite mine area (**a**—uninterpreted and **a**′—interpreted) shows in a sub-vertical section a normal faulting in red clay bed. The fault dips 58° towards N085° at location-6. Hammer (30 cm) as scale

Fig. 9 Field photographs of Giral lignite mine of Northern slope shows parallel tension cracks of different stages at locations 2 and 3: **a** uninterpreted, and **a**′ interpreted. Tension cracks have developed. Few fractured (cracked) blocks have rotated

Fig. 10 Field photographs of tension cracks on plan view at location-2. **a** uninterpreted; **a′** interpreted: curved fractures probably show left lateral shear (left hand side N295°); **b** detached irregular block is rotated and dips 10° towards N146°; **c** possibly left laterally sheared curved fractures (left hand side N308); Pen (12 cm) as scale

3 Conclusion

This field work was carried out in Giral lignite mine. During this study following findings are observed;

- Two sets of normal faults are mapped and documented first time in the area. These sets can be divided into two faults group as F1 and F2.
- F1 faults have dips 31–47° and dip direction towards N160 to 172°. F2 faults show NNW (N340–350°) dip direction and dips vary from 50 to 58°. Due to the presence of two sets of normal faults, graben structures are well exposed at the northern portion of the Giral Mine.
- F1 faults generate due to the NE–SW extension. These are probably related to initiation of strike-slip movement between India and Madagascar. F2 probably resulted due to the main Barmer rifting.
- Along these normal faults, secondary precipitation of salt is observed as moderately hard filling.

Fig. 11 Field photograph of Giral lignite mine at location-3 shows parallel tension cracks, fractured planes dip at different angles; **a** detached plane shows 0.7 m throw and fractured plane dips 20° towards N210°; **b** 30–40 cm throw observed along the slip plane, this plane dips at 10° towards N112°. Another slip plane shows 8 cm throw and dips 20° towards N115°. The third plane shows a maximum inclination of 35° towards N110°. Pen (12 cm) as scale

- The faulted northern portion creates a slope stability problem in the mine. Tension cracks are the result of the slope failure and can be observed at northern side of the mine.
- Tension cracks are sub-circular and show multiple stages of slope failure. Fractured blocks show horizontal and vertical movement. The inclination of these blocks varies 10–30° with N112° to N210° orientation.

Acknowledgements The work was funded by a seed project (grant number: RD/0120-PSUCE19-001) awarded to SM by the agency: Center of Excellence in Oil, Gas and Energy (CoE-OGE, IIT Bombay). We are thankful to the Sh. SC Sharma (Head, CEU-Lignite RSMML, Barmer) for the mine visit permission. Alexis Vizcatrone and the proofreading team (Springer) are thanked for assistance. Paramita Haldar (BITS Pilani, Goa assisted the fieldwork). Mukherjee (2023) summarized this chapter.

References

Biswas, M., Gogoi, M. P, Mondal, B., Sivasankar, T., Mukherjee, S., & Dasgupta, S. (2022a). Geomorphic assessment of active tectonics in Jaisalmer basin (western Rajasthan, India). Geocarto International. https://doi.org/10.1080/10106049.2022.2066726

Biswas, M., Puniya, M. K., Gogoi, M. P., Dasgupta, S., Mukherjee, S., & Kar, N. R. (2022b). Morphotectonic analysis of petroliferous Barmer rift basin (Rajasthan, India). *Journal of Earth System Science, 131*, 140.

Compton, P. M. (2009). The geology of the Barmer Basin, Rajasthan, India, and the origins of its major oil reservoir, the Fatehgarh Formation. *Petroleum Geoscience, 15*, 117–130.

Dasgupta, S., & Mukherjee, S. (2017). Brittle shear tectonics in a narrow continental rift: asymmetric nonvolcanic Barmer Basin (Rajasthan, India). *Journal of Geology, 125*(5), 561–591.

Dolson, J., Burley, S. D., Sunder, V. R., Kothari, V., Naidu, B., Whiteley, N. P., Farrimond, P., Taylor, A., Direen, N., & Ananthakrishnan, B. (2015). The discovery of the Barmer Basin, Rajasthan, India, and its petroleum geology. *American Association of Petroleum Geology Bulletin, 99*, 433–465.

Hughes, N. C., Myrow, P. M., McKenzie, N. R., Xiao, S. H., Banerjee, D. M., & Tang, Q. (2015). Age and implications of the phosphatic Birmania Formation, Rajasthan, India. *Precambrian Research, 267*, 164–173.

Kar, N. K., Mani, D., Mukherjee, S., Dasgupta, S., Puniya, M. K., Kaushik, A. K., Biswas, M., & Babu, E. V. S. S. K. (2022). Source rock properties and kerogen decomposition kinetics of Eocene shales from petroliferous Barmer basin, western Rajasthan, India. *Journal of Natural Gas Science and Engineering, 100*, 104497.

Kumar, M., Rana, S., Pant, P. D., & Patel, R. C. (2017). Slope stability analysis of balianala landslide, Kumaun Lesser Himalaya, Nainital, Uttarakhand, India. *Journal of Rock Mechanics and Geotechnical Engineering, 9*(1), 150–158.

Kumar, A., Singh, A. K., Paul, D., & Kumar, A. (2020). Evolution of hydrocarbon potential with insight into climate and environment present during deposition of the Sonari lignite, Barmer Basin Rajasthan. *Energy and Climate Change, 1*, 100006.

Mukherjee, S. (2023). Introduction to structural geology and tectonics field guidebook—Volume 2. In S. Mukherjee (Ed.), *Structural geology and tectonics field guidebook* (vol. 2) (pp xi–xiv). Springer Nature Switzerland AG. ISBN: 978-3-031-19575-4.

Puniya, M. K, Kaushik, A. K., Mukherjee, S., Dasgupta, S., Kar, N. R., Biswas, M., Choudhary, R. (2022). (this volume). New structural geological input from the Barmer Basin, Rajasthan (India). In S. Mukherjee (Ed.), *Structural geology and tectonics field guidebook* (vol. 2). Springer.

Rajak, P. K., Singh, V. K., Singh, P. K., Singh, A. L., Kumar, N., Kumar, O. P., Singh, V., & Kumar, A. (2018). Geochemical implications of minerals environmentally sensitive elements of Giralliginite, Barmer Basin Rajasthan (India). *Environmental Earth Sciences, 77*, 698.

Roy, A. B., & Jakhar, S. R. (2002). *Geology of Rajasthan (Northwest India) precambrian to recent* (pp. 1–421). Scientific Publishers.

Sharma, K. K. (2007). K-T magmatism and basin tectonism in western Rajasthan, India, results from extensional tectonics and not from reunion plume activity. In G. R. Foulger, D. M. Jurdy

(Eds.), *Plates, plumes and planetary processes. Geological Society of America Special Paper, 430*, pp 775–784
Singh, P. K., Rajak, P. K., Singh, M. P., Singh, V. K., Naik, A. S., & Singh, A. K. (2016). Peat swaps at Giral lignite field of Barmer basin, Rajasthan Western India; Understanding the evolution through petrological modelling. *International Journal of Coal Science Technology, 3*(2), 148–164.
Sisodia, M. S., & Singh, U. K. (2000). Depositional environment and hydrocarbon prospects of the Barmer Basin, Rajasthan, India. *Nafta (Zagreb), 51*, 309–326.
Tang, H. M., Liu, X., Xin-Li, H., & Griffiths, D. V. (2015). Evaluation of landslide mechanisms characterized by high speed mass ejection and long-run-out based on events following the Wenchuan earthquake. *Engineering Geology, 194*, 12–24. https://doi.org/10.1016/j.enggeo.2015.01.004
Xue, D. M., Li, T. B., Zhang, S., Ma, C. C., Gao, M. B., & Liu, J. (2018). Failure mechanism and stabilization of a basalt rock slide with weak layers. *Engineering Geology, 233*, 213–224.

Relationship Between Deformation Structures and Rock Mass Rating: A Case Study of Underground Power House, Andhra Pradesh—India

Mohit Kumar Puniya, Sohan Kumar, Ashish Kumar Kaushik, and Nikhil Puniya

Abstract Polavaram multipurpose project, is an under construction power plant on Godavari River in the West Godavari District and East Godavari District in Andhra Pradesh. This study focuses on deformation structures in the powerhouse area and their relationship with rock mass rating (RMR). Garnetiferous gneiss, suites of khondalite, pegmatite and charnockite are the prime rock types of the Eastern Ghat Mobile Belt (EGMB). The power house has two faces viz., the upstream and the downstream faces. During surface mapping, different structural data was collected e.g., joints, shear zones, faults, spacing of discontinuity, condition of discontinuity, uniaxial compressive strength by Schmidt hammer, rock quality designation (RQD), groundwater condition, weathering condition etc. Four joint sets were identified with one random joint; those give the blocky appearance to the rockmass. Pegmatite occur in three episode of intrusion one intruded along foliation with orientation 40–55° dipping towards 220–275° (SW), dipping at 75–80° angle towards 145° (SE), third set of pegmatite intruded obliquely to the foliation i.e. 60–70 towards SW (240°). Three sets of ductile- brittle shear zones were identified. Based on cross-cut relationship they were arranged in chronological order. Oldest shear zones in the area show parallel to the regional foliation in the area and older one dips at 60° towards NE (110°) and youngest one dips moderately (25–40°) towards NW (290°). At some locations anastomosing shear zone also identified and these structure bring the RMR values varies 20–30 and bring the rock mass into poor to very poor class. Based on the surface mapping weathering grade also varies from highly weathered at the top

M. K. Puniya (✉)
National Geotechnical Facility, 11-C, Acharya Narender Dev Marg, DalanwalaDehradun, Survey of IndiaUttarakhand 248001, India
e-mail: puniyamohit@gmail.com

S. Kumar
Wadia Institute of Himalayan Geology, 33 GMS Road, Dehradun, Uttarakhand 248001, India

A. K. Kaushik
Kurukshetra University, Kurukshetra, Haryana 136119, India

N. Puniya
Department of Petroleum Engineering & Earth Sciences, UPES, Dehradun, Uttarakhand 248001, India

S. Mukherjee (ed.), *Structural Geology and Tectonics Field Guidebook–Volume 2*,
Springer Geology, https://doi.org/10.1007/978-3-031-19576-1_11

of the hill to slightly weathered towards the foot of the hill at the cutting surface. Overall RMR class varies from very poor at the deformed/sheared zone to fair class in relatively less deformed area.

Keywords Deformation · Shear zone · Joint · Pegmatites · Polawaram

1 Introduction

Deformation structures like shear zones, folds, faults and joints weakens rockmass. The frequency of these structures cumulatively can be defined as damage zone or discontinuities. In such a case, damage zone can be defined as '*deformation confined in a zone around a fault surface and may be resulted due to the tectonic activity*' (Caine et al., 1996; Childs et al., 2009; Choi et al., 2016; Cowie & Scholz, 1992; Kim & Hopke, 2004; Kumar et al., 2019; McGrath & Davison, 1995; Mukherjee, 2014; Peacock et al., 2017). Numerous authors identify the effect of deformation structures on engineering structures viz. underground caverns, tunnels and slopes (e.g., Figueiredo & Assis, 2018; Hoek & Bray, 1981; Kumar et al., 2017, 2019; Naji et al., 2018; Paul & Mahajan, 1999; Puniya et al., 2013; Stead and Eberhardt, 2013). For example, Barton et al. calculated the modulus of deformation (E_m) for sheared rock mass. Singh and Goel (2002) used the modulus of deformation (E_m), mean value of rock mass quality (Q_m) and joint roughness number (J_{rm}) to design support systems for shear zones or damage zones with semi- empirical method. Barton (1984) describes the effect of rock mass deformation on tunnel in seismic region and defines the role of joints and water on tunnel stability. Figueiredo and Assis (2018) describe the stability of underground excavation stability in Brazil. Kumar et al. (2017) described the relationship between different discontinuities to the natural slope in Balia Nala (Nainital). Sah et al. (2018) analysed the vulnerability of Nainital City based on the deformation structures. Puniya et al. (2013) and Kumar et al. (2019) describe the slope stability along Nainital Bypass and damage zone relationship with engineering slopes along Shri Kedarnath highway from Kund to Sonparyag in Himalayan region respectively. Naji et al. (2018) mention the impact of shear zone on rockbrust in the deep Neelum-Jhelum Hydropower tunnel and describe the rockbrust problem due to shear zones. Bhasin et al. (1995) calculate Q-value based on the thickness of the shear zones.

In the present study, several shear zones/damage zone are identified by preliminary surface mapping for underground powerhouse section of Polavaram Multipurpose Project. Rock mass is classified on the basis of rock mass rating (RMR) and rock quality designation (RQD) and is correlated with the ground conditions of the strata.

2 Study Site and Geology

Polavaram multipurpose project is an under construction hydropower plant on Godavari River in the West Godavari District and East Godavari District of Andhra Pradesh. This will be utilized for irrigation, drinking water and power generation with 12*80 MW capacity. Main constituents of the project are 1057.40 m long concrete spillway, stilling basin (110 m), earth dam (564 m long) in Gap-I, earth cum rock fill dam in Gap-II, earth dam (140 m long) in Gap-III and underground power house.

The study area is located around 17°13′N latitude and 81°41′E longitude and geologically is a part of the Eastern Ghat Mobile Belt (EGMB) of Achaean age. Lithologicaly, the EGMB consists of granulite-facies rocks including garnet-sillimanite gneiss (khondalite), garnet- biotite bearing quartzofeldspathic gneiss (leptynite), basic granulites, migmatite gneiss, granite, anorthosite and alkaline igneous rocks (Rajasekhara, 2019; Yamamoto et al., 1998). Mainly charnockites, migmatites and a few pegmatitic dykes are observed during mapping undertaken in this study. This mapping work has been carried out with the help of total station (Figs. 1, 2).

3 Methodology

Geological mapping has been carried out in the investigated area on 1:1000 scale (Fig. 3). This work is primarily focused on the identification of shear zones/damage zones and classification of rock mass on the near these shear zones. To calculate the RMR, various characteristics of the rock mass are recorded. These include the orientation of different discontinuities, uniaxial compressive strength (UCS) measured by using Schmidt hammer as per ISRM (1978, 1981, 2007), Bieniawski (1989), and Brencich et al. (2013), description of discontinuities (persistence, spacing, aperture, roughness, filling and alteration of discontinuities) and rockmass description (degree of weathering, groundwater conditions and rock quality designation). Rock quality designation (RQD) is calculated following Singh and Goel (1999) in the field. RQD is calculated using number of joints per unit volume Jv and equal to 115–3.3 Jv. Other available structures such as foliation, folds faults and different shear sense indicators are also recorded during the mapping (Tables 1, 2).

4 Result and Discussion

The study area is divided into two segments, one is the upstream and second is the downstream portion of the powerhouse. Both segments are divided by a central line (Fig. 3). In the upstream portion charnockites and migmatites are the main rock types with intrusions of orthoclase bearing pegmatite dykes. Four prominent set of joints are identified as J1 17–79°/270–290°, J2 60–80°/075–110°, J3 15–70°/330–025°

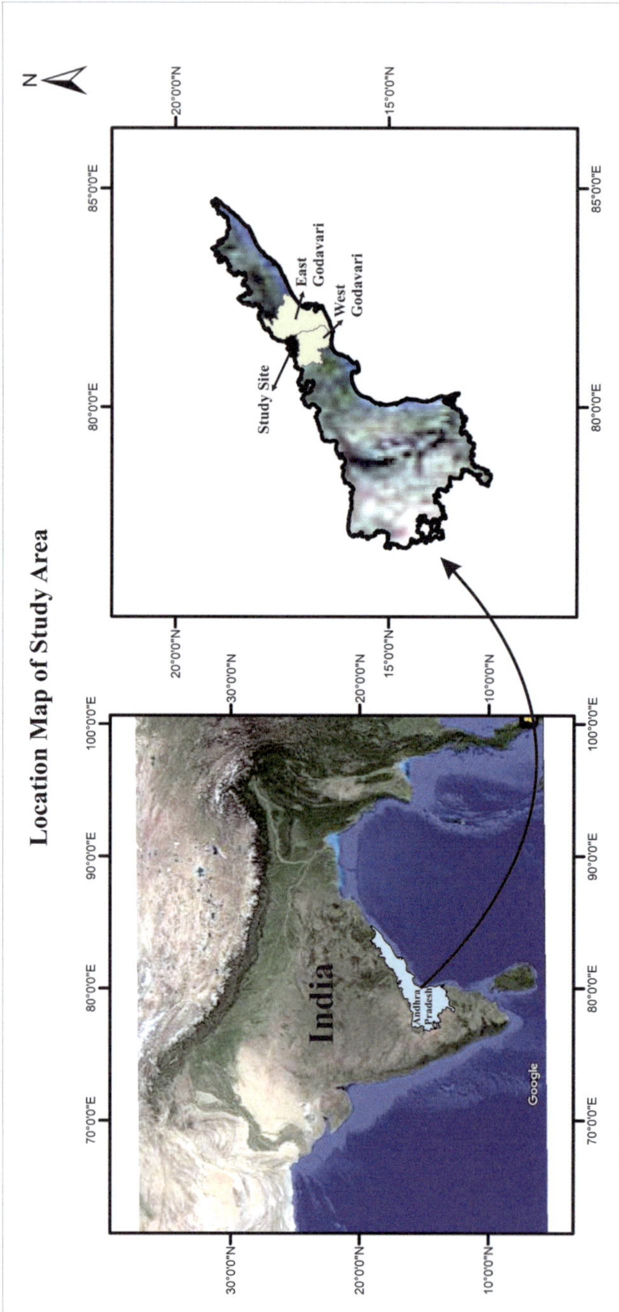

Fig. 1 Location map of the study area

Fig. 2 Field photographs are showing the lithologies; **a** Charnockite xenolith embedded in migmatite; **b** highly weathered migmatite intruded by 2–3 m thick pegmatite dyke

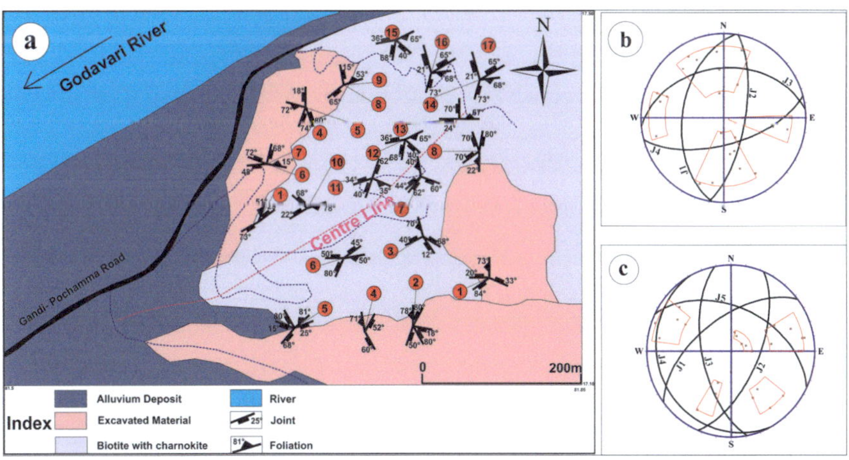

Fig. 3 **a** Geologic map of the study area is representing lithologic units and various discontinuities; **b** stereo-plot of major joint sets in upstream domain of powerhouse; poles and great circles are plotted for dip direction and dip amount of various joint sets; **c** stereo-plot of major joint sets in downstream domain of powerhouse

Table 1 Rock mass classification and prominent structure at different locations in upstream domain of the powerhouse

Outcrop no	RQD		RMR		Geological structure
	Value	Class	Value	Class	
1–2	30–40	Poor	31–55	Poor to fair	Pegmatitte dyke
4–5	50–60	Fair	60–67	Good	20–30 cm thick pegmatite dyke
6–7	50–60	Fair	60–65	Good	Jointed
8–9	30–40	Poor	25–36	Poor	4–6 m thick shear zone
10	40–50	Poor	39–53	Poor to fair	10–20 cm thick shear zone
11	35–45	Poor	40–58	Fair	30–60 cm thick shear zone
12	30–40	Poor	36–54	Poor to fair	2–4 m thick shear zone
13	45–55	Poor to fair	52–60	Fair	0.5–1 m thick shear zone
14–16	50–65	Fair	61–67	Good	Jointed

Table 2 Rock mass classification and prominent structure at different locations in downstream domain of the powerhouse

Outcrop no	RQD		RMR		Geological structure
	Value	Class	Value	Class	
1	45–55	Poor to fair	42–67	Fair to good	Shearing along the joints
2	50–55	Fair	37–60	Poor to fair	1–3 m thick pegmatite dyke
3	48–55	Poor to fair	52–55	Fair	Jointed
4	50–55	Fair	49–65	Fair to good	Jointed
5	50–60	Fair	45–52	Fair	Jointed
6	48–55	Poor to fair	39–60	Poor to fair	40 cm thick shear zone
7	30–34	Poor	32–45	Poor to fair	3–5 m thick sheared pegmatite dyke
8	28–35	Poor	33–46	Poor to fair	Multiple shear zones with cross cutting relationships

and J4 25–70°/140–200°. Fair conditions of rockmass are observed in the region, however, poor rockmass is also encountered at places with RMR value falling up to 25. Shear zones with varying thickness and pegmatite intrusions are the prime structure features observed during the mapping. Pegmatite dykes are observed at location 1 and 4, intruded along joint dipping towards west. Shearing is observed along these intrusions. Approximately 5 m thick shear zone occurs parallel to the foliation at location no. 8. This shear zone is responsible for deformed strata with RMR values of 25. Water drain is observed along this narrow zone of concentrated deformation (Fig. 4). Another 2–4 m thick shear zone was observed at location no. 12 (Fig. 5). In addition, small shear zones upto 50 cm width are also observed at different locations. Along these shear zones, maximum value of RMR reached upto

40, while adjoining strata has higher values going up to 67. At places, charnockite xenoliths are also found enclosed within the migmatite.

In the downstream domain of powerhouse, the lithology is similar to upstream domain with charnockites and migmatites as the main rock types with pegmatite dykes intruded within. Five set of joints are recorded as J1 41–70°/307–335°, J2 52–84°/097–129°, J3 38–76°/239–270°, J4 12–20°/192–270° and J5 16–31°/032–068°. Detailed mapping in downstream domain represents three episode of intrusion pegmatite intrusion; one intruded along foliation with orientation 40–55° dipping towards 220–275° (SW), second dipping at 75–80° angle towards 145° (SE), third set intruded obliquely to the foliation i.e. 60–70 towards SW (240°). At some locations, pegmatite dykes show shearing and convert the rock into sheared material. Three sets of ductile–brittle shear zones were identified at location no. 8. Based on the cross-cut relationship, oldest shear zones are observed parallel to the regional foliation and the other one dips at 60° towards SE (110°).The youngest set of shear zones dip moderately (25–40°) towards NW (290°) (Fig. 6). The array of these shear zones

Fig. 4 Outcrop photograph is showing ~5 m thick shear zone at location 8 in upstream domain of powerhouse

Fig. 5 Outcrop photo showing a ductile brittle shear zone at location 12 in migmatite

is responsible for deformed strata leading to poor RMR values upto 33. Group of small shear zones occur in location 1, where these are parallel to the steep joint set oriented at 145/84° (Fig. 7). Other narrow shear zones with small openings are also encountered at numerous outcrops. Similar to the upstream domain of powerhouse, RMR values are much lower along the sheared zones than the adjacent strata.

5 Conclusions

The powerhouse area is highly deformed with four to five joint sets. These joints give the strata a blocky appearance and is also responsible for the poor rock mass at various locations. Pegmatite dykes are intruded within the prime rock types of charnockite and migmatite. These intruded dykes signify various episodes of deformation. Shear can be observed along these dykes. In addition to these dykes, numerous shear zones are also present in the area. These shear zones are the prime locations of poor rock mass with rock mass rating falling down upto 25.

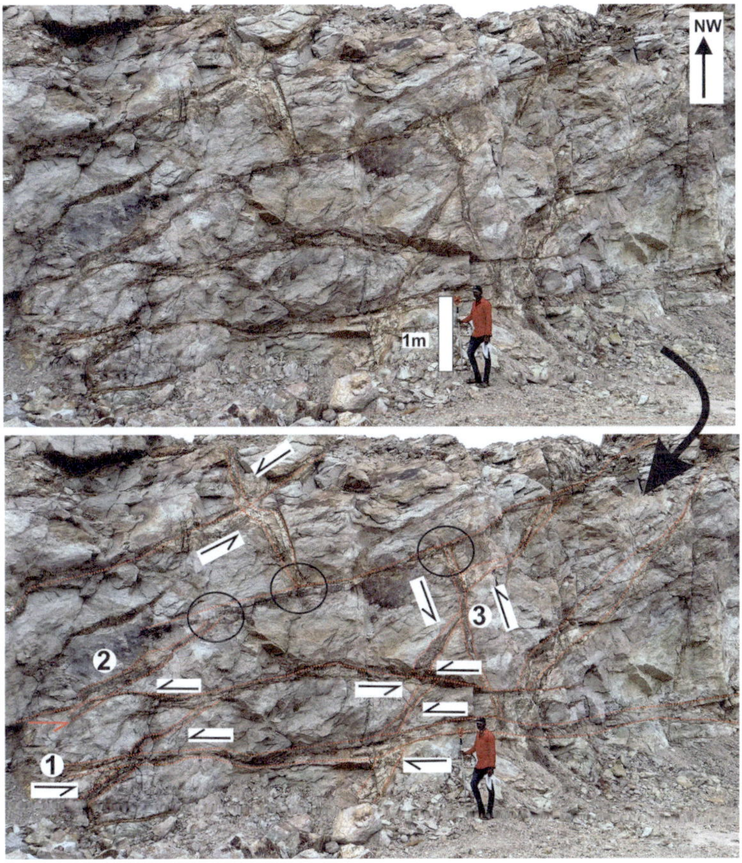

Fig. 6 Moderately weathered migmatite showing 3 sets of shear zones, showing three phases of deformation. Location 8 in downstream domain of the powerhouse

Fig. 7 Outcrop photograph showing sheared pegmatite dyke along vertical joint at location 1 in downstream domain of the powerhouse

Acknowledgements Soumyajit Mukherjee (IIT Bombay) invited and reviewed this article. Mukherjee (2023) summarized this chapter.

References

Barton, N. (1984). Effects of rock mass deformation on tunnel performance in seismic regions. *Advances in Tunnelling Technology and Subsurface Use, 4*(3), 89–99.

Bhasin, R., Barton, N., Grimstad, E., & Chryssanthakis, P. (1995). Engineering geological characterization of low strength anisotropic rocks in the Himalayan region for assessment of tunnel support. *Engineering Geology, 40*(3–4), 169–193.

Bieniawski, Z. T. (1989). *Engineering rock mass classifications: A complete manual for engineers and geologists in mining, civil, and petroleum engineering*. Wiley.

Brencich, A., Cassini, G., Pera, D., & Riotto, G. (2013). Calibration and reliability of the rebound (Schmidt) hammer test. *Civil Engineering and Architecture, 1*(3), 66–78.

Caine, J. S., Evans, J. P., & Forster, C. B. (1996). Fault zone architecture and permeability structure. *Geology, 24*(11), 1025–1028.

Childs, C., Manzocchi, T., Walsh, J. J., Bonson, C. G., Nicol, A., & Schöpfer, M. P. (2009). A geometric model of fault zone and fault rock thickness variations. *Journal of Structural Geology, 31*(2), 117–127.

Choi, J. H., Edwards, P., Ko, K., & Kim, Y. S. (2016). Definition and classification of fault damage zones: A review and a new methodological approach. *Earth-Science Reviews, 152*, 70–87.

Cowie, P. A., & Scholz, C. H. (1992). Displacement-length scaling relationship for faults: Data synthesis and discussion. *Journal of Structural Geology, 14*(10), 1149–1156.

Figueiredo, L. T., & Assis, A. P. (2018). Case study: Stability assessment in underground excavations at Vazante Mine-Brazil. *Soils and Rocks, 41*(2), 203–216.

Hoek, E., & Bray, J. D. (1981). *Rock slope engineering*. CRC Press.

ISRM (International Society for Rock Mechanics). (1978). Suggested methods for the quantitative description of discontinuities in rock masses. *International Journal of Rock Mechanics and Mining Sciences and Geomechanics Abstracts, 15*, 319–368.
ISRM. (1981). Rock characterization testing and monitoring. In E. Brown (Ed.), Pergamon Press, Oxford, 211 p.
ISRM. (2007). The complete ISRM suggested methods for rock characterization, testing and monitoring: 1974–2006. Suggested methods prepared by the commission on testing methods, International Society for Rock Mechanics, Compilation Arranged by the ISRM Turkish National Group Ankara, Turkey, 628 p.
Kim, E., & Hopke, P. K. (2004). Comparison between conditional probability function and nonparametric regression for fine particle source directions. *Atmospheric Environment, 38*(28), 4667–4673.
Kumar, M., Joshi, R. C., & Pant, P. D. (2019). Impact of structural damage zones on slope stability: A case study from Mandakini Valley, Uttarakhand state (India). In S. Mukherjee (Ed.), *Tectonics and structural geology: Indian context* (pp. 397–410). Springer.
Kumar, M., Rana, S., Pant, P. D., & Patel, R. C. (2017). Slope stability analysis of Balia Nala landslide, Kumaun Lesser Himalaya, Nainital, Uttarakhand, India. *Journal of Rock Mechanics and Geotechnical Engineering, 9*(1), 150–158.
McGrath, A. G., & Davison, I. (1995). Damage zone geometry around fault tips. *Journal of Structural Geology, 17*(7), 1011–1024.
Mukherjee, S. (2014). Review of flanking structures in meso-and micro-scales. *Geological Magazine, 151*(6), 957–974.
Mukherjee, S. (2023). Introduction to structural geology and tectonics field guidebook—Volume 2. In: S. Mukherjee (Ed.) *Structural geology and tectonics field guidebook—Volume 2* (pp. xi–xiv). Springer Nature Switzerland AG. Cham. ISBN 978-3-031-19575-4.
Naji, A. M., Rehman, H., Emad, M. Z., & Yoo, H. (2018). Impact of shear zone on rockburst in the deep Neelum-Jehlum hydropower tunnel: A numerical modeling approach. *Energies, 11*(8), 1935.
Paul, S. K., & Mahajan, A. K. (1999). Malparockfall disaster, Kali valley Kumaun Himalaya. *Current Science, 76*(4), 485–487.
Peacock, D. C. P., Dimmen, V., Rotevatn, A., & Sanderson, D. J. (2017). A broader classification of damage zones. *Journal of Structural Geology, 102*, 179–192.
Puniya M.K., Joshi P., Pant P.D. (2013). Geological investigation of Nainital Bypass; A special emphasis on slope stability analysis, Kumaon Lesser Himalaya, Himalayan Vulnerability p. 65.
Rajasekhara R. K. (2019). A study on partial replacement of feldspars with charnockite rocks of Eleswaram, East Godavari District, Andhra Pradesh, India in Red Body Glazed Ceramic Tiles, International Journal of Recent Technology and Engineering (IJRTE) ISSN: 2277-3878, Volume-7, Issue-6C2
Sah, N., Kumar, M., Upadhyay, R., & Dutt, S. (2018). Hill slope instability of Nainital City, Kumaun Lesser Himalaya, Uttarakhand, India. *Journal of Rock Mechanics and Geotechnical Engineering, 10*(2), 280–289.
Singh, B., & Goel, R. K. (1999). Rock mass classification: A practical approach in civil engineering (Vol. 46). Elsevier.
Singh, B., & Goel, R. K. (2002). *Software for engineering control of landslide and tunnelling hazards*. CRC Press.
Stead, D. O. U. G., & Eberhardt, E. R. I. K. (2013). Understanding the mechanics of large landslides. *Italian Journal Engineering Geology Environment Book Series, 6*, 85–112.
Yamamoto, T., Tani, Y., Miyashita, Y., Rao, A. T., & Yoshida, M. (1998). Migmatite and granulites in the Patapatnam-Tekkali area, Eastern Ghats, India. *Journal of Geosciences Osaka City University, 41*, 123–142.

Microstructures Mimic Meso-Scale Structures

Harsh Bhu, Ritesh Purohit, and Riya Dutta

Abstract The chapter deals with the imitation of meso-scale as microstructures in photomicrographs from a complexly deformed Precambrian terrane of the Aravalli craton from northwest India. These structures are mostly seen in low to medium-grade metamorphosed metasediments e.g., calc-silicates, schists, and gneisses.

Keywords South Delhi Terrane · Mount Abu · Pindwara-Abu Road Belt · West Marginal Fault · Todgarh Formation · Basantgarh Formation

1 Introduction

The mesoscale structures present in the rocks are also referred to as small-scale structures by some workers. They can be decoded for regional strain distribution (e.g., Mukherjee, 2021). The imprints of small-scale structures found in thin sections are referred to as microstructures. They can be foliations, cleavages, rotated blasts, and shear-related features.

The study area is located to the east of Mount Abu batholith; along the eastern margin of West Marginal Fault (WMF) in the southern swathe of South Delhi Terrane between Pindwara-AbuRoad Belt (Fig. 1). The stratigraphic status of this area is assigned to the Todgarh Formation and Basantgarh Formation of the Kumbalgarh Group of South Delhi Terrane.

2 Foliations and Cleavages

Foliations can be used to obtain information on the strain, metamorphic conditions, and overprinting relations (Passchier & Trouw, 2005). During metamorphism, when the rocks undergo heating and squeezing, the changing temperature can form new

H. Bhu · R. Purohit (✉) · R. Dutta
Department of Geology, M.L.Sukhadia University, Udaipur, Rajasthan, India
e-mail: riteshpurohit72@gmail.com

S. Mukherjee (ed.), *Structural Geology and Tectonics Field Guidebook–Volume 2*,
Springer Geology, https://doi.org/10.1007/978-3-031-19576-1_12

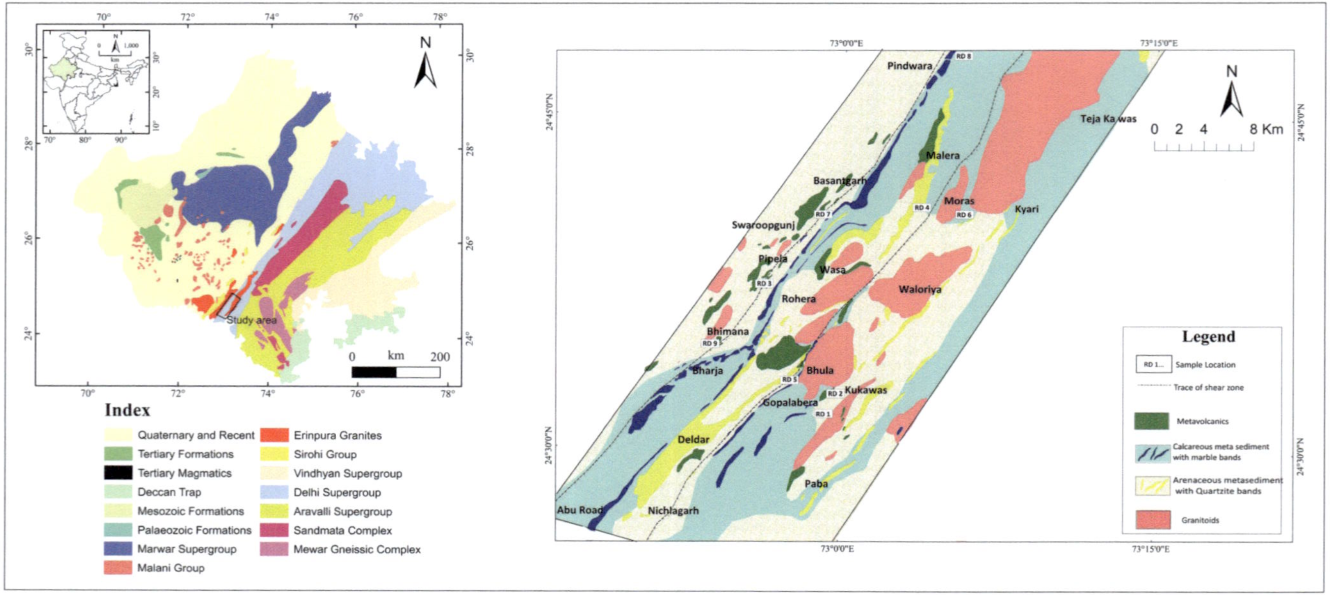

Fig. 1 Location of the study area in South Delhi Terrane between Pindwara-Abu Road Belt

minerals. The newly formed minerals are usually elongated perpendicular to the direction of compression, and the bedding in a rock becomes obscured by foliation.

In Fig. 2a, the outcrop may be described in such a way that the layered sedimentary rock (shale)had undergone metamorphism through pressure, temperature changes, and foliations are developed by lengthening and squeezing of minerals by deformation in the direction perpendicular to the maximum stress (Ramsay & Huber, 1987a, b). The rock has undergone a relatively low degree of metamorphism, for what the bedding is still visible. The photomicrograph (Fig. 2b), is the small-scale manifestation of bedding parallel foliation in mesoscale. Figure 3a shows foliated amphibolite with the preferred alignment of flaky amphiboles that are derived from metamorphism mafic igneous rocks. Figures 2b and 3b show the continuous foliation with a non-layered homogeneous distribution of flaky and platy minerals viz., mica, amphibole along with elongated quartz and other minerals.

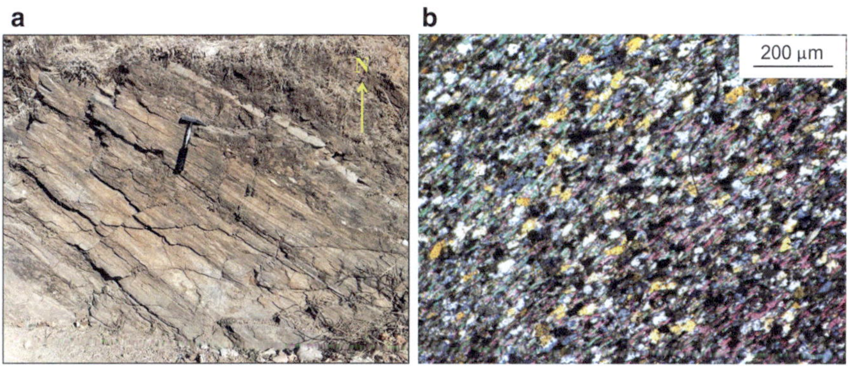

Fig. 2 **a** Foliated quartz-mica schist formed by the preferred alignment of flaky grains (length of hammer: 30 cm) (sample location. RD 1, near Gopalabera); **b** foliation developed by preferred alignment of elongated quartz and platy mica in quartz mica schist (cross-polar light)

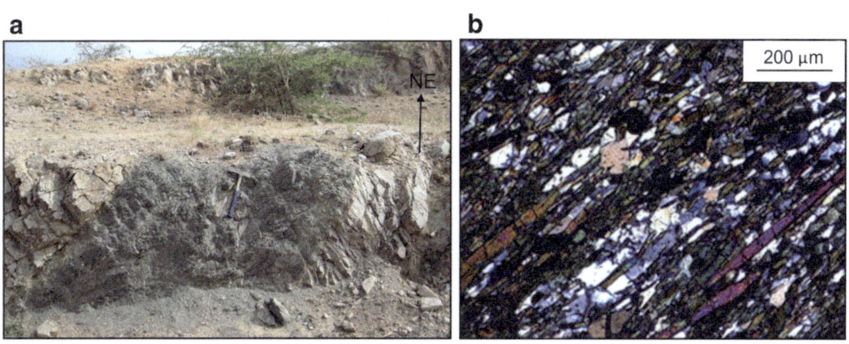

Fig. 3 **a** Green-colored foliated amphibolites (length of hammer: 30 cm) (sample location: RD 2, near Kukawas); **b** the parallel orientation of amphibole minerals and elongated quartz forming a significant foliation in amphibolite (cross-polar light)

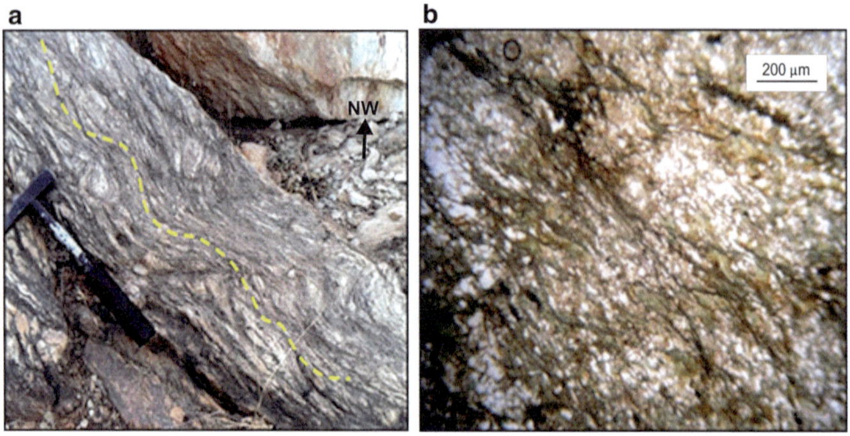

Fig. 4 **a** Field photograph of warping of foliation in foliated quartzite (length of hammer: 30 cm) (sample location: RD 3, Southwest of Pipela); **b** photomicrograph of warping of foliation in foliated quartzite (plane-polarized light)

The warping of foliation in the outcrop-scale indicates the more advanced stage of deformation where the later foliation tends to form distinct foliation plane cross-cutting the earlier one by later deformation at some angle to the previous one and may involve warping of the earlier phase of crenulation formation. That type of feature not only gives the clue for stress pattern but also shows the orientation of earlier foliation (Fig. 4a, b).

3 Shear Related Features

(a) Rotated Porphyroclasts: Fine-grained soft mantled porphyroblasts show crystaloblastic deformation and dislocation *tangles* in response to flow in the matrix. The fine-grained soft mantle distorts from its parallelism. The wings are stretched and change in shape while the core of porphyroblasts remains rigid. Few of the core minerals are symmetric (Mukherjee, 2017) and the others are asymmetric (Mukherjee, 2013). The asymmetric shapes of the wings in the outcrop scale are a useful indicator of shear sense (Fig. 5a, b) but must be used with caution (e.g., Dutta & Mukherjee, 2019; Mulchrone & Mukherjeee, 2020).

(b) Shear Cleavages: These are the results of mylonites, which are fine-grained, banded, and cohesive rocks formed by localized plastic flow and dynamic recrystallization and are generally characterized by very high finite strain. In the outcrops, mylonites are identified as unusually regular and planar strongly developed foliation. Due to intra-crystalline deformation or recrystallization in the ductile regime, the coarser-grained parental rocks alter to form layers or lenses of fine-grained material. The elongation of long axes of individual crystal is stretched (Fig. 6a) and quartz is

Fig. 5 **a** Parallel arrangement of calc-silicate layer distorted by porphyroblasts (length of pencil: 14.5 cm) (sample location: RD 4, South of Malera); **b** rotated porphyroblasts of quartz with undulose extinction bounded by fine-grained elongated quartz and calcareous grains (cross-polar light)

drawn out as long narrow ribbons (Fig. 6b). Both meso-scale and microscale features are in the section parallel to the stretching lineation and perpendicular to the mylonitic foliation.

On the increase of temperature large porphyroblasts are seen embedded in a fine-grained matrix (Fig. 7a) and in thin sections, fine elongated grains with elongated porphyroblasts are seen (Fig. 7b).

The shear cleavage formation in serpentinized mafic rocks develops a strong preferred orientation on a meso-scale (Fig. 8a) and the serpentine mineral fibres are perfectly aligned along the shear plane (Fig. 8b).

(c) Pinch and swell structures: These are developed in the rocks with significant competency differences, where more competent layers within the weaker matrix are subjected to layer parallel extension. In the field, pinch and swell structures have

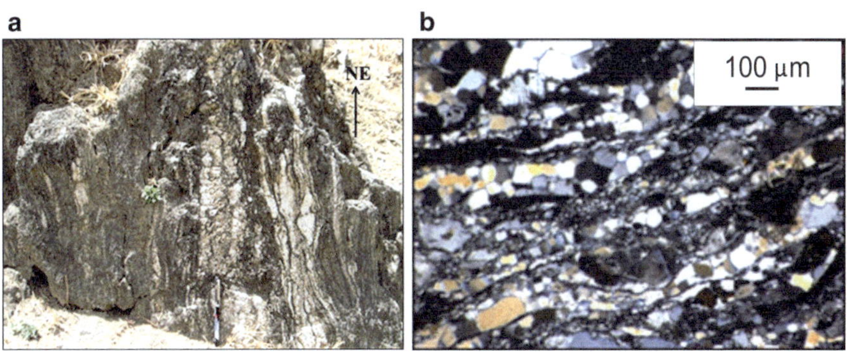

Fig. 6 **a** Mylonite in schistose metasediments. Low strain lenses of feldspar occur around which planar-shaped fabric anastomoses occur as mylonitic foliation (length of the pen: 14 cm) (sample location: RD 5, southwest of Bhula granite); **b** mylonitized metasediments with porphyroblasts of plagioclase, orthopyroxene, and amphibole. The stretched quartz are recrystallized into angular grains (cross-polar light)

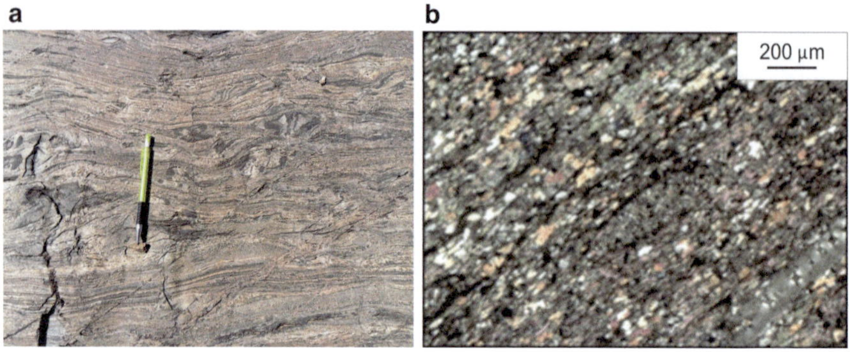

Fig. 7 **a** Closely spaced mylonitic foliation in calc silicates (length of pencil: 14.5 cm) (sample location: RD 6, south of Moras); **b** mylonitic foliation is formed by the parallel arrangement of fine-grained elongated quartz and calcite (cross-polar light)

Fig. 8 **a** Sheared partially serpentinized metavolcanics (marker: Riya Dutta, geologist; height: 5ft) (sample location: RD 7, near Basantgarh); **b** the preferred orientation of serpentine minerals in partially serpentinized metavolcanics (cross-polar light)

great significance for determining strain localization and flow behavior. Pinch and swell structures in finely foliated rocks show elongation along a direction (Fig. 9a) and the phenomena can also be observed in thin sections (Fig. 9b).

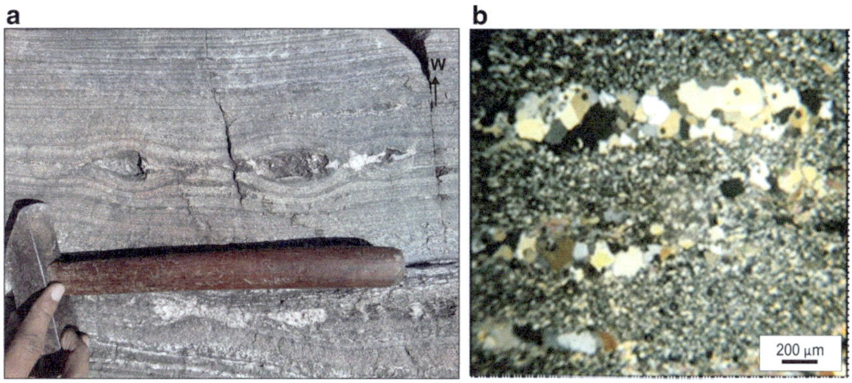

Fig. 9 **a** Pinch and Swell structures in calc silicates (length of hammer: 30 cm) (sample location: RD 8, near Pindwara); **b** lenses of coarse-grained quartz embedded in a very fine-grained groundmass of quartz, calcite, and mafics (cross polar light)

4 Faults and Displacement

Faults are fracture discontinuities in a rock along which a significant differential displacement has taken place. They transect and displace lithological layers. Faults are generally formed during the brittle failure of rock under stress, but transitions exist between brittle and ductile regimes. The plane along which displacement has taken place is generally filled by fault scrap intruded by veins (Fig. 10a). On the outcrop scale, the fault can be identified by several criteria among which the simplest way is to identify the break in the continuity of strata with corresponding cut-off lines, piercing points, and by dragging. In Fig. 10a, dragging is considered to divulge the relative motion along the fault plane. In thin sections, the displacement can be observed on the minerals where they are faulted (Fig. 10b, c).

5 Summary

Meso-scale structures in various forms are depicted as microstructures in more or less similar fashion and pattern. The fabric on the meso-scale is imitated on the microscale and can be deciphered by texture and mineral identification. On the meso-scale, mineral identification is not so pronounced but can be distinctly identified on the microscale. Such studies on meso-scale and micro-scale when done simultaneously can be taken up as perfect field guides by structural geologists.

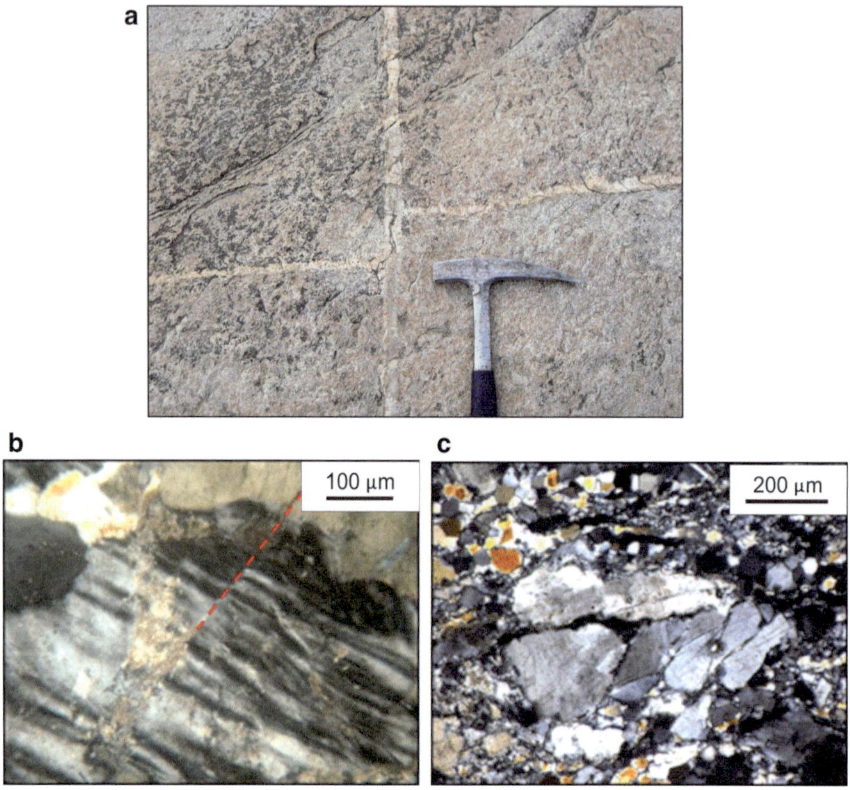

Fig. 10 **a** Displaced vein of quartz in granite (length of hammer: 30 cm) (sample location: RD 9, near Bhimana); **b** displaced twin lamellae in plagioclase (cross-polar light); **c** irregular fracture zone formed by displacement in orthoclase grain. The fracture planes are filled with mafics (cross-polar light)

Acknowledgements We thank Soumyajit Mukherjee (IIT Bombay) for inviting, editing and reviewing this article. Mukherjee (2023) summarized this chapter. The authors wish to acknowledge the funding agency DST, GOI for their support in this work through the WOS-A scheme to RD.

References

Dutta, D., & Mukherjee, S. (2019). Opposite shear senses: Geneses, global occurrences, numerical simulations and a case study from the Indian Western Himalaya. *Journal of Structural Geology, 126*, 357–392.

Mukherjee, S. (2013). *Deformation microstructures in rocks* (pp. 1–111). Springer Geochemistry/Mineralogy. ISBN: 978-3-642-25608-0.

Mukherjee, S. (2017). Review on symmetric structures in ductile shear zones. *International Journal of Earth Sciences, 106*, 1453–1468.

Mukherjee, S. (2021). *Atlas of structural geology* (2nd edn, pp. 1–260). Elsevier. ISBN: 978012816802.

Mukherjee, S. (2023). Introduction to structural geology and tectonics field guidebook, vol 2. In S. Mukherjee, (Ed.) *Structural geology and tectonics field guidebook* (vol. 2) (pp. xi–xiv). Springer Nature Switzerland AG. ISBN 978-3-031-19575-4.

Mulchrone, K. F., & Mukherjee, S. (2020). Numerical modelling and comparison of the temporal evolution of mantle and tails surrounding rigid elliptical objects in simple shear regime under stick and slip boundary conditions. *Journal of Structural Geology, 132*, 103968.

Passchier, C. W., & Trouw, R. A. J. (2005). *Microtectonics* (pp. 1–366). Springer. ISBN: 3-540-64003.

Ramsay, J. G., & Huber, M. I. (1987a). *The techniques of modern structural geology: Strain analysis* (vol. 1, pp. 1–307). Academic Press. ISBN: 0-12-576901-06.

Ramsay, J. G., & Huber, M. I. (1987b). *The techniques of modern structural geology: Folds and fractures* (vol. 2, pp. 308–700). Academic Press. ISBN: 0-12-576902-4.

Review on Role of Multi-Constellation Global Navigation Satellite System-Reflectometry (GNSS-R) for Real-Time Sea-Level Measurements

Kutubuddin Ansari

Abstract The coast is a unique environment where land, sea and atmosphere interact and interplay continuously influencing a strip of spatial zone defined as coastal zone. In other words, coastal zones are the areas having the influence of both marine and terrestrial processes. This article reviews Global Navigation Satellites System (GNSS) based reflectometry technique using various GNSS constellation systems to predict sea-level rise. The study summarized various kinds of reflectometry studies, which involves three global navigation satellites systems (GPS, GLONASS and Galileo) and three regional navigation satellite systems (BeiDou, IRNSS and QZSS) worldwide. The GNSS constellations and individual and combined performance including their comparative analysis are included. It is expected that when a field-worker works on the coastal zone, this kind of study will be useful in understanding the tectonic (as well as oceanic) activities.

Keywords GNSS-Reflectometry · GPS · GLONASS · Galileo · BeiDou · IRNSS and QZSS

1 Introduction

A Multi-constellation Global Navigation Satellite System (GNSS) receiver is the system capable to estimate the velocity, position, and time by receiving the multi navigation signals broadcasted by multiple satellites (Ansari & Park, 2018; Ansari et al., 2021; Carlin et al., 2021; Li et al., 2015a, 2015b). Previously, Global positioning system (GPS) was the representative satellite navigation system operated by the United States. Presently other positioning systems e.g., Globalnaya Navigatsionnaya Sputnikovaya Sistema (GLONASS) operated by Russia, Galileo operated by Europe, BeiDou (Compass) operated by China, Indian Regional Satellites System (IRNSS) operated by India and the Quasi-Zenith Satellite System (QZSS) operated by Japan are available (Fig. 1). Moreover, Satellite Based Augmentation System (SBAS)

K. Ansari (✉)
Integrated Geoinformation (IntGeo) Solution Private Limited, New Delhi 110025, India
e-mail: kdansarix@gmail.com

S. Mukherjee (ed.), *Structural Geology and Tectonics Field Guidebook–Volume 2*,
Springer Geology, https://doi.org/10.1007/978-3-031-19576-1_13

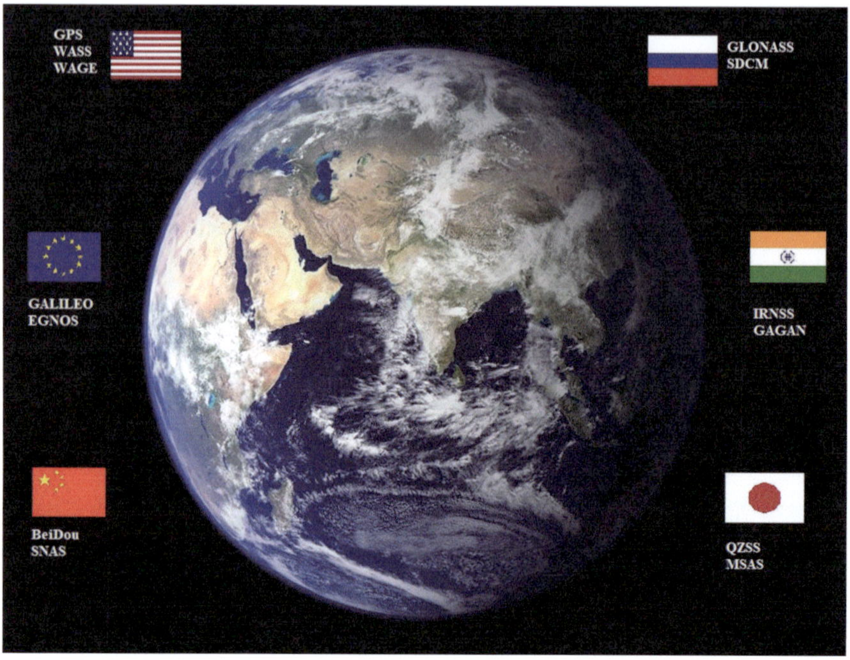

Fig. 1 The world of Multi-constellation Global Navigation Satellite System (GNSS) and Satellite Based Augmentation System (SBAS). Here GNSS are Global positioning system (GPS), operated by the United States, Globalnaya Navigatsionnaya Sputnikovaya Sistema (GLONASS) operated by Russia, Galileo operated by Europe, BeiDou operated by China and Indian Regional Satellites System (IRNSS) operated by India and Quasi-Zenith Satellite System (QZSS) operated by Japan. The SBAS are Wide Area Augmentation System (WAAS) and Wide Area GPS Enhancement (WAGE) of United States, System for Differential Corrections and Monitoring (SDCM) from Russia, European Geostationary Navigation Overlay Service (EGNOS) of Europe, Satellite Navigation Augmentation System (SNAS) from China, GPS-aided GEO augmented navigation (GAGAN) from India, Multi-functional Satellite Augmentation System (MSAS) of Japan (https://en.wikipedia.org/wiki/GNSS_augmentation)

is a network of the geostationary satellites e.g., Wide Area Augmentation System (WAAS) and Wide Area GPS Enhancement (WAGE) from United States, System for Differential Corrections and Monitoring (SDCM) from Russia, European Geostationary Navigation Overlay Service (EGNOS) from Europe, Satellite Navigation Augmentation System (SNAS) from China, GPS-aided GEO augmented navigation (GAGAN) from India and Multi-functional Satellite Augmentation System (MSAS) from Japan are providing quality services. The Southern Positioning Augmentation Network (SouthPAN) established by Australia and New Zealand planned to be in operation by 2023 (Wikipedia, Retrieved on 04-10-2021). Few private industries have been operating their own SBAS networks. For example, the commercial Starfix DGPS System and the OmniSTAR system are worked by Fugro, the commercial

StarFire navigation system is worked by John Deere and C-Nav Positioning Solutions, the commercial Atlas GNSS Global L-Band Correction Service system worked by Hemisphere GNSS. The GPS Correction (GPS.C) was a differential GPS data source maintained by the Canadian Active Control System (part of Natural Resources Canada).

The impact of sea level rising is specially interest for the population residing in coastal areas or islands, which does not only impact the public safety but also their infrastructure developments, protection of relocation, sustainability and heath ecosystem services and rising and fall of economical situations (Löfgren et al., 2011; Sasaki et al., 2014; Oke et al., 2015; Ansari et al., 2020).For example, a strong earthquake with Mw = 9.1 occurred in Japan on 11-March-2011, known as Tohoku-Oki (TO) earthquake, and caused a giant tsunami in Japan. Occurrence of these extreme events has been increasing worldwide with an anticipated rise of sea-level (Parker, 2019; Vitousek et al., 2017; Vousdoukas et al., 2018). Global warming and change of climate are also expected cause of sea level rising. Thus, an effective and accurate modelling of sea level variation and its impact on economic growth is an important mission (Cardellach et al., 2019; Chen et al., 2019; Liu et al., 2020; Montillet et al., 2018; Tabibi et al., 2020; Wang et al., 2018).

The Global Navigation Satellite System-Reflectometry (GNSS-R) have comparable accuracy for monitoring sea-level (see Table 1). In the following years GNSS-R technique not only used in oceanographic field but also in several other purposes (see Table 2). The GNSS-R technique developed to monitor the sea level measurement, uses satellite reflected signals data from nearby water surface in contrast to the standard GNSS applications for high-precession positioning that use direct satellite signal data received by GNSS sites (Larson et al., 2017). These reflected signals oscillate after interference from the multipath surface and can be recorded in the form of signal-to-noise ratio (SNR) data (Fig. 2). Later, vertical distance from the water surface is estimated by analyzing the frequency of the SNR oscillations. GNSS-R can sense larger areas (hundreds of square meters) and may capture extreme sea level changes at some distance along the coast (Peng et al., 2019).

The GNSS-R technique has been kept on used to measure the global sea level changes by providing the relative measurement with respect to land, offer high-precision topography of the sea surface (Feng et al., 2013), but often has poor-precision near to the coastal areas because of the fast and complex dynamical changes of the ocean in the same region (Bouffard et al., 2011; Roussel et al., 2015). Moreover, sometimes it may not provide mesoscale level phenomena/information such as swell and waves with real tide amplitudes, needed better spatiotemporal level of sampling (Martín-Neira et al. 2011; Wang et al., 2018). At these situations, the use of multi-GNSS reflected signals is an optimal method to enhance the precision and sampling rates by taking the advantage of more satellites improved signals and extensive sensing coverages (Larson et al., 2017; Tabibi et al., 2017; Wang et al., 2019).Several kinds of errors can be there, e.g., ionospheric and tropospheric errors, satellite and clock error can be reduced when more constellations track more signals. Several researchers have been used GNSS-R algorithm by using multi-constellation systems and presented comparative solutions (Hobiger et al., 2014; Tabibi et al., 2020; Wang

Table 1 List of few GNSS-R studies for monitoring sea-level (references with their key input, key output and terrains of the country)

Authors	Key input	Key output	Terrains of the country
Löfgren et al. (2011)	The GNSS-derived time series of local sea level is compared with independent data	A high degree of agreement is found in the time domain, with correlation coefficients of up to 0.96	Data were collected during three months in 2010 from September 16 to December 19 located at 18 km south and 33 km north of OSO, Sweden respectively
Larson et al. (2013)	Tide gauge data from single frequency GPS receiver	Best suitable results of GPS-R measurements	Friday Harbor GPS site, PBO, USA and OSO, Sweden
Roussel et al. (2015)	GLONASS-R along with GPS-R and worked on Sea level measurements by using a single geodetic receiver	Combined GPS and GLONASS constellation result better than individual GPS and GLONASS	A geodetic antenna set up at ~ 60 m above the surface of the Atlantic ocean, at Cordouan lighthouse. Data were acquired from 3 March 2013 to 31 May 2013
Santamaría-Gómez and Watson (2017)	A new approach based on optimization of the unknown receiver bandwidth and the estimation of frequency changes in the SNR oscillation through an extended Kalman filter/smoother algorithm is developed. The geometric bending of the GNSS signals due to tropospheric refraction using local meteorological observations is corrected	Possible to estimate small but significant changes in the frequency of the SNR oscillation by elevation angle	From January 7–30, 2015, a field experiment was carried out at the TG station at Spring Bay, Tasmania, Australia
Larson et al. (2017)	A geodetic GPS receiver Aquatrak tide gauge, collocated are used	Individual water level estimates showed an RMS error of about 12 cm. The errors are slightly reduced at lower water levels and slightly raised at higher levels	The tide gauge, collocated at Friday Harbor, Washington, are used to assess the quality of 10 years (2006–2015)

(continued)

Table 1 (continued)

Authors	Key input	Key output	Terrains of the country
Wang et al. (2019)	A multi-GNSS combination algorithm is developed to formulate SNR data of four constellations (GPS, GLONASS, Galileo, and BeiDou)	The combined sea level retrieval time series had an approximately 40%–75% accuracy improvement compared to individual signal sea level retrievals	BRST station located in Brest harbor on the west coast of France. The SNR data in DOY (day of year) 141–347 of year 2017 are collected in this work
Zhang et al. (2019)	SNR signal decomposition method based on empirical mode decomposition (EMD) is presented	The EMD-based improved algorithm results are more accurate and can be used to obtain the change trend of local mean sea level	SNR sequence of PRN31 on Day of Year (DOY) 1 in 2013, which was obtained from Friday Harbor in the US and Rosslyn Bay in Australia
Chen et al. (2019)	GLONASS (L1, L2) signals SNR including to the other GNSS (GPS and BeiDou) system	GLONASS-R estimated observation have mean error about 0.02 m and standard deviation about 0.32. The histogram plot of sea surface height error estimation approximated a normal distribution	MAYG station located in Mayotte near the Indian Ocean
Geremia-Nievinski et al. (2020)	The GPS SNR data for L1-C/A signal was processed by four groups, in Germany, Sweden, Luxembourg and United Kingdom	All groups processed tide gauge results showed good agreement	Onsala Space Observatory for one-year period (2015 to 2016)
Ansari et al. (2020)	GPS SNR measurement predicted with ARMA and SSA methods	Significant improvement and perfect suitability obtained of GPS-R after ARMA and SSA method	South Beach, Oregon, United States

et al., 2019). This article reviews multi-constellation GNSS-R (GPS, GLONASS, Galileo, BeiDou, QZSS and IRNSS) for real-time sea-level measurements.

The interaction amongst tectonic plates frequently results in intra- and inter-plate seismicity (Ansari & Park, 2019; Sagiya et al., 2000). Primary mechanisms of post-seismic land motions are generally considered as after-slip and viscoelastic relaxation. Tide gauge recorded observation from active tectonic zones are normally recognized to be biased by seismic events connected process (Klos et al., 2019). Estimations of relative sea-level fall and rise resulted from tide gauge observations

Table 2 List of few GNSS-R studies used for various kind of purposes (references with their key input, key output and terrains of the country)

Authors	Key Input	Key Output	Terrains of the Country
Zhou et al. (2019)	GPS triple frequency (L1, L2 and L5) signals are utilized for snow depth estimation and a new approach utilizing the SNR Combination is used	The proposed approach presents a high correlation of 0.95 and an accuracy in terms of RMS improvement of over 30%	KIRU (Sweden) from July 2015 to May 2016 and YEL2 (Canada) from August 2015 to June 2016
Rodriguez-Alvarez et al. (2019)	Sea ice multi-step classification approach based on bistatic radar reflections applied	GNSS-R has the potential to be an effective measurement for classifying different types of sea ice	October 2015 in the Beaufort and Chukchi seas region
Peng et al. (2019)	GPS (L1, L2C and L5) signal SNR data at under normal and abnormal weather conditions	GPS-R is successfully able to produce sea-level variability during the normal and strong wind too	HKQT site from Hong Kong
Jia et al. (2019)	GNSS-R soil moisture Retrieved using XGboost and Machine Learning Aided Methods	Statistical analysis of ground-truth measurements showed a good correlation. The study provides some experimental insights into the behavior of the GNSS-R soil moisture retrieval	Grugliasco and Agliano, Italy
Edokossi et al. (2020)	Detailed review on the current soil moisture content measurement techniques, retrieval approaches, products, and applications are presented	GNSS-R techniques based on different models that can estimate soil moisture have been examined	Northern Argentina and Uruguay with soil moisture content in March 2007 and other places
Martín et al. (2020)	Multi-constellation solution using GPS, GLONASS, and GALILEO satellites are used for soil moisture	GNSS solutions, including the three constellations and the two sensors (geodetic and mass market), were highly correlated, with a correlation coefficient between 0.7 and 0.85	3 December 2018 to 6 February 2019 at Cajamar Centre of Experiences, Paiporta, Valencia, Spain

(continued)

Table 2 (continued)

Authors	Key Input	Key Output	Terrains of the Country
Yan et al. (2020)	An effective schematic is developed for estimating soil moisture from CYclone GNSS data	The use of CYclone GNSS soil moisture significantly enhances the pan-tropical coverage of soil moisture active passive by about 22% on average	The data examined in this work span the year of 2018 and over the full region covered by CYclone GNSS (within ± 37° latitudes)
Hammond et al. (2020)	GNSS-R ocean wind speed retrieval performed	The accurate attitude knowledge and a good characterization of GNSS-R nadir antenna patterns should be prioritized for future GNSS-R missions of wind speed	Three years of the UK TechDemoSat-1 (TDS-1) mission. TDS-1 was launched in July 2014
Balasubramaniam and Ruf (2020)	GNSS-R used for L-Band navigation signals that can penetrate through clouds and rain	This work binds together several rain-related phenomena and enhances our overall understanding of rain effects on GNSS-R measurements	8 Cyclone GNSS observatories over a period of 200 days between DOY 77 and DOY 276 of 2017

is affected by absolute rise of sea level as well as vertical land motions due to the ocean warming (Wöppelmann and Marcos, 2016). Generally, most of the government/private organizations provide data of the land and elevations from the mean sea level. They plan to establish GNSS-based control stations for precise positional measurements where signals from the GNSS satellites are observed continuously. The GNSS network that was designed by them is used to study the continental and oceanic tectonic activities along the coastal areas. The GNSS site primarily mounted on bedrock to measure the high-precision surface displacements and thus provide the opportunity to study the sea level measurement. Each GNSS site is occupied with such type of geodetic quality of GNSS antenna and receiver so that they can access most of the multi-constellation GNSS signals. Fieldworkers working on the coastal zones need to collect GNNS data in the Rinex format. The Rinex files contain SNR measurements. After following the proper methodology (Sect. 2), vertical rise of sea level can be quantified.

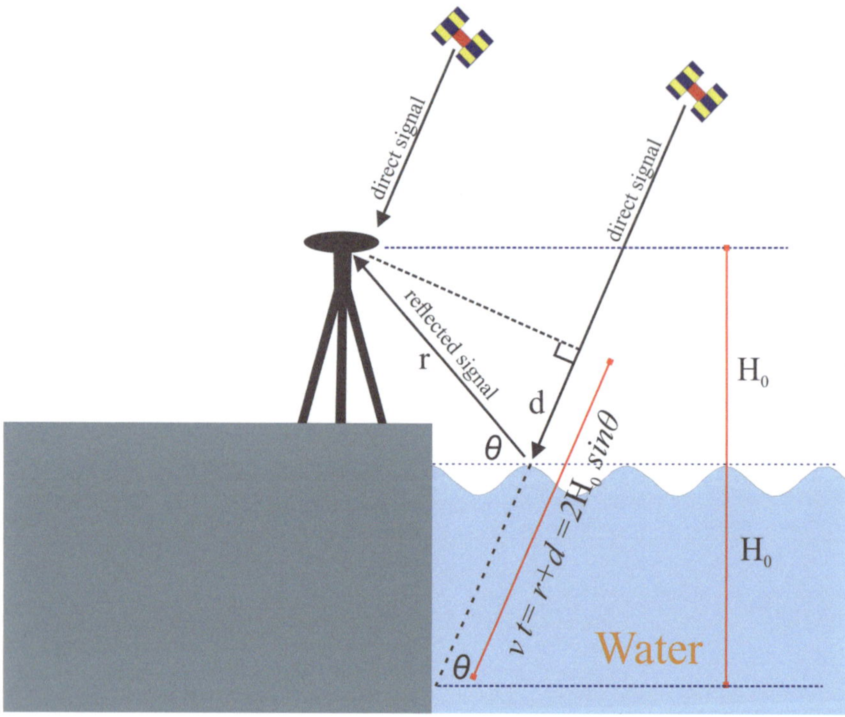

Fig. 2 The Global navigation satellite system-reflectometry (GNSS-R) technique where the land-based receiver receives the GNSS signal electromagnetic wave phase. The GNSS direct signal reaches the satellite directly while the reflected signal arrives after reflection from the nearby water surface (Figure archived from Ansari et al., 2020)

2 Mathematical Background of GNSS-R

The study of GNSS-R is based on the interference of the direct and reflected satellites signal or the multipath signal from surface area (Roesler & Larson, 2018). The oscillation waves of the reflected signals directly influenced to the signal-noise-ration (SNR) measurements. The SNR equations for the signal wave can be presented by most basic format like this:

$$SNR = A\cos(\omega\, t + \phi), \tag{1}$$

Here A is amplitude (in meter) functionally deviated from zero, ω stands for angular frequency given in radians/sec, and ϕ is phase in radians and t is time in seconds. The ordinary frequency f equals the number of oscillations per second. One can write:

$$SNR = A\cos(2f\pi\ t + \phi), \tag{2}$$

Say wavelength of the signals is given by λ (in meter) and velocity by v (in meter/s), then the Eq. (2) can be written like this:

$$SNR = A\cos\left(\frac{2\pi}{\lambda} v\,t + \phi\right). \tag{3}$$

Referring to Fig. 2, travelling path *vt* can be transferred in terms of priori reflector height (H_0). This H_0 is called vertical distance between GPS antenna phase center and surface of reflection.

$$SNR = A\cos\left(\frac{4\pi}{\lambda} H_0\ \sin\theta + \phi\right). \tag{4}$$

Here elevation angle is θ (degree), and the reflector height is (H_0) in meter. From Eqs. (1) and (4):

$$\omega\,t = \frac{4\pi}{\lambda} H_0\ \sin\theta. \tag{5}$$

The left side of the Eq. (5) varies with time and right side shows the phase delay of reflected signals in terms of H_0. According to Axelrad et al. (2005), the affect multipath error, which varies with time and can be eliminated by directly considering the rate of change in phase delay with respect to the sine of elevation angle. Differentiating Eq. (5):

$$\frac{d(\omega\,t)}{d(\sin\theta)} = \frac{4\pi}{\lambda} H_0 = 2\pi\frac{2H_0}{\lambda}. \tag{6}$$

Here $2H_0/\lambda$ is called spatial frequency of the multipath measurement denoted by f_M (Axelrad et al., 2005), in terms of cycles per full satellites arc from 0 to 90°. Now the Eq. (6) will be from of like this:

$$f_M = \frac{2H_0}{\lambda}. \tag{7}$$

In order to record the only interference pattern and remove its influence, the SNR interferogram is detrended in a polynomial of low order (here second order), estimated in linear scale (volt/volt) for each tracking satellite of the selected day (Chew et al., 2016):

$$SNR_{V/V} = 10^{SNR_{dB-Hz}/20}, \tag{8}$$

Here, $SNR_{db\text{-}Hz}$ presents the SNR data in hertz and $SNR_{V/V}$ denotes the SNR data in linear volt/volt scale. The SNR value from Eq. (8) is utilized to calculate the f_M by using Lomb-Scargle periodogram (Press et al., 1996), called the dominant frequency ($f_{m,t}$). After implementing LSP, the simple reflector planar distance will change in an effective reflector height (H_{eff}) (Chew et al., 2016; Larson et al., 2013; Peng et al., 2019)

$$f_{m,t} = \frac{2H_{eff}}{\lambda} \quad or \ H_{eff} = \frac{1}{2} f_{m,t}\lambda \tag{9}$$

Here unit of dominant frequency ($f_{m,t}$) is dimensionless. After wavelength (m) multiplication, the reflector height (H_{eff}) will be become in meter (m).

3 Application of GNSS-R for Real-Time Sea-Level Measurement

3.1 *Global Positioning System-Reflectometry (GPS-R)*

These days a series of reflectometry reports have been investigated to examine how a geodetic GPS receiver with standard quality, situated at the coast with an open view of the sea, can be used as an accidental tide indicator (e.g., Ansari et al., 2020; Geremia-Nievinski et al., 2020; Larson et al., 2013; Peng et al., 2019). Larson et al. (2013) presented a methodology to investigate sea level changes by utilizing the data from single geodetic quality of GPS receiver. They analyzed GPS tide gauge data from the Friday Harbor GPS site of the Earth Scope Plate Boundary Observatory (PBO, USA) and the Onsala Space Observatory (OSO, Sweden). The obtained results of sea level measurements are validated with independent sea level elevation data from closest tide gauge data. It has been found that root mean square (RMS) value between both data for Friday Harbor was about 10 cm and for OSO was 5 cm. Both sites showed correlation coefficient of more than 0.97. These results indicating the best suitability of GPS-R measurements. In another study, Peng et al. (2019) studied the seal-level measurement performance from GPS signal SNR data at HKQT site from Hong Kong under normal weather conditions. They used GPS SNR data from L1, L2C and L5 signals individually and demonstrated the best agreement of tide gauges with seal level changes. In Second attempt Peng et al. (2019) chose abnormal conditions-period of Super Typhoons Hato (2017) and Mangkut (2018) hitting to Hong Kong. Observed results verified that GPS-R is successfully able to produce sea-level variability during the strong wind too. These authors covered tide gauge observation from HKQT co-located Quarry Bay and approved the reality of surge measurement.

There are more studies available in literature based on GPS-R and provided best estimates of sea level measurements. Ansari et al. (2020) used the GPS SNR data from the GPS station of South Beach, Oregon, United States. They estimated tide gauge measurements by using seasonal observation. These GPS-R measurements are compared with closest traditional tide gauge sentinel station (Station ID: 9435380) from National Water Level Observation Network (NWLON) for validation purpose

(Ansari et al., 2020). The obtained clarified the correlation of 0.942 and RMS of 12.90 cm between GPS tide gage and NWLON tide gauge. In the same study GPS tide gauge are predicted by auto-regressive moving average (ARMA) and singular spectrum analysis (SSA) methods to improve the precision of predicted results. ARMA predicted tide gauge results provide correlation coefficient of ~0.981 and RMS of ~4.80 cm while SSA gives correlation coefficient of ~0.998 and RMS of ~0.88 cm. Significant improvement and perfect suitability of GPS-R can be imagined from these results. Geremia-Nievinski et al. (2020) collected GPS SNR data from Onsala Space Observatory for one-year period (2015–2016) and compared the results with tide gauge stations located at nearest station. The main goal of this research was to process the data from independent research group and cross validate the outcomes obtained at under comparable conditions. The GPS SNR data for L1-C/A signal was processed by four groups, in Germany, Sweden, Luxembourg and United Kingdom. All groups processed tide gauge results showed good agreement. There are large numbers of papers based on only GPS-R and including to the other GNSS constellations and are published with the comparison of various prediction methods (Table 3). Some of them research paper will be discussed in next sections. The results showed that GPS constellation provide perfect estimation of tide gauge measurements and indicates its suitability of tide gauge measurement.

3.2 Globalnaya Navigatsionnaya Sputnikovaya Sistema-Reflectometry (GLONASS-R)

The GLONASS system of Russia has potential to allow and handle the direct and reflected signal such as the bistatic radar (Zhang et al., 2019). Implication of GLONASS R system in various remote sensing application make it cheap and valuable (Hobiger et al., 2014). There are many investigations has been done by using GLONASS-R measurement with the inclusion of other constellation systems (Chen et al., 2019; Hobiger et al., 2014; Löfgren and Haas, 2014; Wang et al., 2019). Hobiger et al. (2014) used such type of system data located at Onsala Space Observatory, Sweden that has an operational GLONASS-R system including to the GPS-R system. They investigated sea level monitoring over a period of two weeks in the month of October 2013.The results by using GLONASS constellations including of GPS and noticed that the accuracy and precision of the GLONASS based GNSS-R solutions are comparable to, or even better than conventional GPS-based GNSS-R solutions. Löfgren and Haas (2014) used GPS and GLONASS constellations of L1 and L2 signals and estimated tide gauge recorded data. They included SNR and phase delay analysis using dual-frequency L-band signals of GPS and GLONASS during 1 month from September 29, DOY 273, to October, DOY 303, 2012 located at Onsala Space Observatory the west coast of Sweden (57.4° N, 11.9° E). The obtained measurements were compared to the independent sea level observations from co-located tide station. The results showed a high correlation coefficient of

Table 3 List of few GNSS-R studies used for comparison of various prediction methods (references with their key input, key output and terrains of the country)

Authors	Key input	Key output	Terrains of the country
Hobiger et al. (2014)	An operational GLONASS-R system including to the GPS-R system is used	The results by using GLONASS constellations including of GPS and noticed that the accuracy and precision of the GLONASS based GNSS-R solutions are comparable to, or even better than conventional GPS-based GNSS-R solutions	Onsala Space Observatory, Sweden for period of two weeks in the month of October 2013
Löfgren and Haas (2014)	GPS and GLONASS constellations of L1 and L2 signals and estimated tide gauge recorded data	A better agreement has been noticed with tide gauge compared to SNR data with the RMS of 3/3.2 cm (GLONASS L1/L2) and of 3.5 cm (GPS L1and L2 bands) compared to 4.7/8.9 cm (GLONASS L1/L2) and 4.0/9.0 cm (GPS L1/L2 bands)	DOY 273, to October, DOY 303, 2012 located at Onsala Space Observatory the west coast of Sweden
Pascual et al. (2014)	GPS L1 signals and Galileo E1 signals are used, and their performance is compared for the determination of sea state and altimetry	The Galileo E1 signals are able to provide up to 27% better accuracy than the GPS L1signal in height precision, when the optimum bandwidths are utilized	–
Roussel et al. (2015)	GLONASS-R along with GPS-R and worked on Sea level measurements by using a single geodetic receiver	Combined GPS and GLONASS constellation result better than individual GPS and GLONASS	A geodetic antenna set up at ~60 m above the surface of the Atlantic ocean, at Cordouan lighthouse. Data were acquired from 3 March 2013 to 31 May 2013

(continued)

Table 3 (continued)

Authors	Key input	Key output	Terrains of the country
Zhang et al. (2015)	BeiDou SNR measurements with two coastal experiments	GEO satellites monitoring accuracy is much better than geosynchronous orbit satellites	Two coastal experiments were performed. The Lake experiment was at Dishui Lake, Shanghai, on August 21, 2014. The ocean experiment was at Dayangshan port on October 18–19, 2014
Jin et al. (2017)	BeiDou constellation system with three SNR frequencies of L2, L6 and L7 signals	The obtained result from BDS-R is compared with GPS-R. The GPS SNR measurements of L5 signals showed better results compared to BDS-R	MAYG station located in Mayotte near the Indian Ocean
Wang et al. (2019)	A multi-GNSS combination algorithm is developed to formulate SNR data of four constellations (GPS, GLONASS, Galileo, and BeiDou)	The combined sea level retrieval time series had an approximately 40–75% accuracy improvement compared to individual signal sea level retrievals	BRST station located in Brest harbor on the west coast of France. The SNR data in DOY (day of year) 141–347 of year 2017 are collected in this work
Chen et al. (2019)	GLONASS (L1, L2) signals SNR including to the other GNSS (GPS and BeiDou) system	GLONASS-R estimated observation have mean error about 0.02 m and standard deviation about 0.32. The histogram plot of sea surface height error estimation approximated a normal distribution	MAYG station located in Mayotte near the Indian Ocean
Peng et al. (2019)	GPS (L1, L2C and L5) signal SNR data at under normal and abnormal weather conditions	GPS-R is successfully able to produce sea-level variability during the normal and strong wind too	HKQT site from Hong Kong
Geremia-Nievinski et al. (2020)	The GPS SNR data for L1-C/A signal was processed by four groups, in Germany, Sweden, Luxembourg and United Kingdom	All groups processed tide gauge results showed good agreement	Onsala Space Observatory for one-year period (2015–2016)

(continued)

Table 3 (continued)

Authors	Key input	Key output	Terrains of the country
Rover and Vitti (2019)	Galileo SNR data along with GPS and GLONASS signals with sample of direct and reflected components of SNR measurements	The outcomes of the study revealed the viability of GNSS-R even with non-geodetic-grade tools. The results opened a way towards diffuse of GNSS-R targeted applications	March 2018 and February 2019 at Province of Trento, Italy
Martín et al. (2020)	Multi-constellation solution using GPS, GLONASS, and GALILEO satellites are used for soil moisture	GNSS solutions, including the three constellations and the two sensors (geodetic and mass market), were highly correlated, with a correlation coefficient between 0.7 and 0.85	3 December 2018 to 6 February 2019 at Cajamar Centre of Experiences, Paiporta, Valencia, Spain
Wang et al. (2020)	Galileo signals data and other three constellations (GPS, GLONASS and BeiDou) from three GNSS sites and snow depth retrieval is demonstrated	The multi-GNSS signals provide a higher number of redundant retrievals. These redundant retrievals subsequently provide the possibility of a multi-GNSS retrieval combination	Station PRDS is located in Alberta, Canada
Ansari (2022)	QZSS-R with L1, L2 and L5 signals	QZSS-R are comparable with GPS and GLONASS-R	P109 GNSS site located at Sado Island of Japan

0.86 to 0.97 for both systems and frequency bands, respectively. The phase delay outcome showed a better agreement with tide gauge compared to SNR data with the RMS of 3/3.2 cm (GLONASS L1/L2) and of 3.5 cm (GPS L1 and L2 bands) compared to 4.7/8.9 cm (GLONASS L1/L2) and 4.0/9.0 cm (GPS L1/L2 bands).

Roussel et al. (2015) studied GLONASS-R along with GPS-R and worked on Sea level measurements by using a single geodetic receiver. An inversion approach based on least square method is applied in their study for both constellations. They obtained best results with linear correlation up to 0.96 when both GPS and GLONASS constellation are applied. By considering only GLONASS constellation the correlation decreases up to 0.03. RMSE reduced approximately 6 cm when only GPS constellation used, and it reduced approximately 4 cm when combined constellations is used. There were bias almost equal to zero when both constellations are used, while bias value was about 0.22 m in case of only GLONASS and about 0.05 m with GPS constellation. In conclusion we can say, because the use of both constellations SNR data increase number of observations, it allows to enhance the temporal resolution. Their dynamic SNR method permits very nice estimate results of the main tide periods and allows to detect swell and waves with real amplitudes.

Chen et al. (2019) presented a new surface height by using GLONASS (L1, L2) signals SNR including to the other GNSS (GPS and BeiDou) system at MAYG station located in in Mayotte, France near the Indian Ocean. The GLONASS-R results for L1 and L2 signals observation data over a period of seven days (DOY 190 to DOY 196 in 2017) showed the RMS of correlation coefficient of 0.80 and 0.81 respectively and their RMS of about 31 and 33 cm, respectively. In their study, they found GLONASS-R estimated observation have mean error about 0.02 m and standard deviation about 0.32. Moreover, the histogram plot of sea surface height error estimation approximated a normal distribution. Since the mean error was very close to zero, the authors concluded that the result is unbiased. Peng et al. (2019) examined the GLONASS-R and compared them with GPS-R. They noticed that power of GLONASS reflected signal was generally smaller than to the power of GPS reflected signals. The reflector height produced by GLONASS-R had more outliers for example non-physical reflector height compared to GPS-R. Peng et al. (2019) studied these points and showed such kind of drawbacks in GLONASS-R. There are several authors who used GLONAS-R, cleared that the above points are not true for each case. Wang et al. (2019) used BRST station of Multi-GNSS Experiment (MGEX) network and studied GLONASS SNR along with the constellations of GPS, Galileo and BeiDou system. Multi-GNSS combined algorithm was developed by them to express a 10-min sea level time-series. They noticed that GLONASS SNR data has power levels differing by up to ~20 dB. GLONASS-R estimated sea level measurement root mean square error (RMSE) showed small difference of ~2 cm compared to other constellations. Here it is notable, although the orbit of GLONASS satellites is lower than the orbit of GPS satellites, still they showed similar performance. It means developing of GLONASS system works as a complementary system and playing an important role in remotes sensing application such a reflectometry.

3.3 Galileo-Reflectometry (Galileo-R)

The Galileo consistsof constellation of 30 available satellites developed with the aim to provide more precise and flexible measured positioning service at global level. Likewise, GPS and GLNASS there is Galileo SNR observation are also utilized for reflectometry. Pascual et al. (2014) used GPS L1 signals and Galileo E1 signals and compared the performance of both signals in space and airborne situations for the determination of sea state and altimetry. They noticed that the Galileo E1 signals are able to provide up to 27% better accuracy than the GPS L1signal in height precision, when the optimum bandwidths are utilized. Wang et al. (2019) used BRST (48.380N; – 4.497E, France) station of Multi-GNSS Experiment (MGEX) network and studied Galileo SNR along with the constellations of GPS, GLONASS and BeiDou system. They found that Galileo signals S5X SNR type had optimal precisions with the number of initial used SNR arcs around 60 for Galileo per day. In their study they provided an example of PRN 04 Galileo to show the performance of Lomb-scale periodogram (LSP) on 310 DOY of year 2017. The study showed that the SNR for all Galileo signals have power levels differing by up to ~8 dB. The LSP peak of all other constellation (GPS, GLONASS, and BeiDou) were almost same while Galileo has an outlier belongs to S1C, differ from the peaks of S5X, S7X, and S8X, corresponds to the peak-to-noise below 3. They found that a lower frequency always corresponds to an SNR type with better quality specifically, 1176.45 for the Galileo S5X. Rover and Vitti (2019) used Galileo SNR data along with GPS and GLONASS signals with sample of direct and reflected components of SNR measurements. The outcomes of the study revealed the viability of GNSS-R even with non-geodetic-grade tools. The results opened a way towards diffuse of GNSS-R targeted applications.

Martín et al. (2020) studied multi-constellation solution using Galileo system along with GPS and GLONASS (excluding to BeiDou) systems and presented a comparative analysis of combine GNSS-R methodology of the daily soil moisture. A geodetic GNSS receiver located at the Polytechnic University of Valencia, Span were utilized to gain multi-constellation SNR observations (Galileo, GPS and GLONASS). There are E1 and E5 two main frequencies of the emitted signals from the Galileo satellites but Martín et al. (2020) used only E1 frequency in the experiment. In the required result for the Galileo constellation was least favorable results in terms of RMSE and standard deviation compared to others. They concluded one of the possible causes was that Galileo does not have as many satellites in the constellation as the GPS and GLONASS constellations do. Wang et al. (2020) used Galileo signals data and other three constellations (GPS, GLONASS and BeiDou) from three GNSS sites and snow depth retrieval is demonstrated. It is worth noting that the power spectral density grid for Galileo signals has fewer gaps because of the repeatability lack of Galileo revisits. Although there are many research papers based on GNSS-R investigated, research on Galileo-R is limited. Therefore, no consensus on the signal performance of Galileo constellation has been reached. Thus, more work is needed in that direction.

3.4 BeiDou Navigation Satellite System-Reflectometry (BDS-R)

The BeiDou Navigation Satellite System (BDS) was established in early 1990, currently operating more than 19 BDS satellites. These satellites included Geostationary Earth Orbit (GEO) satellites, Medium Earth Orbit (MEO) satellites and Inclined Geosynchronous Orbit (IGSO) satellites. The BDS provides regional positioning, navigation, and timing (PNT) service and planned to make it globally in future. Jin et al. (2017) used BeiDou constellation system with three SNR frequencies of L2, L6 and L7 signals to estimate the sea level changes. In this work, first time they used BDS-R to estimate the sea level changes based on SNR measurements and triple-frequency phase and code combinations, which are compared to tide gauge observations. The five available BDS satellites (PRN6, 7, 8, 9, and 10) are used estimate the changes in sea level from January 2015 to June 2015. To perform the spectral analysis in the study azimuths between 20°–80° and 110°–170° are used during the complete evaluation of the maps around the stations and reliability of the observations. An interpolation method is used to interpolate the co-located tide gauge observations to get the observations at one time corresponding to the antenna height. It was notable that sea level changes observation derived from BDS remains relative to the BDS site while sea level tide gauge observations remain relative to the tide gauge benchmark. Therefore, a mean of sea level was retrieved by comparing the difference between tide gauge observations and BDS-derived results. The outcomes of the study point out BDS SNR and phase combination sea level changes have a perfect agreement with correlation coefficients of 0.83–0.91and RMSE of less than 0.6 m. The BDS code combination sea level does not show good comparison like other measurements. In the same investigation, Jin et al. (2017) analyzed a new negative linear model between phase and code peak frequencies and tide gauge measurements. This linear model improved the results from the combinations of three-frequency phase and code observations with RMSE of about 10 and 18 cm. The obtained result from BDS-R is compared with GPS-R with three frequencies observations from PRN 1, 3, 6, 9, 24, 25, 27 and 30 satellites. The GPS SNR measurements of L5 signals showed better results compared to the other with the correlation coefficient of 0.87 and RMSE of 0.43 m.

Zhang et al. (2015) investigated the sea-level altimetry using BeiDou SNR measurements with two coastal experiments (ocean and lake in China). The lake study was performed by taking reflected left-handed circularly polarized (R-LHCP) antenna and direct right-handed circularly polarized (D-RHCP) antenna, established potential availability of water surface. They noticed that the obtained error (0.11 m) from the BeiDou GEO satellite signals was smaller than IGSO satellite signals obtained error (1.61 m) This verified that that the suitability of R-LHCP signals from high-elevation satellites for altimetry. The ocean surface altimetry study was investigated on East China Sea for 28 h observation data. The predicted results using R-LHCP BeiDou GEO satellites signals showed a good agreement with the measured data with height RMS of 0.37 m. In the whole analysis, Zhang et al. (2015) concluded that since geostationary Earth orbit satellites monitoring accuracy is much better than geosynchronous orbit satellites, they point out that the high-altitude satellites reflected signals are more appropriate for tide gauge measurements.

Wang et al. (2019) studied four constellation system (GPS, GLONASS, Galileo and BeiDou; Fig. 3), and concluded that the combined result of GNSS-R provided improved accuracy of 40–75% compared to individual performance. These kinds of combine results are beneficial in terms of precision and sampling rate. Wang et al. (2020) used three sites in their work in which XJJC site recorded the SNR data of S2I, S6I, S7I from BeiDou and studied the snow depth retrieval. They showed the trajectories of BeiDou satellites at XJJC site on DOY30, 2017. They noticed that all three SNR (S2I, S6I, and S7I) data can be used to extract frequencies in significantly single peak. BeiDou SNR showed a greater accuracy compared to GPS S1C while they have lower accuracy compared to GPS S2X/S5X data. In terms of combined

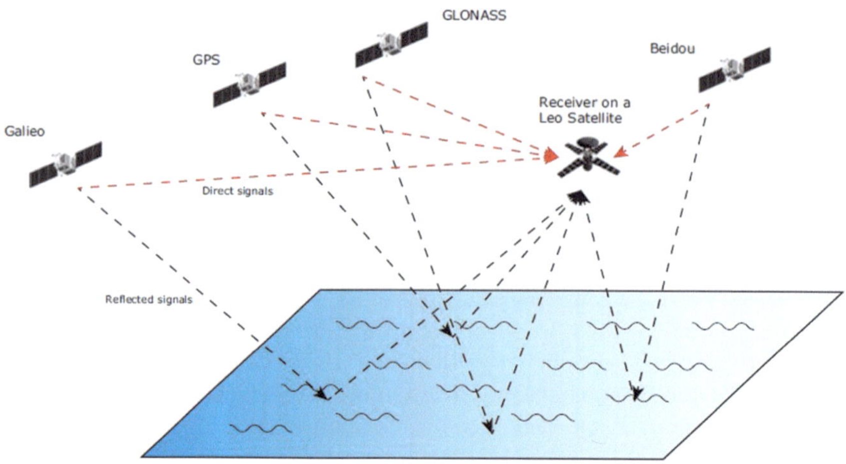

Fig. 3 Multi-GNSS (GPS, GLONASS, Galileo and BeiDou) reflected signals

multi-GNSS, RMSE improved about 50%. In conclusion we can say that with the development of BeiDou, the applications of BeiDou-R will perform a key role in GNSS-R signals.

3.5 Indian Regional Navigation Satellite System-Reflectometry (IRNSS-R)

The Indian Regional Navigation Satellite System (IRNSS) is an autonomous regional satellite navigation system with an operating name of Navigation with Indian Constellation (NavIC). The system provides real-time PNT services covering a region of India and 1500 km extension of around it (Fig. 4). Currently the system consists of a constellation of eight satellites with few additional satellites on ground as stand-by. The constellation is in orbit as of 2018 and is planned to available in consumer mobile phones in the first half of 2020. There is planning to expand IRNSS constellation size from 8 to 11. Indian peninsula surrounded by sea from three sides, in East it has Bay of Bengal, west Indian Ocean and in the north Arabian sea. Although there is no research has been conducted on IRNSS-R, but rapid development of IRNSS system must be beneficiary in future to measure the seal level activity in the Indian coast areas. In conjunction with the known tectonics of the Indian coast (e.g., Misra et al., 2014; Misra and Mukherjee, 2017), more robust tectonic models from such areas can be constructed.

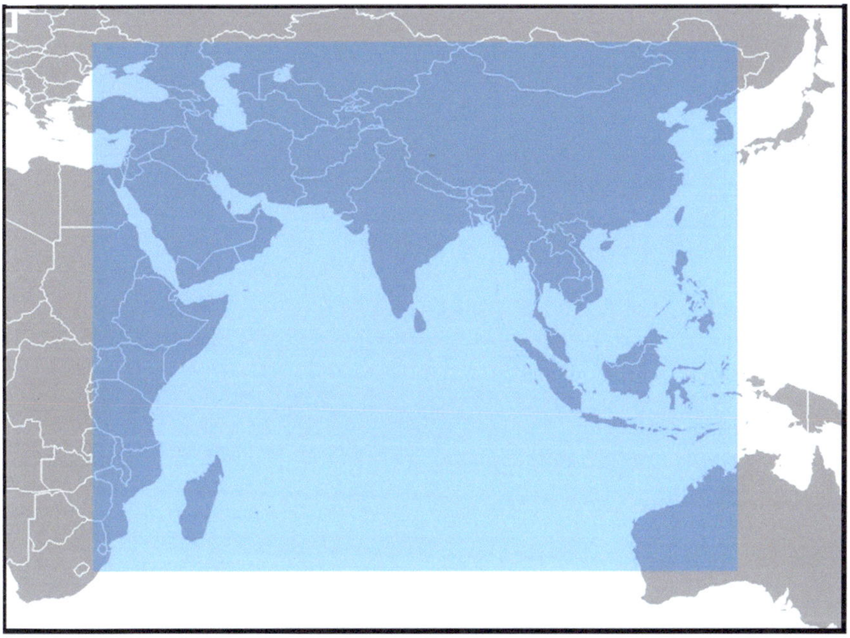

Fig. 4 Coverage area of Indian Regional Navigation Satellite System (https://en.wikipedia.org/wiki/Indian_Regional_Navigation_Satellite_System)

3.6 Quasi Zenith Satellite System-Reflectometry (QZSS-R)

The QZSS establish a GPS complementary system to improve several aspects of positioning, for example, integrity, accessibility, accuracy and reliability, over Asia–Pacific region in conjunction with the particular orbits that are designed to increase the high-elevation signals in Japan (QZSS, 2018; Wu et al., 2004). The complete QZSS constellation includes three satellites in inclined elliptical geosynchronous orbits, which are named as quasi-zenith orbits (QZO) satellites. The QZO satellites broadcast signals compatible with the GPS L1C/A signals, along with modernized GPS L1C, L2C signals and L5 signals. The combination of GPS and QZSS systems delivers an enhanced performance of positioning via data ranging correction method.

An initial study was done by Ansari (2022) to present the sea level measurements by using QZSS-R. His measurement retrieved the SNR data of QZSS (L1, L2 and L5) at P109 site of GNSS Earth Observation Network (37.815°N; 138.281°E; 44.70 m elevation height) located at Sado Island of Japan in the month of November 2019 (Fig. 5). The results of QZSS observation data for L1, L2 and L5 signals with tide gauge observation over a period of one month showed the average correlation coefficient of 0.8158, and their average RMSE was 0.0445 m. These results were comparable with the other GNSS-R observations of GPS-R (L1, L2, and L5 signals), and GLONASS-R (L1 and L2 signals). The GPS-R observation for L1, L2 and L5 signals

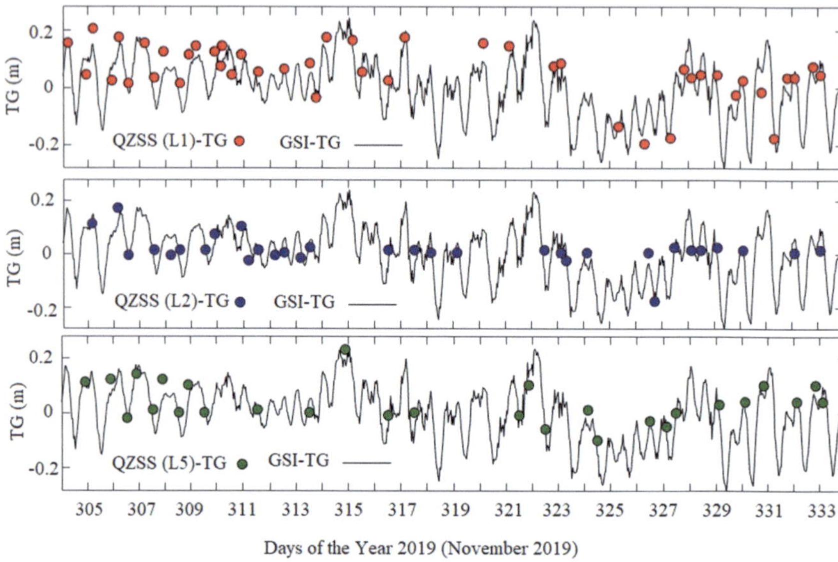

Fig. 5 Daily sea level changes (unit: meter) obtained from QZSS L1, L2 and L5 SNR observations (namely, QZSS-TG) as well as GSI tide gauge (namely, GSI-TG) observations during the month of November 2019 (30 days) (Ansari 2022)

showed the average correlation coefficient of 0.8432, and their average RMSE was 0.0445 m while GLONASS-R observation for L1 and L2 signals showed the average correlation coefficients of 0.8696 and their average RMSE was 0.0485 m. He reconstructed sea level measurements by utilizing the kernel extreme learning machine (KELM) technique based on variational mode decomposition. During the comparison of the sea level measurements with the KLEM reconstructed outcomes; the studied results demonstrate that the average correlation coefficient of 0.8764 for QZSS-R L1, L2 and L5 signals. The mean errors for QZSS-R, GPS-R, and GLONASS-R are estimated and to be noted very close to zero or equivalent to zero. This indicates that the estimated results can be considered as unbiased estimations. No other studies involved QZSS-R; hence it can be said that the outcomes of the study showed that the rapid development of QZSS may provide a new possible opportunity to monitor sea level changes with the available frequencies.

4 Conclusions

GNSS-R technique is a remote sensing application widely used for multipurpose such as oceanography, ice and snow monitoring, soil moisture determination and ocean wind speed. This article reviews multi-constellation GNSS (GPS, GLONASS,

Galileo, BeiDou, IRNSS and QZSS) reflectometry. Although the individual measurement of reflectometry has been measuring the global sea level changes by providing the relative measurement with respect to land while offering high-precision topography of the sea surface, sometimes it has poor-precision near to the coastal areas because of the fast and complex dynamical changes of the ocean in the same region. Sometimes it may not provide meso-scale phenomena/information e.g., swell and waves with real tide amplitudes. In these situations, the use of multi-GNSS reflected signals is an optimal method to enhance the precision and sampling rates by taking the advantage of more satellites improved signals and extensive sensing coverages. These kinds of combine results are beneficial in terms of precision and sampling rate.

Acknowledgements Self-financed research. Soumyajit Mukherjee (IIT Bombay) invited and reviewed this chapter. Mukherjee (2023) summarized this chapter.

Appendix

Abbreviations	Meaning
GNSS	Global Navigation Satellites System
GPS	Global Positioning System
GLONASS	Globalnaya Navigatsionnaya Sputnikovaya Sistema
BDS	BeiDou Navigation Satellite System
IRNSS	Indian Regional Satellites System
NavIC	Navigation with Indian Constellation
QZSS	Quasi-Zenith Satellite System
GNSS-R	Global Navigation Satellite System-Reflectometry
GPS-R	Global Positioning System-Reflectometry
GLONASS	Globalnaya Navigatsionnaya Sputnikovaya Sistema-Reflectometry
Galileo-R	Galileo-Reflectometry
BDS-R	BeiDou Navigation Satellite System-Reflectometry
IRNSS-R	Indian Regional Navigation Satellite System -Reflectometry
QZSS-R	Quasi-Zenith Satellite System-Reflectometry
SBAS	Satellite Based Augmentation System
WAAS	Wide Area Augmentation System
WAGE	Wide Area GPS Enhancement
SDCM	System for Differential Corrections and Monitoring
EGNOS	European Geostationary Navigation Overlay Service
SNAS	Satellite Navigation Augmentation System

(continued)

(continued)

Abbreviations	Meaning
GAGAN	GPS-aided GEO Augmented Navigation
MSAS	Multi-functional Satellite Augmentation System
SouthPAN	Southern Positioning Augmentation Network
GPS.C	GPS Correction
SNR	Signal-noise-ration
TO	Tohoku-Oki
GEO	Geostationary Earth Orbit
MEO	Medium Earth Orbit
IGSO	Inclined Geosynchronous Orbit
QZO	Quasi-Zenith Orbits
PNT	Positioning, Navigation, and Timing
R-LHCP	Reflected Left-handed Circularly Polarized
D-RHCP	Direct Right-handed Circularly Polarized
PBO	Plate Boundary Observatory
OSO	Onsala Space Observatory
NWLON	National Water Level Observation Network
GEONET	GNSS Earth Observation Network
MGEX	Multi-GNSS Experiment
RMS	Root Mean Square
ARMA	Auto-regressive Moving Average
SSA	Singular Spectrum Analysis
DOY	Day of Year
LSP	Lomb-scale periodogram
KELM	Kernel Extreme Learning Machine

References

Ansari, K., Bae, T. S., & Inyurt, S. (2020). Global positioning system interferometric reflectometry for accurate tide gauge measurement: Insights from South Beach, Oregon, United States. *Acta Astronautica, 173*, 356–362. https://doi.org/10.1016/j.actaastro.2020.04.060

Ansari, K., Bae, T. S., Seok, H. W., & Kim, M. S. (2021). Multiconstellation global navigation satellite systems signal analysis over the Asia-Pacific region. *International Journal of Satellite Communications and Networking, 39*(3), 280–293. https://doi.org/10.1002/sat.1389

Ansari, K., & Park, K. D. (2018). Multi constellation GNSS precise point positioning and prediction of propagation errors using singular spectrum analysis. *Astrophysics and Space Science, 363*(12), 1–7. https://doi.org/10.1007/s10509-018-3479-7

Ansari, K., & Park, K. D. (2019). Contemporary deformation and seismicity analysis in Southwest Japan during 2010–2018 based on GNSS measurements. *International Journal of Earth Sciences, 108*(7), 2373–2390. https://doi.org/10.1007/s00531-019-01768-w

Ansari, K. (2022). Quasi Zenith Satellite System-Reflectometry for sea level measurement and Implication of Machine learning methodology (An initial study, and complete study need to be done)

Axelrad, P., Larson, K., & Jones, B. (2005, September). Use of the correct satellite repeat period to characterize and reduce site-specific multipath errors. In Proceedings of the 18th international technical meeting of the satellite division of the institute of navigation (ION GNSS 2005) (pp. 2638–2648).

Balasubramaniam, R., & Ruf, C. (2020). Characterization of rain impact on L-Band GNSS-R ocean surface measurements. *Remote Sensing of Environment, 239*, 111607. https://doi.org/10.1016/j.rse.2019.111607

Bouffard, J., Roblou, L., Birol, F., Pascual, A., Fenoglio-Marc, L., Cancet, M., Morrow, R., & Ménard, Y. (2011). Introduction and assessment of improved coastal altimetry strategies: case study over the Northwestern Mediterranean Sea. In Coastal altimetry (pp. 297–330). Springer, Berlin, Heidelberg. https://doi.org/10.1007/978-3-642-12796-0_12

Cardellach, E., Li, W., Rius, A., Semmling, M., Wickert, J., Zus, F., Ruf, C. S., & Buontempo, C. (2019). First precise space borne sea surface altimetry with GNSS reflected signals. *IEEE Journal of Selected Topics in Applied Earth Observations and Remote Sensing, 13*, 102–112. https://doi.org/10.1109/JSTARS.2019.2952694

Carlin, L., Hauschild, A., & Montenbruck, O. (2021). Precise point positioning with GPS and Galileo broadcast ephemerides. *GPS Solutions, 25*(2), 1–13. https://doi.org/10.1007/s10291-021-01111-4

Chen, F., Liu, L., & Guo, F. (2019). Sea surface height estimation with multi-GNSS and wavelet de-noising. *Scientific Reports, 9*(1), 1–10. https://doi.org/10.1038/s41598-019-51802-9

Chew, C., Small, E. E., & Larson, K. M. (2016). An algorithm for soil moisture estimation using GPS-interferometric reflectometry for bare and vegetated soil. *GPS Solutions, 20*(3), 525–537. https://doi.org/10.1007/s10291-015-0462-4

Edokossi, K., Calabia, A., Jin, S., & Molina, I. (2020). GNSS-reflectometry and remote sensing of soil moisture: A review of measurement techniques, methods, and applications. *Remote Sensing, 12*(4), 614. https://doi.org/10.3390/rs12040614

Feng, G., Jin, S., & Zhang, T. (2013). Coastal sea level changes in Europe from GPS, tide gauge, satellite altimetry and GRACE, 1993–2011. *Advances in Space Research, 51*(6), 1019–1028. https://doi.org/10.1016/j.asr.2012.09.011

Geremia-Nievinski, F., Hobiger, T., Haas, R., Liu, W., Strandberg, J., Tabibi, S., Vey, S., Wickert, J., & Williams, S. (2020). SNR-based GNSS reflectometry for coastal sea-level altimetry: Results from the first IAG inter-comparison campaign. *Journal of Geodesy, 94*(8), 1–15. https://doi.org/10.1007/s00190-020-01387-3

Hammond, M. L., Foti, G., Gommenginger, C., & Srokosz, M. (2020). Temporal variability of GNSS-Reflectometry ocean wind speed retrieval performance during the UK TechDemoSat-1 mission. *Remote Sensing of Environment, 242*, 111744. https://doi.org/10.1016/j.rse.2020.111744

Hobiger, T., Haas, R., & Löfgren, J. S. (2014). GLONASS-R: GNSS reflectometry with a frequency division multiple access-based satellite navigation system. *Radio Science, 49*(4), 271–282. https://doi.org/10.1002/2013RS005359

Jia, Y., Jin, S., Savi, P., Gao, Y., Tang, J., Chen, Y., & Li, W. (2019). GNSS-R soil moisture retrieval based on a XGboost machine learning aided method: Performance and validation. *Remote Sensing, 11*(14), 1655. https://doi.org/10.3390/rs11141655

Jin, S., Qian, X., & Wu, X. (2017). Sea level change from BeiDou navigation satellite system-reflectometry (BDS-R): First results and evaluation. *Global and Planetary Change, 149*, 20–25. https://doi.org/10.1016/j.gloplacha.2016.12.010

Klos, A., Kusche, J., Fenoglio-Marc, L., Bos, M. S., & Bogusz, J. (2019). Introducing a vertical land motion model for improving estimates of sea level rates derived from tide gauge records affected by earthquakes. *GPS Solutions, 23*(4), 1–12. https://doi.org/10.1007/s10291-019-0896-1

Larson, K. M., Löfgren, J. S., & Haas, R. (2013). Coastal sea level measurements using a single geodetic GPS receiver. *Advances in Space Research, 51*(8), 1301–1310. https://doi.org/10.1016/j.asr.2012.04.017

Larson, K. M., Ray, R. D., & Williams, S. D. (2017). A 10-year comparison of water levels measured with a geodetic GPS receiver versus a conventional tide gauge. *Journal of Atmospheric and Oceanic Technology, 34*(2), 295–307. https://doi.org/10.1175/JTECH-D-16-0101.1

Li, X., Ge, M., Dai, X., Ren, X., Fritsche, M., Wickert, J., & Schuh, H. (2015b). Accuracy and reliability of multi-GNSS real-time precise positioning: GPS, GLONASS, BeiDou, and Galileo. *Journal of Geodesy, 89*(6), 607–635. https://doi.org/10.1007/s00190-015-0802-8

Li, X., Zhang, X., Ren, X., Fritsche, M., Wickert, J., & Schuh, H. (2015a). Precise positioning with current multi-constellation global navigation satellite systems: GPS, GLONASS Galileo and BeiDou. *Scientific Reports, 5*(1), 1–14. https://doi.org/10.1038/srep08328

Liu, H., Jin, S., & Yan, Q. (2020). Evaluation of the ocean surface wind speed change following the super typhoon from space-borne GNSS-reflectometry. *Remote Sensing, 12*(12), 2034. https://doi.org/10.3390/rs12122034

Löfgren, J. S., & Haas, T. (2014). Sea level measurements using multi-frequency GPS and GLONASS observations. *EURASIP Journal on Advances in Signal Processing, 2014*(1), 50. https://doi.org/10.1186/1687-6180-2014-50

Löfgren, J. S., Haas, R., Scherneck, H. G., & Bos, M. S. (2011). Three months of local sea level derived from reflected GNSS signals. *Radio Science 46*(6). https://doi.org/10.1029/2011RS004693

Martín-Neira, M., D'Addio, S., Buck, C., Floury, N., & Prieto-Cerdeira, R. (2011). The PARIS ocean altimeter in-orbit demonstrator. *IEEE Transactions on Geoscience and Remote Sensing, 49*(6), 2209–2237. https://doi.org/10.1109/TGRS.2010.2092431

Martín, A., Ibáñez, S., Baixauli, C., Blanc, S., & Anquela, A. B. (2020). Multi-constellation GNSS interferometric reflectometry with mass-market sensors as a solution for soil moisture monitoring. *Hydrology and Earth System Sciences, 24*(7), 3573–3582. https://doi.org/10.5194/hess-24-3573-2020

Misra, A. A., Bhattacharya, G., Mukherjee, S., & Bose, N. (2014). Near N-S paleo-extension in the western Deccan region, India: Does it link strike-slip tectonics with India-Seychelles rifting? *International Journal of Earth Sciences, 103*(6), 1645–1680. https://doi.org/10.1007/s00531-014-1021-x

Misra, A. A., & Mukherjee, S. (2017). Dyke–brittle shear relationships in the western Deccan strike-slip zone around Mumbai (Maharashtra, India). In S. Mukherjee, A. A. Misra, G. Calvès, M. Nemčok (Eds.) Tectonics of the Deccan large igneous province. Geological Society, London, Special Publications, 445(1), 269–295. https://doi.org/10.1144/SP445.4

Montillet, J. P., Melbourne, T. I., & Szeliga, W. M. (2018). GPS vertical land motion corrections to sea-level rise estimates in the Pacific Northwest. *Journal of Geophysical Research: Oceans, 123*(2), 1196–1212. https://doi.org/10.1002/2017JC013257

Mukherjee, S. (2023). Introduction to structural geology and tectonics field guidebook— Volume 2. In: Mukherjee S. (Ed.) Structural geology and tectonics field guidebook—Volume 2 (pp. xi–xiv). Springer Nature Switzerland AG, Cham. ISBN 978-3-031-19575-4.

Oke, P. R., Larnicol, G., Fujii, Y., Smith, G. C., Lea, D. J., Guinehut, S., Remy, E., Balmaseda, M. A., Rykova, T., Surcel-Colan, D., & Martin, M. J. (2015). Assessing the impact of observations on ocean forecasts and reanalyses: Part 1, global studies. *Journal of Operational Oceanography, 8*(sup1), s49–s62. https://doi.org/10.1080/1755876X.2015.1022067

Parker, A. (2019). Sea level oscillations in Japan and China since the start of the 20th century and consequences for coastal management-Part 1: Japan. *Ocean & Coastal Management, 169*, 225–238. https://doi.org/10.1016/j.ocecoaman.2018.12.031

Pascual, D., Park, H., Camps, A., Arroyo, A. A., & Onrubia, R. (2014). Simulation and analysis of GNSS-R composite waveforms using GPS and Galileo signals. *IEEE Journal of Selected Topics in Applied Earth Observations and Remote Sensing, 7*(5), 1461–1468. https://doi.org/10.1109/JSTARS.2014.2311116

Peng, D., Hill, E. M., Li, L., Switzer, A. D., & Larson, K. M. (2019). Application of GNSS interferometric reflectometry for detecting storm surges. *GPS Solutions, 23*(2), 1–11. https://doi.org/10.1007/s10291-019-0838-y

Press, W. H., Flannery, B. P., Teukolsky, S. A., & Vetterling, W. T. (1996). Numerical recipes in Fortran 90: The art of parallel scientific computing. Cambridge University Press.

QZSS. (2018). Overview of the Quasi-Zenith satellite system (QZSS). Cabinet Office, National Space Policy Secretariat. http://qzss.go.jp/en/overview/services/sv01_what.html. Accessed April 2018

Rodriguez-Alvarez, N., Holt, B., Jaruwatanadilok, S., Podest, E., & Cavanaugh, K. C. (2019). An Arctic sea ice multi-step classification based on GNSS-R data from the TDS-1 mission. *Remote Sensing of Environment, 230*, 111202. https://doi.org/10.1016/j.rse.2019.05.021

Roesler, C., & Larson, K. M. (2018). Software tools for GNSS interferometric reflectometry (GNSS-IR). *GPS Solutions, 22*(3), 1–10. https://doi.org/10.1007/s10291-018-0744-8

Roussel, N., Ramillien, G., Frappart, F., Darrozes, J., Gay, A., Biancale, R., Striebig, N., Hanquiez, V., Bertin, X., & Allain, D. (2015). Sea level monitoring and sea state estimate using a single geodetic receiver. *Remote Sensing of Environment, 171*, 261–277. https://doi.org/10.1016/j.rse.2015.10.011

Rover, S., & Vitti, A. (2019). GNSS-R with low-cost receivers for retrieval of antenna height from snow surfaces using single-frequency observations. *Sensors, 19*(24), 5536. https://doi.org/10.3390/s19245536

Sagiya, T., Miyazaki, S. I., & Tada, T. (2000). Continuous GPS array and present-day crustal deformation of Japan. *Pure and Applied Geophysics, 157*(11), 2303–2322. https://doi.org/10.1007/PL00022507

Santamaría-Gómez, A., & Watson, C. (2017). Remote leveling of tide gauges using GNSS reflectometry: Case study at Spring Bay. *Australia. GPS Solutions, 21*(2), 451–459. https://doi.org/10.1007/s10291-016-0537-x

Sasaki, Y. N., Minobe, S., & Miura, Y. (2014). Decadal sea-level variability along the coast of Japan in response to ocean circulation changes. *Journal of Geophysical Research: Oceans, 119*(1), 266–275. https://doi.org/10.1002/2013JC009327

Tabibi, S., Geremia-Nievinski, F., & van Dam, T. (2017). Statistical comparison and combination of GPS GLONASS and Multi-GNSS multipath reflectometry applied to snow depth retrieval. *IEEE Transactions on Geoscience and Remote Sensing, 55*(7), 3773–3785. https://doi.org/10.1109/TGRS.2017.2679899

Tabibi, S., Geremia-Nievinski, F., Francis, O., & van Dam, T. (2020). Tidal analysis of GNSS reflectometry applied for coastal sea level sensing in Antarctica and Greenland. *Remote Sensing of Environment, 248*, 111959. https://doi.org/10.1016/j.rse.2020.111959

Vitousek, S., Barnard, P. L., Fletcher, C. H., Frazer, N., Erikson, L., & Storlazzi, C. D. (2017). Doubling of coastal flooding frequency within decades due to sea-level rise. *Scientific Reports, 7*(1), 1–9. https://doi.org/10.1038/s41598-017-01362-7

Vousdoukas, M. I., Mentaschi, L., Voukouvalas, E., Verlaan, M., Jevrejeva, S., Jackson, L. P., & Feyen, L. (2018). Global probabilistic projections of extreme sea levels show intensification of coastal flood hazard. *Nature Communications, 9*(1), 1–12. https://doi.org/10.1038/s41467-018-04692-w

Wang, N., Xu, T., Gao, F., & Xu, G. (2018). Sea level estimation based on GNSS dual-frequency carrier phase linear combinations and SNR. *Remote Sensing, 10*(3), 470. https://doi.org/10.3390/rs10030470

Wang, X., He, X., & Zhang, Q. (2019). Evaluation and combination of quad-constellation multi-GNSS multipath reflectometry applied to sea level retrieval. *Remote Sensing of Environment, 231*, 111229. https://doi.org/10.1016/j.rse.2019.111229

Wang, X., Zhang, S., Wang, L., He, W., & Zhang, Q. (2020). Analysis and combination of multi-GNSS snow depth retrievals in multipath reflectometry. *GPS Solutions, 24*(3), 77. https://doi.org/10.1007/s10291-020-00990-3

Wikipedia. (2021). https://en.wikipedia.org/wiki/GNSS_augmentation Retrieved on 04 October 2021, Original Source Geoscience Australia, Trial of accurate positioning, Retrieved 25 April 2020. https://www.ga.gov.au/scientific-topics/positioning-navigation/positioning-australia/trial-of-accurate-positioning

Wöppelmann, G., & Marcos, M. (2016). Vertical land motion as a key to understanding sea level change and variability. *Reviews of Geophysics, 54*(1), 64–92. https://doi.org/10.1002/2015RG000502

Wu, F., Kubo, N., & Yasuda, A. (2004, April). A study on GPS augmentation using Japanese Quasi-Zenith satellite system. In PLANS 2004. Position location and navigation symposium (IEEE Cat. No. 04CH37556) (pp. 347–356), IEEE. https://doi.org/10.1109/PLANS.2004.1309016

Yan, Q., Huang, W., Jin, S., & Jia, Y. (2020). Pan-tropical soil moisture mapping based on a three-layer model from CYGNSS GNSS-R data. *Remote Sensing of Environment, 247*, 111944. https://doi.org/10.1016/j.rse.2020.111944

Zhang, Y., Tian, L., Meng, W., Gu, Q., Han, Y., & Hon, Z. (2015). Feasibility of code-level altimetry using coastal BeiDou Reflection (BeiDou-R) setups. *IEEE Journal of Selected Topics in Applied Earth Observations and Remote Sensing, 8*(8), 4130–4140. https://doi.org/10.1109/JSTARS.2015.2446684

Zhang, S., Liu, K., Liu, Q., Zhang, C., Zhang, Q., & Nan, Y. (2019). Tide variation monitoring based improved GNSS-MR by empirical mode decomposition. *Advances in Space Research, 63*(10), 3333–3345. https://doi.org/10.1016/j.asr.2019.01.046

Zhou, W., Liu, L., Huang, L., Yao, Y., Chen, J., & Li, S. (2019). A new GPS SNR-based combination approach for land surface snow depth monitoring. *Scientific Reports, 9*(1), 1–20. https://doi.org/10.1038/s41598-019-40456-2

Architecture and Structures of Kiradu Temple (Barmer Region, Rajasthan, India)

Paramita Haldar, Mohit Kumar Puniya, Mery Biswas, Soumyajit Mukherjee, Nihar Ranjan Kar, and Ratna Choudhary

Abstract Kiradu temple, known as the *Khajuraho temple of Rajasthan*, is an epitome of Maru-Gurjara style of architecture. Due to weathering, several sections of the temple complex has been decayed. Though literature refers the architectural style of the Vishnu temple of Kiradu, very few works mention the details of other temples existing within the temple complex. In this study we attempted to understand the architecture and present structural condition of Kiradu. We have performed strength analysis of one of the partially destructed temple to understand the vulnerability of the fractures. We studied different fractures and destructed artifacts to understand the real reason of their decay. This study will help in preserving and restoring the temple and to understand the cultural and social lifestyle of the western India during medieval period.

Keywords Kiradu · Temple restoration

P. Haldar (✉)
Department of Chemical Engineering, BITS Pilani, K.K. Birla Goa Campus, Goa 403726, India
e-mail: paramitah@goa.bits-pilani.ac.in

M. K. Puniya
National Geotechnical Facility, Survey of India, Dehradun, India

M. Biswas
Department of Geography, Presidency University, Kolkata, West Bengal 700073, India

S. Mukherjee
Department of Earth Sciences, Indian Institute of Technology Bombay Powai, Mumbai, Maharashtra 400076, India

N. R. Kar
Council of Scientific and Industrial Research–National Geophysical Research Institute, Uppal Road, Habsiguda, Hyderabad, Telangana 500007, India

R. Choudhary
Banasthali University, Jaipur, Rajasthan 302001, India

S. Mukherjee (ed.), *Structural Geology and Tectonics Field Guidebook–Volume 2*, Springer Geology, https://doi.org/10.1007/978-3-031-19576-1_14

1 Introduction

India is the central peninsula in the south Asia. Due to diversity of climate, variety of races inhabited in the country. India has witnessed a wide-spread cultural diversity over ages. Temples and their architecture are the integral parts of Indian history.

Locations of north Indian temples of medieval period (A.D. 900–1000) is classified into four major zones as the Upper, the Central, the Eastern and the Western. It is observed that the Western zone is the largest (Paranagar near Alwar in Upper Rajasthan to Parol near Mumbai, a north–south stretch of over a thousand km) and longest (west to east from Dewalthatha in Sind to Atru in eastern-most Rajasthan, a distance of ~ 600 km) and most productive amongst these regions (Dhaky, 1967). Rajasthan is considered as one of the most important places in the western India for temple architecture and sculpture. Unfortunately only the ruins of many temples are left in this Indian state. Between seventh and tenth centuries, several temples were constructed and different regional architectural styles developed as temple designs. The temples of late medieval periods in Rajasthan are mostly constructed as per the Maru-Gurjara style.

The Kiradu temple in Barmer (Rajasthan) (Fig. 1a) is a fine example of an eleventh century Maru-Gurjara architectural style. As Kiradu is an arid area, the diurnal temperature variation is extreme especially during the summer and the winter time. This resulted in decay of the temple material. Most of the temples of Kiradu are made up of sandstones (Chanchani, 2014).

We present the architectural design and the present structural condition of the Kiradu temple. In most of the literatures, we obtain information only regarding the Vishnu temple. Though the Maru-Gurjara style of the architecture is elaborated in several literature, the detail design pattern of the outer walls and the Mandovara section of the Kiradu temple complex remained a due. This study provides detail information of the constructional design, which can be further utilised for the preservation and restoration of the temple.

2 Location of Kiradu

The Kiradu temple (Fig. 1) is located ~ 43 km NW from the town Barmer, Rajasthan at the Hathma village, which lies north of Khadon Railway station situated on Barmer–Munabao Railway line.

3 General Description of the Kiradu Temple

The purple polygon in Fig. 1a demarcates ~ 0.16 km^2 of the Kiradu temple complex. Yellow rectangles denote the temples of the Kiradu. From the entry gate of the

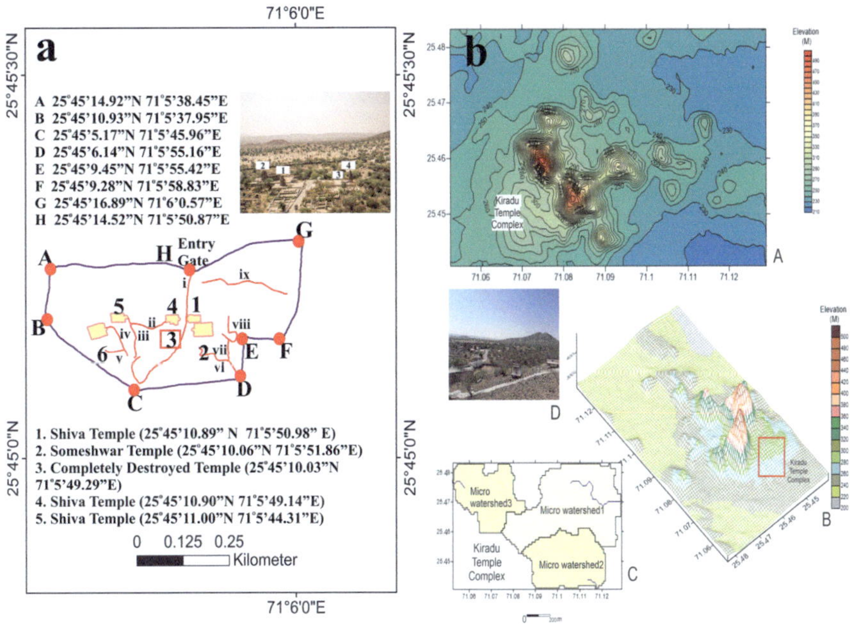

Fig. 1 Google Map of the Kiradu Temple **a** coordinates of longitude and latitude of Kiradu, **b** Kiradu temple and its adjacent area

temple complex, the first temple (temple no. 1) is the Shiva temple. Temple no. 2 is the Someshwar temple. Rectangle 3 denotes the completely ruined temple area. Rectangle 4 and 5 denote temple no. 4 and 5, which are Shiva temples and rectangle 6 is a Vishnu temple. In Fig. 1b describes the Kiradu temple and its surrounding area. From Fig. 1b it is clear that micro watersheds surround the temple complex. These micro watersheds can indicate habitat surrounding the Kiradu temple in the past.

4 Weather of Kiradu

Kiradu is situated in an arid region. The temperature of Kiradu fluctuates much during summer and winter. In summer, the temperature rises upto 48–49 °C in the day time and drops to 27–29 °C at night. In winter, the temperature in day time is usually 26–27 °C and falls to 11–13 °C at night (Singh et al., 2010). The average rainfall in Barmer is 277 mm in a year but remains very dry in the monsoon (Poonia & Rao, 2018). Sometimes Barmer experienced abnormal rainfall due to weather depressions. Due to being in arid region, Kiradu experiences high evaporation rate, wide temperature swing both daily and seasonally and very low precipitation (Saifuddin, 2000).

Fig. 2 Notices by Archeological survey of India (ASI), History of Kiradu-Group of temples

5 History of the Foundation

According to Ghurye (1968), and inscription dated 1153–1178 CE, the temples of Kiradu were built around twelfth century C.E (Ghurye, 1968). Art historian Percy Brown and Madhusudan Dhakay assigned the temples of Kiradu to the eleventh century C.E. (Dhaky, 1998). Inscriptions of 1161 A.D. is found on the pillars. The ancient name of 'Kiradu' is being mentioned as Kirat Koop. It appears from the dispersed rocks that Kiradu was a fully established town. Sindhu Raj was the founder of the Parmer rule of Kiradu town. The Parmer and Chauhan rules of Kiradu were under the regime of Solanki dynasty of Gujarat. The town got ruined due to foreign invasion. An 1178 C.E. Kiradu inscription, issued during the reign of the Chaulukya monarch Bhima II, records repairs to a temple damaged by the Turushkas (Turkic people). These Turushkas are identified with the Ghurids led by Muhammad of Ghor (Chaudhary, 2012). Figure 2 displays the temple history mentioned by the Archeological Survey of India. There is no specific information available regarding who built the temples of Kiradu. Though there are three inscription of twelfth century available, but they do not provide information related to the construction of the temple. The temples of the present Kiradu town are in a ruined state after the desolation of the magnificent Kirat Koop town, yet they are still known for their architectural splendor and testimony of ancient glory.

Fig. 3 Kiradu temple complex

6 Architectural Style of the Kiradu Temple

The Kiradu temples are a group of ruined Hindu temples. Presently five temples are found in remnant state. They are mainly the Someshwar temple, three Shiva temples, and an almost ruined Vishnu temple. Besides these five temples, there are two completely ruined temples. Figure 3 shows the temple complex of Kiradu. In this figure, temple numbers 1 and 4 indicate the Shiva temple and temple number 2 is the Someshwar temple. Point 3 is a completely ruined temple (25°45′10.03″ N Latitude and 71°5′49.25″ E Longitude).

According to M. A. Dhaky the structure of the temple can be studied into two parts. The first one is along the vertical direction from the base foundation of the temple to the pinnacle of the temple top. This section is known as Udharvachhand (Chhand means roof, and Udharva means outer wall of the section where main deity is installed) (Fig. 4).The second one is along the horizontal length from the one end (entry gate of the temple) to the other end of the structure. This portion is called Talchhand (Talchhand means under the roof in the front section of the temple) (Fig. 4).The Udharvachhand can be divided into four main sections viz., Jagati, Pitha or Adhisthan, Mandovara and Sikhara.

Fig. 4 Architecture of the Someshwar temple (Temple No. 2)

Jagati (Figs. 5, 6 and 7): Jagati is the pedestal or platform, which is the moulded base of the structure. The Pitha of the Kiradu temple uniquely represents the early form of the Maru-Gurjara style of architecture. It is mainly constituted by seven or eight mouldings, starting with two successive Bhittas followed by Jadyakumbha, Kumuda, Grasapattika, Gajapitha, Ashvapitha and Narapitha. Rectangular Bhitta is the first moulding above the Jagati. Jadyakumbha is the moulding above the two Bhittas. It is inverted cyma-recta engraved with designs. Kumuda is the torus moulding above the Jadyakumbha. Grasapattika represents the band showing the gorgon heads. Gajapitha is the basal-band above the Grasapattika showing the frontal posture of the elephant figures. Ashvapitha is the band displaying the frontal posture of the horses. Narapitha displays the pitha-course showing human engaged in manifold activities.

Mandovara (Fig. 5): It is the temple's closed hall's portion above the Pitha and below the roof of the temple. The Mandovara consists of the Vedibandha, the Jangha and the Varandika. The Vedibandha is the aggregate of five basal wall-moulding consisting primarily of Khurda, Kumbha, Kalasa, Antarapatta and Kapotapalli. Khurda is the basal plain moulding of Vedibandha. Kumbha is the Vedibandha's 2nd moulding above the Khurda.

Sikhara (Fig. 5): It is the spire of the temple.

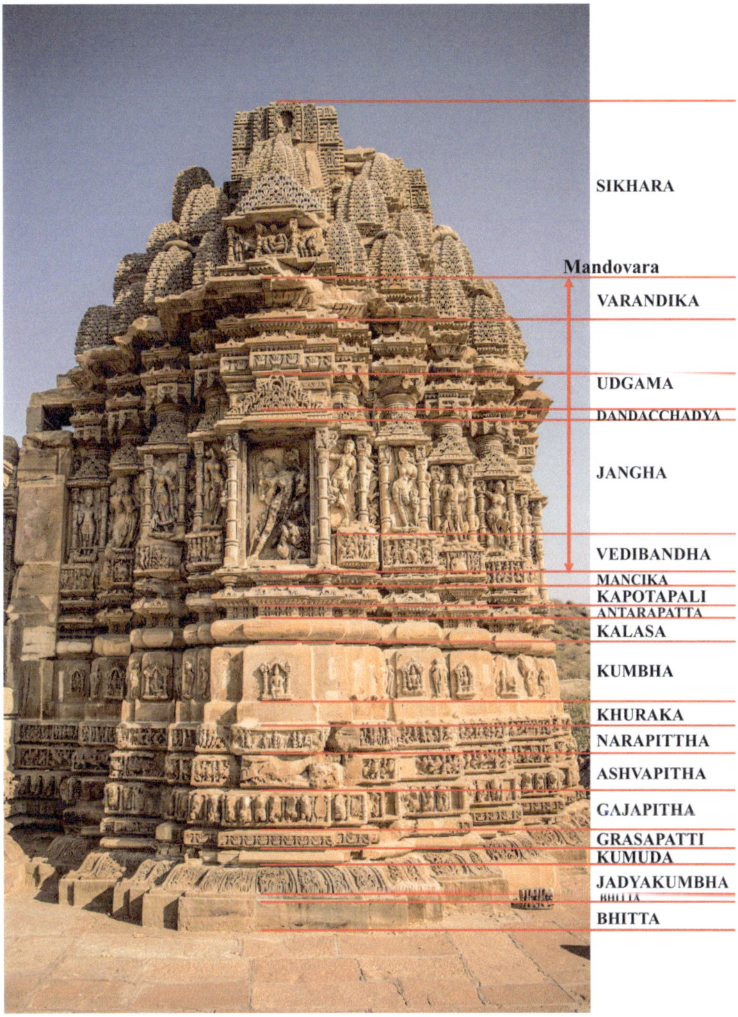

Fig. 5 Architectural details of Udharvachhand section of the Someshwar Temple (Temple No. 2) (Dhaky, 1998)

Figure 4 displays the Talchhand and Udharvachhand sections of the Someshwar temple (Temple No. 2). Architectural details of Udharvachhand section of the Someshwar Temple (Temple No 2) is presented in Fig. 5. A detail description of architecture of Talchhand of the temples of Kiradu was presented by Dhaky (1998) and Chaudhary (2012) about the Vishnu temple (Temple no. 6; Fig. 7) (Chaudhary, 2012; Dhaky, 1998). It is evident from their literature that the temples of Kiradu hold a significant turning point in the history of western architecture. They further demonstrate that these temples are landmarks attesting to the beginning of the fusion of Maha-Gurjara style with Maha-Maru style to transmute into the Maru-Gurjara

Fig. 6 Architectural details of Udharvachhand section of the Shiva Temple (Temple No. 1) (Dhaky, 1998)

style (Bhandarkar, 1912; Agrawala, 1954, 2011; Banerjee, 2008; Dhaky, 1975).The Shiva temple (Temple no. 1; Fig. 6) and the Vishnu temple (Temple no. 6; Fig. 7) have also similar kind of pattern at the wall design. But the ashvapitha slab of pattern is missing from the Vishnu temple. This depicts that the Vishnu temple was built earlier than the Shiva temple and the Someshwar temple because the ashvapitha patten of design was introduced in Maru-Gurjara style.

Fig. 7 Architectural details of Udharvachhand section of the Vishnu Temple (Temple No. 6) (Dhaky, 1998)

7 New Observations and Studies in This Work

The temples of Kiradu is majorly made of sandstone. Most of the temples of Kiradu are in ruined state. Some of them are due to weathering (Figs. 5 and 6) and some of the destructions are man-made (Figs. 26 and 27).

7.1 Temple No. 1 (The Shiva Temple)

The Talchhand section is completely destroyed in the Shiva temple. Only the Garbhagriha section is remaining. We have observed that the wall of the remaining section has been fabricated based on interlocking blocks. Interlocking blocks are suited for wall constructions with loads specially for two or more storeyed constructions. The temples of Kiradu are at a height with heavily sculptures inscribed wall. So this interlocking blocking helps to keep the strength of the wall. Figure 8 shows that lower panel section is attached to the upper panel with block to block locking. Some part of the temple is under restoration (Fig. 9). Figure 9a shows that the due to weathering, the Jadyakumbha portion of the wall is damaged and is restored (Fig. 9b). Adhesive (calcareous cementing material) are usually used to attach one damaged section to the main construction (Fig. 10a, b). In Fig. 10b, observe that the Narapitha block is broken and is then attached with the Khuraka section using calcareous cement. Figure 11 presents a wind-induced erosion in the temple rock.

Figure 11a presents damages in the gajaptha, ashvapitha and narapitha slabs. Figure 11b shows destruction of the frontal part of the sculpture. Figure 11c displays that the trunks of the elephants are damaged, and Fig. 11d is clearly shows that the middle sculpture is completely destroyed. These sections are the elevated portion of the sculptures. Due to weathering these elevated parts were affected most and got damaged. Due to weathering the top part of the Sikhara is also ruined. As Sikhara is the upper part of the temple, this section is completely exposed to the nature and we observe damages at different sections of the Sikhara. Figure 12a–c show that the frontal part of the sculpture is ruined. Figure 12d displays the damaged part at the back end of the temple.

Restoration work done in this temple seems to be a continuous process. As restoration is performed at different time with different materials, color differences at different restoration sections are clearly evident (Fig. 13). In Fig. 13a, the dark

Fig. 8 Block to block locking system of rocks in the Shiva temple (Temple No. 1) (Pen of length 13 cm as marker)

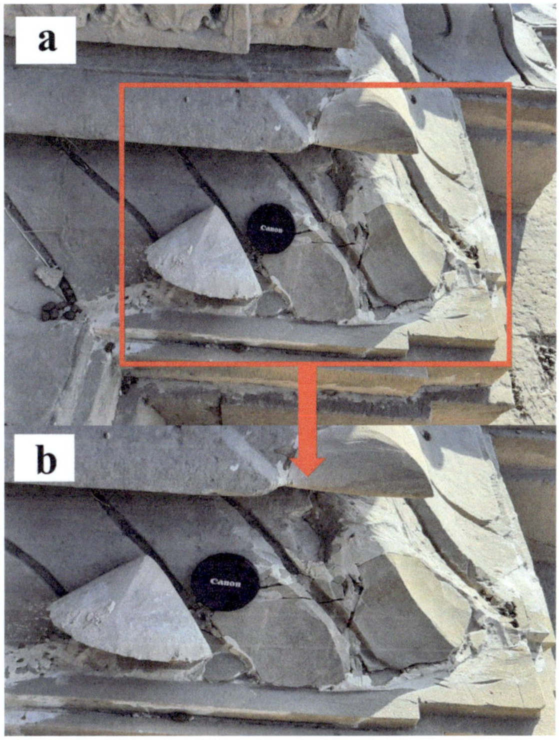

Fig. 9 Restoration of the Shiva Temple (Temple no. 1) (Camera cover marker, diameter 5.4 cm). **a** Restoration work is done in the damaged Jadyakumbha section; **b** restoration done using calcareous cement

yellow color at the bhitta section and pale yellow color at the restored section are observed. Similarly Fig. 13b shows that the color of the restored part is pale yellow, and the unrestored portion is dark yellow at the Jadyakumbha.

In the Shiva temple, several natural fractures are detected (e.g., Fig. 14). Arms, legs and breasts of the female idols show human-induced damages (Fig. 14b), and this is a persistent observation in the Kiradu temples.

Figure 15 shows restoration work within the sandstone using calcareous cementing material. We have seen a drain of width 8.1 cm and length 20 cm in north–south direction within the garbhagriha of the Shiva temple. Figure 16 indicates the water outlet along north–south direction.

Due to either weak metamorphism or some deformation, foliation planes developed within the sandstone, as observed in few places. The foliation is defined by the preferred concentration of micasalong some planar zones (Fig. 17). Such foliation planes act as planes of preferential breakage (Ghosh, 1993), and we note that such pieces of sandstones were not used in any vulnerable portion of the temple. Figure 17b shows a black coloured mineral weakly defining foliation in sandstone. Detail stress-strength analyses, and microstructural studies under an optical microscope would be important to understand the rheology and structures of the sandstone slabs that were used in constructing the Kiradu temples. This can give important constrain on the restoration process.

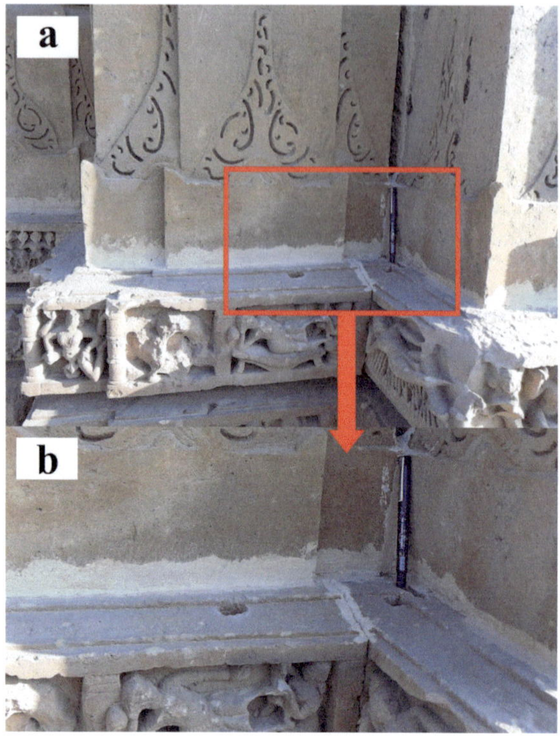

Fig. 10 Adhesive (calcareous cementing material) at the restoration site of the Shiva Temple (Temple no. 1) (Pen marker of length 13 cm) **a** Broken narapitha slab attached with the Khuraka portion with cementing, **b** A broken narapitha slab attached with another such slab with (calcareous cementing material)

Microorganisms and microchiropterans are very important for damaging any heritage monument. Six species of microchiropterans (Pipistrillus tenuis, Rhinopoma hardwickii, Taphozous perforates, Rhinopoma microphyllum kinneri, Taphozous melanopogon, and Taphozous nudiventris), most commonly known as bats are sited in the temples of Kiradu (Purohit, 2013). These bats induce moisture leading to leaching at the ceiling of the Shiva temple. Figure 18 shows leaching of water in sandstone that was produced by the increased moisture content of the wall created by bats.

7.2 Temple No. 2 (The Someshwar Temple)

The Someshwar Temple (Temple no. 2) is the only temple where the Mahamandap and the Garbhagriha are not destroyed. In fact, this temple gives an idea about the architectural details of the temples of Kiradu. At the entry of the Udharvachhand section of the temple, partially preserved inscriptions are observed at the wall (Fig. 19). A decorative door inscribed with several male and female sculptures are observed at the garbhagriha (Rep. Fig. 1). On this roof, there exists a design of lock and key within the beam to provide a rigid support to the roof (Fig. 20a, b). This type

Fig. 11 Wind erosion affected in parts of the Shiva temple (Temple no. 1), **a** Ruined portions within three different slabs, **b** Frontal part is destroyed, **c** Trunks of the elephants are destroyed and **d** a sculpture is completely damaged

of architectural pattern shows how different stone slabs are stacked from floor to the ceiling of the temple.

We have observed different types of fractures on the connecting slabs of the pillars forming the Mahamandap (Fig. 21). These connecting slabs are directly exposed to the nature since the Mahamandap has no roof. Natural weathering has created fractures in them. The fractures are quit deep and the slabs are presently in a vulnerable condition. These connective slabs require immediate restoration. Figure 22 exhibits different types of fractures at the top part of the horizontal slabs connecting the pillars. Natural fractures are also observed at the temple wall (Fig. 23a). Water-induced erosion has created ~2 cm deep cavities in sandstones, giving rise to finally honeycomb appearance (Fig. 23b). In Rep. Fig. 2, the Talchhand section (From the entry gate to the gate of Garbhagriha) is illustrated.

Natural erosion was also observed at the wall of the Someshwar temple. Figure 24 presents the erosion of the wall. In this figure, the top portion of the left side of the slab is more damaged whereas the right side is more intact. This can occur due to direction of wind flow along a specific direction. In the lower section of this slab, observe that the sculptures are decayed due to the action of water. The natural decay at the wall of the Someshwar temple clearly indicates that we need a detail defect and stress distribution study of the temple rocks. The sandstones used to construct the

Fig. 12 Top part of the Sikhara of the Shiva temple (Temple no. 1) is completely ruined (Soumyajit Mukherjee as marker), **a** Several part of the Sikhara are damaged, **b, c** Frontal sculpture of the Sikhara is damaged, **d** Part of the back of the Sikhara is damaged

Someshwar temple shows grain size variation at few places (Fig. 25a). In Fig. 25b, the fine-grained softer particles are pink, whereas the harder coarse grains are yellow.

We have collected one sample foundation stone from the ruins of this temple and performed uniaxial compressive strength (UCS), transverse strength, porosity, density and abrasive strength calculation to understand the physical properties of

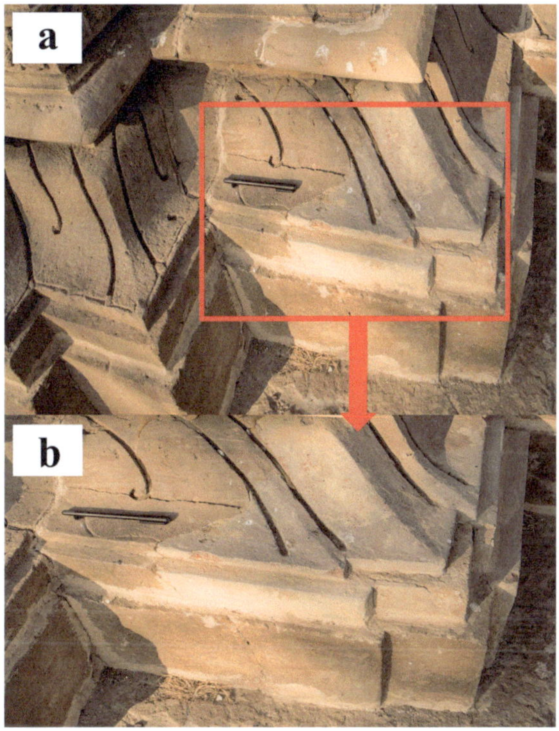

Fig. 13 Color difference in the restored section (black pen marker of length 15 cm) (Temple no. 1). **a** Dark yellow color at the Bhitta section and pale yellow color at the restored section, **b** The color of the restored part is pale yellow and not restored portion is dark yellow at the Jadyakumbha

the samples. Repository Figure 3 displays the sample. These physical properties are engineering properties of the building stone (Deere & Miller, 1966).

Indian standard code, as per the present day norm, define the engineering criteria to select the building stone. Rock blocks used in building construction should have compressive strength between 60 and 200 N/mm^2 (IS9143, 1973). Similarly, specific gravity of the building stone can vary 2.4–2.8. Highly porous stone is not good for the moisture full environment condition, so, absorption should not be more than 5% and porosity should be <25% [IS 13030 (1991 reaffirmed 1996 and 2001)]. The porosity of sandstone varies 10–40%. In this study, sandstone has porosity 35–37% in the dry and deserted Barmer area (as per our lab testing). UCS values lies between 12 and 17 N/mm^2. It shows week rock type (IS9143, 1973). But the samples of the Kiradu Temple are weathered; it might be possible, when this temple was constructed, rocks were fresh and had higher UCS values.

7.3 *Temple No. 3 (A Completely Ruined Temple)*

We have observed that there is no intense weathering related features but yet it is broken. (The fragmented sections of the temple clearly indicate that this temple is

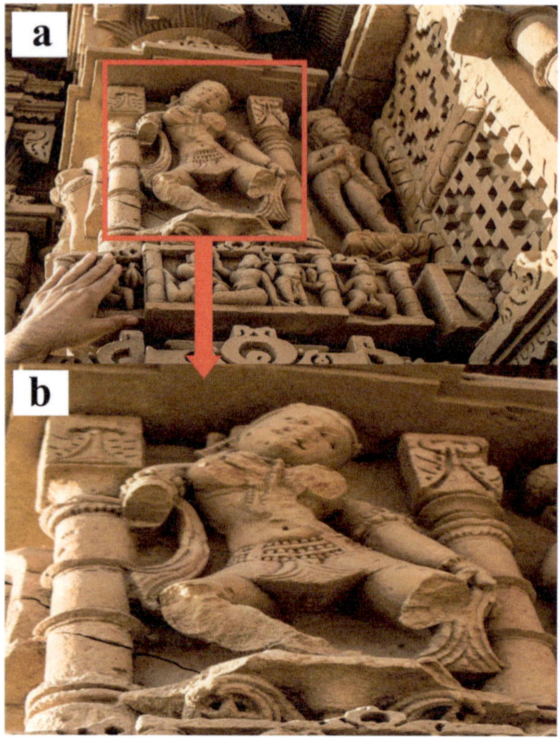

Fig. 14 Natural fracture at the Shiva temple (Temple no. 1) (**a**) and **b** arm, legs and breasts are damaged

not destroyed by natural process. So this might be man-made (Fig. 26a). Figures 26b and 27 present a broken pillar of the Mahamandap. Repository Figure 4 represents a fallen idol (Temple no. 3). We have observed that the idol had worn beaded broad head-band, where the beads are designed as small pots fixed downward direction. The idol had worn a beaded choker and a kundan bead necklace. There are bangles at the arms and wrists of the statue.

7.4 Temple No. 4 (The Shiva Temple)

In this temple, only the garbhagriha is remaining. The remainder is completely ruined. We have measured tensile strength variation at different sections of this temple using smith hammer. The first foundation stone from the base is of 25 cm thick. The following strengths are measured in four different basement stones (measured from bottom to top), 55 N mm^{-2}, 55 N mm^{-2}, 48 N mm^{-2} and 48 N mm^{-2}, respectively. Strength of the pillar foundation base is 46 N mm^{-2}. The height of the upper base of pillar is 90 cm. Figure 28 shows strength measurement with Smith Hammer at (a) 1st Foundation stone, (b) base of the pillar, (c) base of the Garbhagriha, from Bhitta to Grasapattika. An average of the following strengths measured in three different sides

Fig. 15 Restoration of sandstone block (black pen marker) (Temple no. 1) with calcareous cementing material

of the pillar are 44 N mm^{-2}, 62 N mm^{-2}, and 51 N mm^{-2} respectively. The strength of the pillar without restoration is evaluated as 56 N mm^{-2}.The strength of the pillar just above 200 cm height from the base is measured of 40 N mm^{-2} (Fig. 28d). The same measurements are performed on the second pillar both on vertical and horizontal plane and strengths are 56–57 N mm^{-2} and N mm^{-2}, respectively. Figure 29 presents the strength measurements at the pillars of the Shiva Temple in (a) vertical and (b) horizontal directions. The base of the Jagati is made up of basalt (Fig. 29d).

7.5 Temple No. 5 (The Shiva Temple)

Temple no. 5 is also a partially preserved Shiva temple (Rep. Fig. 5). In this figure it is clearly evident that the temple sanctum without Mahamandap where the Sikhara is damaged and some parts of the pillar sections have been restored.

Fig. 16 Water outlet along north–south direction of width 8.1 cm and length 20 cm (Temple no. 1) (Canon cap marker of diameter 5.4 cm)

7.6 Temple No. 6 (The Vishnu Temple)

Only the Mahamandap and the damaged wall of the garbhagriha are preserved (Figs. 30 and 31). Within this temple area, several broken slabs inscribed with male and female sculptures are scattered on the ground. Figure 32a, b exhibit the scattered broken slabs of the Vishnu temple. We have visualized male "(Ganesha?)" and female sculptures in those slabs. Decorations (A warrior in the jaw of Makara) at the partially damaged pillars of the Mahamandap are also observed (Rep. Fig. 6). Detail floral motif decorations at the partially damaged pillars of the Mahamandap of the Vishnu temple are still present (Rep. Fig. 7). At the inside wall of the damaged garbhagriha, brown patches are spotted within the sandstone (Fig. 33a, b) A chemical analysis is needed to fully understand the composition of the sandstone.

7.7 Selective Destruction of Female Statues

We have also detected that mostly the female statues of the Kiradu temple complex are destroyed. After a detail observation, it is clearly evident that most of the destruction

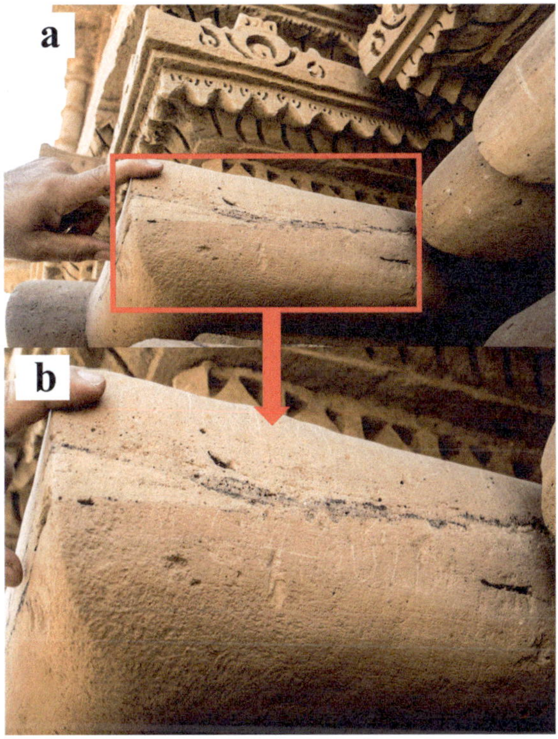

Fig. 17 **a** Black mineral grains marking weak foliation in sandstone (finger marker) (Temple no. 1), **b** A line of black material is observed at the surface of the sandstone

is due to act of vandalism. An evidence of selectively female idols being vandalized is shown in Fig. 34 (Temple no. 5). In this figure we can clearly see from figure (b) and (d) that men sculptures are intact but figure (c) shows that hands, legs and breasts of the female idol is damaged. Similarly Fig. 35 also exhibits that only female statues are defaced at the Shiva temple (Temple No. 5).

Repository Figure 8 represents vandalized female statues in the Shiva temple (Temple no. 4). In figure Rep. Fig. 8a hands and breasts are defaced. Repository Figure 8b shows that head and leg part of the female sculpture are ruined. Similarly Rep. Fig. 8c manifests that head, breast, hand and leg part of the female idol are completely broken.

Repository Figure 9 shows another female statue vandalized at the Shiva temple (Temple no. 4). Repository Figure 9b, c and d show the vandalized sections of the female body in details. The head, hands and leg part are defaced. Repository Figures 10 and 11 exhibit that only the female statues at the wall of the Shiva temple (Temple no. 4 and 5 respectively) are damaged. From Rep. Fig. 12, the same conclusion can be reached. (Temple no. 5).

Fig. 18 Leaching of water in sandstone at the ceiling of the Shiva temple (Temple no. 1)

7.8 Miscellaneous

In the Kiradu temple complex, we have found male and female idols involving in various activities, which highlight a picture of the cultural and social life of Indian civilization during the period the temple was built. They look like isolated sculptures, but at times they are in a row indicating elephant ride and celebration.

Figure 36 displays miniatures in playing and dancing mode. From the left side of this panel, we can see idols playing musical instruments e.g., flute and sarangi. Figure 37 shows female miniatures in various dancing poses. Man playing/fighting with elephants is also sculpted (Fig. 38). Statue in yoga posture (padmasana) is observed (Fig. 39). We have found female idol holding some scriptures in hand (Fig. 40). Warriors moving a chariot is also inscribed at the wall (Fig. 41). Panels of warriors fighting with swords are also observed (Fig. 42). From Fig. 43, we can see that sculpture ornamented with various jewellery like heavily designed kamar-bandh, baju-bandh, choker, necklace. We can also get an idea regarding the lifestyle and economic condition of the people of that time from these sculptures.

Besides, we have also observed different types of fractures at the panels of the temple wall. Figure 44 displays fracture cutting all the idols. Almost all almost sculptures eroded naturally (Fig. 45).

Fig. 19 Partly destroyed inscription at the wall of the Someshwar temple (Temple no. 2)

Indian temples sometimes tell a story like the story of the Ramayana or Mahabharata or the life of Buddha. In future, we need to check whether any coherent stories are told in any temple, or are they just designs.

8 Importance of Study and Future Scope

Despite few restoration works visible, some sections of the Kiradu temple are in a vulnerable state and restorations are urgently required. We need to investigate specific gravity, porosity, permeability, moisture content and chemical composition of the constituent material of the temple, i.e., the sandstone. Analysis of physical properties, pore structures and a more detail strength test of the sandstones used to construct this temple would be the future task. We also should perform defect study at the surface to understand the diffusion, interface behavior and deformation of the constructional material.

Fig. 20 Lock and key design atop the garbhagriha of the Someshwar temple (Temple no. 2). **a** Inside part of the roof over the door of the garbhagriha, **b** Lock and key design within the slab to hold the roof

9 Conclusions

In this study we have focused on the present status of the Kiradu temple. We have looked into different temples and tried to understand fractures and their origin and vulnerability. We present the architectural design of the Shiva temple and the Someshwar temple and tried to make a connection between the architecture of the Vishnu temple as mentioned in the literature. We have measured the strength of various sections of one of the partially damaged Shiva temple. We investigated several destructed male and female idols to understand reason of their decay. We have observed the foliation and weathering at the outer wall of the temples. We noted that few fractures are actually in a very vulnerable situation and an immediate restoration is required. We have also documented several female idols which are clear indication of man-made destruction.

Rep. Fig. 1 Decorative door inscribed with several male and female sculptures are observed at the garbhagriha (Temple no. 2)

Fig. 21 Fracture at the Mahamandap of the Someshwar temple (Temple no. 2). **a** Fracture at the top part of the connecting slab of the pillars. **b** The fracture is deep and vulnerable

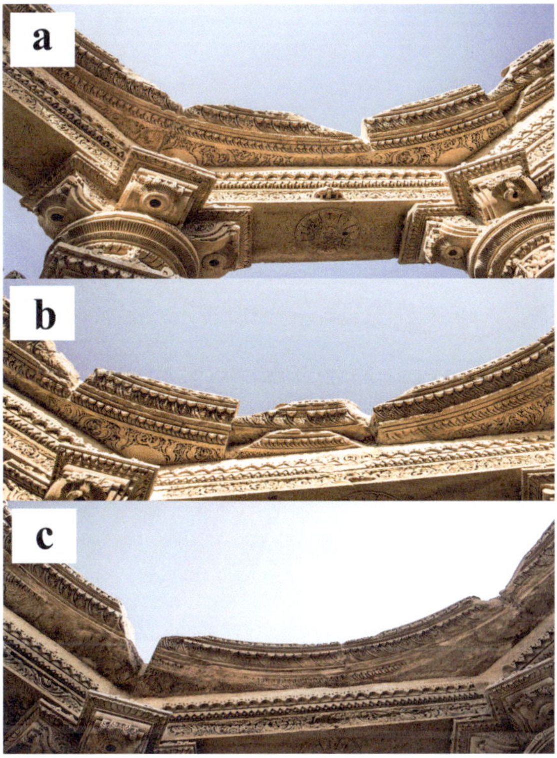

Fig. 22 Different types of fracture at Mahamandap of the Someshwar temple (Temple no. 2). **a–c** Fractures at the top part of the ligands connecting the pillar of Mahamandap

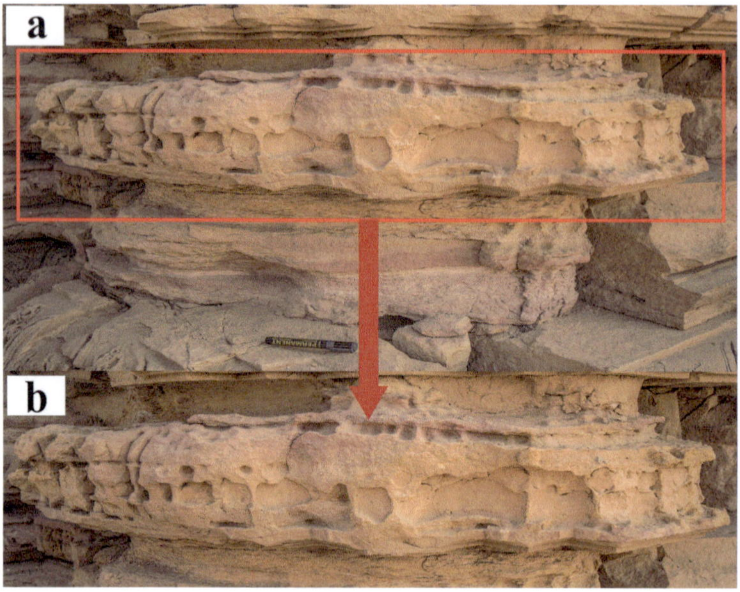

Fig. 23 **a** Natural fracture at the wall of the Someshwar temple (Temple no. 2) (black pen of length 14 cm), **b** Weathering of the wall producing a honeycomb geometry in sandstone

Rep. Fig. 2 Decorative pillars of Mahamandap and Talchhand (Temple no. 2). Soumyajit Mukherjee and Ritojit Mukherjee as markers

Fig. 24 Weathering of the Someshwar temple wall (Temple no. 2) (black pen of length 14 cm)

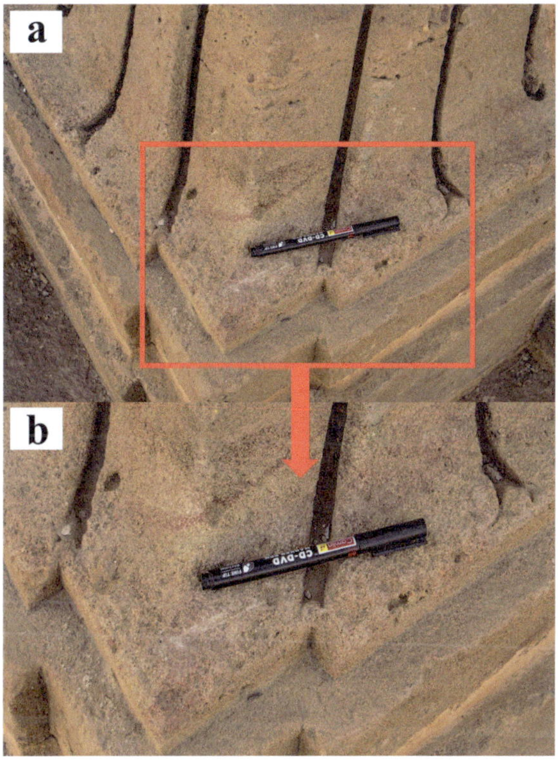

Fig. 25 **a** Grain size difference noted at the wall of the Someshwar temple (Temple no. 2) (black pen marker of length 14 cm); **b** coarse and fine grain within the sandstone are clearly distinguished

Rep. Fig. 3 Sample collected for testing from Temple no. 2

Fig. 26 **a** Completely destroyed temple (Temple no. 3). **b** Fragmented pillar of Mahamandap (black pen maker kept in between the two vandalised statues (Temple no. 3))

Fig. 27 **a** Scattered part of completely destroyed temple (Temple no. 3), **b–d** Demolished part of the pillars (Temple no. 3)

Rep. Fig. 4 A fallen idol ornamented with various jewellery (Temple no. 3)

Fig. 28 Strength measurement with Smith Hammer at the Shiva Temple (Temple no. 4) for **a** first foundation stone, **b** base of the pillar, **c** Height (from the base of the pillar to the section where the smith hammer is kept) and strength measurement at the base of the Garbhagriha (From Bhitta to Grasapattika), and **d** strength measurement at the pillar

Fig. 29 Strength measurement with Smith Hammer at the pillars of the Shiva Temple (Temple no. 4) in **a** vertical direction, **b** horizontal direction, **c** Basalt base of Jagati (Smith hammer marker of length 24 cm)

Rep. Fig. 5 Frontal image of the partially damaged Shiva temple (Temple no. 5)

Fig. 30 Vishnu Temple (Temple no. 6). Only Mahamandap is remaining and Garbhagriha is also in damaged condition

Fig. 31 Vishnu Temple (Temple no. 6). **a** Only Mahamandap is remaining, **b** Garbhagriha is completely ruined

Fig. 32 **a, b** Display the broken slab of the Vishnu temple (Temple no. 6) inscribed with male "(Ganesha?)" and female sculptures

Rep. Fig. 6 Decorations (A warrior in the jaw of Makara) at the partially damaged pillars of the Mahamandap of the Vishnu temple (Temple no. 6)

Rep. Fig. 7 Detail Decorations (Floral motif) at the partially damaged pillars of the Mahamandap of the Vishnu temple (Temple no. 6)

Fig. 33 **a** Presence of iron within the sandstone brown in color (black pen marker), **b** brown patches, possibly indicating iron (finger marker) at the wall of the Garbhagriha of Vishnu temple (temple no. 6)

Fig. 34 **a** Only female statues are vandalized at the Shiva temple (Temple no. 5) (hand as marker), **b** sculpture of man idol is intact, **c** hand, legs of the female idol are damaged. **d** An intact idol

Fig. 35 Only female statues are vandalized at the Shiva temple (Temple no. 5) (black pen marker of length 14 cm)

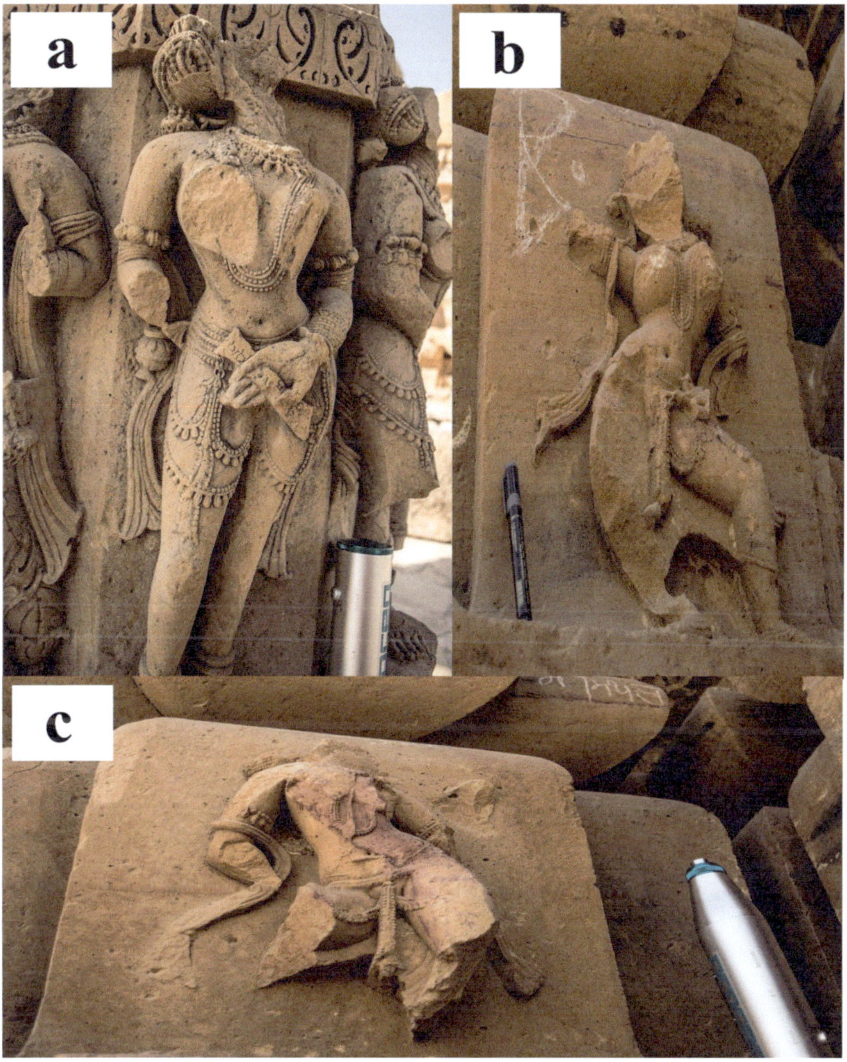

Rep. Fig. 8 Female statues are damaged due to act of vandalism at the Shiva temple (Temple no. 4) **a** Head and breasts destroyed (Smith hammer marker), **b** Head and leg of the female idol are ruined (black pen marker of length 14 cm), **c** Head, breasts, hands and legs of the female idol are broken (Smith hammer marker)

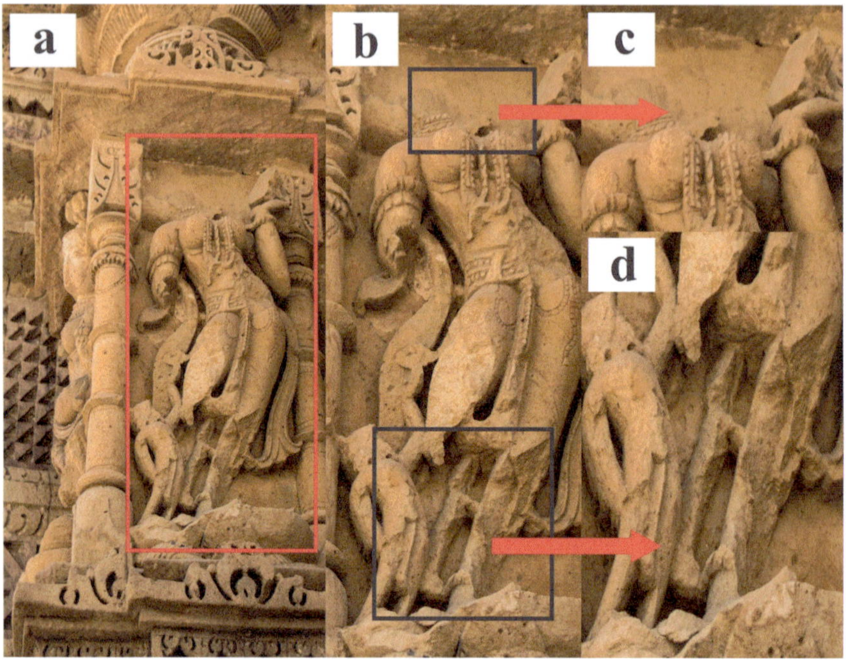

Rep. Fig. 9 **a** Female statue is damaged due to act of vandalism at the Shiva temple (Temple no. 4), **b** Head, hands and leg parts destroyed

Rep. Fig. 10 a, b Female statues are vandalized at the Shiva temple (Temple no. 4)

Rep. Fig. 11 **a** Only the female statues are vandalized at the Shiva temple (Temple no. 5), **b** Broken female idol, **c** Readers are encouraged to interpret!

Rep. Fig. 12 At the Shiva temple (Temple no. 5) (black pen marker of length 14 cm)

Fig. 36 Statues playing music and dancing

Fig. 37 Female idols in dancing pose

Fig. 38 Wall decoration where a man is playing/fighting with elephants

Fig. 39 Statue in a yoga posture (Padmasana)

Fig. 40 Female idol holding some scripture

Fig. 41 Moving a chariot

Fig. 42 Statues fighting with swords

Fig. 43 An idol ornamented with various jewellery

Fig. 44 Through going fracture cutting all the idols

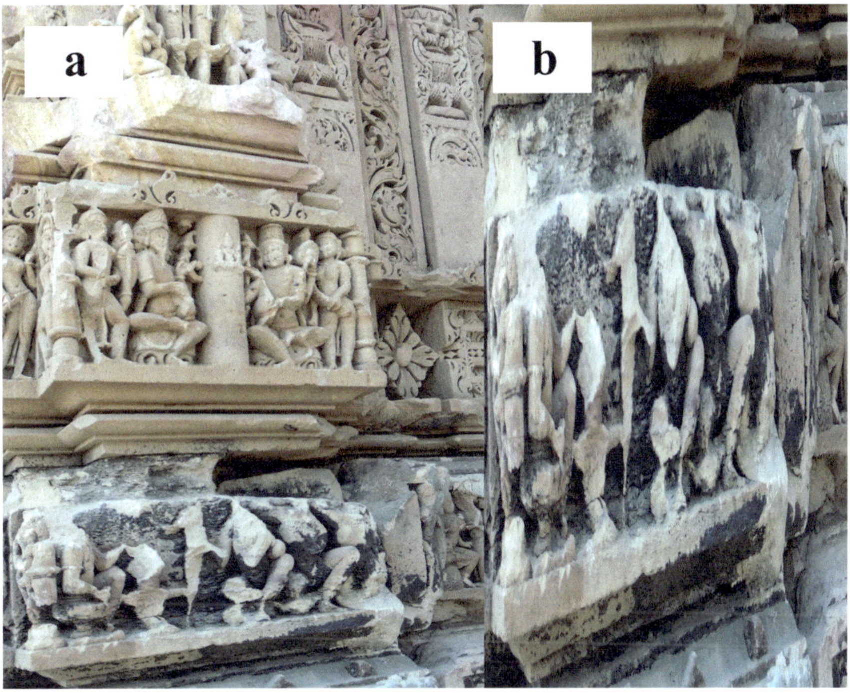

Fig. 45 Almost destructed sculpture by natural erosion

Acknowledgements CPDA Grant (IIT Bombay) supported S Mukherjee. Mukherjee (2023) summarized this article.

References

Agrawala, R. C. (1954). Kṛṣṇa and Balarāma in Rajasthan Sculptures and Epigraphs. *The Indian Historical Quarterly, 30*(4), 339–353.

Agrawala, R. C. (2011). Rajasthanme Krsnabhakti Pradarshan. *Shodhapatrika*, *5*(4), 1–12.

Banerjee, U. K. (2008). The subtle art of story telling. *Indian Literature, 52*(4), 147–152.

Bhandarkar, D. R. (1912). *The temples of Osia, annual report of archaeological survey of India 1908–1909* (pp. 100–115). Superintendent, Government Printing.

Chanchani, N. (2014). From Asoda to Almora, the roads less taken: Māru-Gurjara architecture in the central Himalayas. *Arts Asiatiques, 69*, 3–16.

Chaudhary, M. (2012). *Kiradu mandir samuh ki sthpatya kala evam murtikala.* Literary Circle.

Deere, D. U., & Miller, R. P. (1966). Engineering classification and index properties for intact rock, Technical report No. AFNL-TR-65-116.

Dhaky, M. A. (1967). Kiradu and the Maru-Gurjara style of temple architecture. *Bulletin of the American Academy of Benares, 1*, 35–45

Dhaky, M. A. (1975). The genesis and development of Maru-Guraja temple architecture. In P. Chandra (Ed.), *Studies in Indian temple architecture* (pp. 114–165). American Institute of Indian Studies.

Dhaky, M. A. (1998). *Encyclopedia of Indian temple architecture—North India beginnings of medieval idiom c. A.D. 900–1000*, Vol. 2, Issue 3. American Institute of Indian Studies and Oxford University Press.

Ghosh, S. K. (1993). *Structural geology: Fundamentals and modern developments* (p. 10). Pergamon Press.

Ghurye, G. S. (1968). *Rajput architecture.* Popular Prakashan, ISBN 978-81-7154-446-2

IS 13030. (1991 reaffirmed 1996 and 2001). Method of test for laboratory determination of water content, porosity, density and related properties of rock material [CED 48: Rock Mechanics].

IS 9143. (1973). Methods for the determination of unconfined compressive strength of rock material, Indian Standard Institution [CED 48: Rock Mechanics].

Mukherjee, S. (2023). Introduction to structural geology and tectonics field guidebook, vol 2. In S. Mukherjee, (Ed.) *Structural geology and tectonics field guidebook—Volume 2.* (pp. xi-xiv). Springer Nature Switzerland AG. ISBN 978-3-031-19575-4.

Poonia, S., & Rao, A. S. (2018). *Climate and climate change scenarios in the Indian Thar Region.* Springer Nature Switzerland, Handbook of Climate Change Resilience. https://doi.org/10.1007/978-3-319-71025-9_12-1

Purohit, A., Soni, P., Kaur, A., & Ram, H. (2013). Eco-status of Chiropteran fauna in and around Barmer, India. *International Journal of Conservation Science, 4*(1), 119–123. ISSN: 2067–533X

Saifuddin, I. (2000). Quaternary signatures of paleo-humidity in arid zone, Rajasthan, India. *Journal of Arid Environments, 45*, 151–158. https://doi.org/10.1006/jare.1999.0629

Singh, V. S., Pandey, D. N., Gupta, A. K., & Ravindranath, N. H. (2010). *Climate change impacts, mitigation and adaptation: Science for generating policy options in Rajasthan, India.* Rajasthan State Pollution Control Board, Jaipur, Rajasthan, India.

MIX
Papier aus verantwortungsvollen Quellen
Paper from responsible sources
FSC® C105338

If you have any concerns about our products,
you can contact us on
ProductSafety@springernature.com

In case Publisher is established outside the EU,
the EU authorized representative is:
Springer Nature Customer Service Center GmbH
Europaplatz 3, 69115 Heidelberg, Germany

Printed by Libri Plureos GmbH
in Hamburg, Germany